Springer Proceedings in Complexity

Springer Proceedings in Complexity publishes proceedings from scholarly meetings on all topics relating to the interdisciplinary studies of complex systems science. Springer welcomes book ideas from authors. The series is indexed in Scopus.

Proposals must include the following:

- name, place and date of the scientific meeting
- a link to the committees (local organization, international advisors etc.)
- scientific description of the meeting
- list of invited/plenary speakers
- an estimate of the planned proceedings book parameters (number of pages/articles, requested number of bulk copies, submission deadline)

Submit your proposals to: Hisako.Niko@springer.com.

Tassos Bountis · Filippos Vallianatos ·
Astero Provata · Dimitris Kugiumtzis ·
Yannis Kominis
Editors

Chaos, Fractals and Complexity

Springer

Editors
Tassos Bountis
Mathematics
University of Patras
Patras, Greece

Astero Provata
Institute of Nanoscience
and Nanotechnology
National Center for Scientific Research
Demokritos
Agia Paraskevi, Athens, Greece

Yannis Kominis
School of Applied Mathematical
and Physical Sciences
National Technical University of Athens
Athens, Greece

Filippos Vallianatos
Geology and Geoenvironment
National and Kapodistrian University
of Athens
Athens, Greece

Dimitris Kugiumtzis
Electrical and Computer Engineering
Aristotle University of Thessaloniki
Thessaloniki, Greece

ISSN 2213-8684 ISSN 2213-8692 (electronic)
Springer Proceedings in Complexity
ISBN 978-3-031-37403-6 ISBN 978-3-031-37404-3 (eBook)
https://doi.org/10.1007/978-3-031-37404-3

This Springer imprint is published by the registered company Springer Nature Switzerland AG
The registered company address is: Gewerbestrasse 11, 6330 Cham, Switzerland

This volume is dedicated to the 70th birthday of Athanassios S. Fokas, Professor of the Department of Applied Mathematics and Theoretical Physics of Cambridge University.

Organization

The 28th Summer School-Conference was organized by the COSA (Complex Systems and Applications) group of scientists of the National Center of Scientific Research "Demokritos" of Athens and the Laboratory of Geophysics and Seismology, Department of Natural Resources and Environment, Technological Educational Institute of Crete, Chania, Greece.

Organizing Committee

Conference Chair: Tassos Bountis (University of Patras, Greece)
Program Chair: Philippos Vallianatos (University of Athens, Greece)
Organizing Chair: Vassilios Constantoudis (NCSR Demokritos, Athens, Greece)
Assistant Chair: Astero Provata, (NCSR Demokritos, Athens, Greece)
Members: Yannis Kominis (National Polytechnic School of Athens, Greece); Dimitris Kugiumtzis (University of Thessaloniki, Greece); Theodore Karakasidis (University of Thessaly, Greece); Vassilios Basios (Free University of Brussels, Belgium)
Tutorials: George Tsironis (University of Heraklion, Crete, Greece)

Sponsors

Local Facilities: Deputy Regional Governor of the Chania Region of Crete
Other Sources: ADMIE Independent Power Transmission Operator, S.A. for the Hellenic Electricity Transmission System; GAPSTI: Gianna Angelopoulos Programme for Science, Technology and Innovation; Cambridge University, Cambridge UK; ACCADEMIS Company, Athens, Greece

Sponsoring Institutions

Academy of Athens, Athens, Greece
National Center of Scientific Research "Demokritos", Athens, Greece
Technological Educational Institute of Chania, Crete, Greece
University of Patras, Patras, Greece

Preface

Introduction

As is by now widely recognized, the study of nonlinear dynamical systems and chaos was founded by the great French Mathematician and Physicist Henri Poincaré at the end of the nineteenth century. He was the first to appreciate the vital importance of geometric methods as a means of understanding *qualitatively* the solutions of differential equations describing, for example, the motion of celestial bodies, that appeared analytically intractable. In the years that followed, Poincaré's work inspired many scientists, but it took nearly 60 years before the advent of the Kolmogorov-Arnold-Moser theory and the progress in computer technology revealed the great importance of Poincaré's contributions.

In the 1960's and 70's, the science of Chaos, revealed the secrets of unpredictability in the solutions of a great majority of nonlinear dynamical systems of physical, biological, economic, and technological significance. Moreover, the appreciation of the *spatial complexity* of objects and shapes that surround us led to the discovery of Fractal Geometry and by the end of 1980's Chaos and Fractals formed the foundations of what we call today Complexity Science.

In Greece, we were fortunate to realize early the significance of these developments. After organizing an international meeting on "Nonlinear Dynamics and Chaos in Classical and Quantum Systems", in August 1986, at Thessaloniki, we decided to start an annual series of Summer Schools and Conferences on these topics, aiming to bring together experienced researchers with a new generation of aspiring "nonlinear scientists". It was a successful endeavor, which has lasted to this day, adopting Complexity in the title of annual events that continued uninterrupted for 33 years, leading to 28 Summer Schools and 5 international Ph.D.Schools. The interested reader can find the history of these activities in a book called "The Meaning of Education", at the site http://cosa.inn.demokritos.gr/ of the National Center of Scientific Research of Athens "Demokritos".

In 2022, the Organizing Committee responsible for the 28th Summer School-Conference on "Dynamical Systems and Complexity" decided that the event would take place at the Cultural Center of Chania, Crete, 18–26 July and be online, except for 22–25/7, when it would also be live. The first two of these dates were devoted to celebrating the 70th birthday of Athanassios S. Fokas, Professor of Cambridge University, while on July 25 thirty selected Greek graduate and undergraduate students followed an intensive full-day training seminar on Machine Learning.

The remaining days were devoted to introductory talks, as well as more specialized lectures, on a wide variety of topics of Nonlinear Science. The invited speakers presented fundamental theoretical, experimental, and computational advances in chaos, fractals, and complexity. More details can be found at the site http://cosa.inn.demokritos.gr/28th-summer-school-dynamical-systems-and-complexity/.

Contents of the Volume

Our Organizing Committee appealed to all the speakers and received a significant number of contributions on a wide range of theoretical and experimental topics. They are listed in Parts I, II, and III corresponding to Chaos, Fractals, and Complexity, respectively. Some authors presented new results and reviewed recent achievements obtained in collaboration with Professor Athanassios S. Fokas. As these are primarily based on mathematical advances, they are contained in Part IV of the volume entitled "Fokas and Mathematics". Finally, we included as Part V a paper by Professor George Dassios describing the full spectrum of Professor A. Fokas' achievements to date in several sciences.

Chaos

The paper by S. Aubry, "Diffusion Without Spreading of a Wave Packet in Nonlinear Random Models", offers a remarkable account of what is known to date regarding a very interesting ongoing debate on the long- time behavior of energy wave packets in 1one-dimensional Hamiltonian lattices in the presence of nonlinearity and disorder.

P. A. Patsis, in his article on "Nonlinear Phenomena Shaping the Structure of Spiral Galaxies", reviews dynamical models, which successfully reproduce the morphology of disck galaxies and their evolution in time. He then shows how the interplay of order and chaos can explain the presence of bars and spiral arms in the disks.

M. Katsanikas and S. Wiggins, in their paper "Phase Space Transport and Dynamical Matching in a Caldera-Type Hamiltonian System", discuss phase space mechanisms by which a Caldera-type potential energy surface exhibits the phenomenon of dynamical matching and determine the conditions under which it occurs in their system.

In the paper "The Building Blocks of Spiral Arms in Galaxies", M. Harsoula reviews present- day theories regarding the kind of orbits that support spiral arms in various types of galaxies. The author explains the role of stable periodic orbits of spiral galactic models in creating spiral density waves similar to those observed in real galaxies.

The paper by A. C. Tzemos on "Ordered and Chaotic Bohmian Trajectories" reviews recent results on the emergence of chaos in arbitrary two-dimensional Bohmian dynamical systems. He examines the relation between chaos and entanglement and discusses its role in establishing Born's rule for arbitrary initial distributions of Bohmian particles.

Next, M. Robnik in his paper "A Brief Introduction to Quantum Chaos of Generic Systems" summarizes the remarkable progress of the last 40 years in the field of quantum chaos. He analyzes regular eigenstates associated with invariant tori as well as chaotic eigenstates through their corresponding energy spectra.

T. Bountis, K. Kaloudis, and H. Christodoulidi, in their paper "Dynamics and Statistics of Weak Chaos in a 4-D Symplectic Map", study the nature of chaos near an unstable fixed point of two coupled two-dimensional (2D) MacMillan maps. They discover "weakly" chaotic states characterized by $1 < q < 3$ Gaussian probability distributions, as well as cases of "strong chaos" obeying purely Gaussian ($q = 1$) statistics.

Fractals

In the paper by M. Chatzigeorgiou, V. Constantoudis, M. Katsiotis, and N. Boukos, "Multifractal Analysis of SEM Images of Multiphase Materials: The Case of OPC Clinker", the authors study properties of multiphase materials affected by the spatial distribution of their phases and the geometry of their interfaces. Using multifractal analysis they provide a novel quantification of phase distributions.

F. Minicucci, F. D. Oikonomou, and A. De Sanctis, in their paper "Fractal Dimensional Analysis for Retinal Vascularization Images in Retinitis Pigmentosa: A Pilot Study", regard retinal blood vessels as forming a fractal pattern. They thus show that retinal vascularization (RV) images can help doctors achieve an early diagnosis of retinitis pigmentosa.

Finally, V. Basios, in "Extending the Bayesian Framework from Information to Action", examines Bayesian inference methods and their connection in biological processes, where fractals and chaos play a crucial role. In the former, probability space is contracted, while in the latter it is extended to include latent and observable variables that elucidate the differences between artificial and biological information processing.

Complexity

In the paper by N. E. Protonotarios, K. Kalimeris, and G. A. Kastis, "Fokas on Medical Imaging: Analytic Reconstructions for Emission Tomography", the authors summarize the seminal work of Fokas in the area of mathematical image reconstruction, based on modern methods of complex analysis. They also review the mathematical theory of emission tomography, focusing on the inversion of non-attenuated and attenuated Radon transforms.

Authored by G. Paraskevopoulou, A. S. Fokas, A. Charalambopoulos, and S. Perantonis, the article "Inverse EEG Problem, Minimization and Numerical Solutions" focuses on brain activation, in the form of neuronal electric currents generating electric fields. In this work, the authors present a novel numerical formulation for computing the current, which includes a crucial boundary term that was missing in earlier papers.

The paper "Traveling Waves in Flowing Sand: The Dynamical Systems Approach", by Ko van der Weele, D. Razis, and G. Kanellopoulos, addresses the complex problem of travelling surface waves in a shallow sheet of granular matter. They simplify the problem by deriving a dynamical system that captures the transition from a monoclinal shock wave to a periodic train of roll waves.

In the paper by T. Dogkas, M. Eleftheriou, G. D. Barmparis, and G. P. Tsironis, "Identifying Discrete Breathers Using Convolutional Neural Networks", the authors study physical phenomena related to the time evolution of localized periodic oscillations called Discrete Breathers (DB) in one-dimensional nonlinear chains. They use Convolutional Neural Networks to differentiate between DB and linearized phonon modes.

In the next paper, "Subthreshold Oscillations in Multiplex Leaky Integrate-and–Fire Networks with Nonlocal Interactions", by K. Anesiadis, J. Hizanidis, and A. Provata, the authors investigate the complex dynamics of identical Leaky Integrate-and-Fire (LIF) neurons on a multiplex consisting of two identical ring networks. They show that inter-ring coupling favors in-phase synchronization and determine the relevant parameter region.

J. Courson, Th. Manos, and M. Quoy, in their paper "Networks' Modulation: How Different Structural Network Properties Affect the Global Synchronization of Coupled Kuramoto Oscillators", study how synchronization arises, when different oscillating objects tune their rhythm of interaction. They investigate different network architectures on coupled Kuramoto phase oscillators and measure the global degree of synchrony when different fractions of nodes receive the stimulus.

Finally, in "Neural Correlates of Human-Machine Trust in Autonomous Vehicles Context", A. Dragomir, I. Lazarou, M. S. Seet, S. Nikolopoulos, I. Kompatsiaris, and A. Bezerianos, examine how driver state monitoring systems may be used to support interfacing between human drivers and automated driving systems to enhance road safety. They point out that recent progress has revealed promising results that can be implemented on future intelligent vehicles.

Fokas and Mathematics

In the paper by Y. Cao, A. S. Fokas, and J. He, "High-Order Localized Wave Solutions of the New (3+1)-Dimensional Kadomtsev-Petviashvili Equation", the authors study an integrable extension of the Kadomtsev-Petviashvili equation in three-spatial dimensions and use Hirota's bilinear method to construct smooth multi-solitons and high-order rational and semi-rational solutions.

Alexandrou Himonas, in "Progress in Initial-Boundary Value Problems for Nonlinear Evolution Equations and the Fokas Method", focuses on what he calls the unified transform method, often called the Fokas method, and its success in solving initial-boundary value problems (IBVP) for linear and integrable nonlinear PDEs. The author describes how, using this method, one can derive linear estimates of solutions in Sobolev, Hadamard, and Bourgain spaces.

In the paper by A. Chatziafratis, L. Grafakos, S. Kamvissis, and I. G. Stratis, "Instabilities of Linear Evolution PDEs via the Fokas Method", the authors use a formula provided by the Fokas method for initial-boundary-value problems to study the linearized KdV equation on the half-line for $t > 0$. Depending on the sign of the dispersive term, they discuss how long-range asymptotics can depend very sensitively on the behavior of the data at the point (0, 0).

D. A. Smith, in his paper on "Fokas Diagonalization", discusses an approach for solving linear IBVP formulated as a spectral transform method, through which the underlying spatial differential operator can be diagonalized in two-point initial-boundary-value problems on networks of finite intervals. Here, the author extends these results to problems involving semi-infinite domains, nonlocal boundary conditions, and PDEs with mixed derivatives.

In their paper "A Novel Difference-Integral Equation Satisfied Asymptotically by the Riemann Zeta Function", A. S. Fokas, K. Kalimeris, and J. Lenells first review some basic results by A. S. Fokas, regarding a novel difference-integral equation satisfied asymptotically by the Riemann zeta function, $\zeta(1/2 + it)$. This equation is obtained starting with a singular integral equation presented for the first time in 2019 and using a finite Fourier transform representation of the Riemann zeta function.

B. Pelloni and D.A. Smith, in their paper "The Role of Periodicity in the Solution of Third Order Boundary Value Problems", first explain how the solution of certain boundary value problems connected with Airy's equation can be expressed as a perturbation of the solution of the periodic problem. They explain that their motivation is to understand the role of boundary conditions in the analysis of linear dispersive problems with discontinuous initial data.

Finally, in the paper by D. Mantzavinos on "The Fokas Method for the Well-posedness of Nonlinear Dispersive Equations in Domains with a Boundary", the author discusses the Fokas transform method and elucidates its analogy with the Fourier transform for both linear and nonlinear dispersive equations. He also discusses a novel approach for proving the well-posedness of initial-boundary-value problems for general nonlinear dispersive equations.

Athanassios S. Fokas: A Renaissance Scientist

As the final paper of this volume, we have included a contribution by George Dassios, Emeritus Professor of the University of Patras, entitled "Athanassios S. Fokas: A Renaissance Scientist". Professor Dassios has co-authored many papers and a book with Prof. Fokas. His paper in this volume offers a comprehensive account of Prof. Fokas' remarkable breakthroughs in many sciences, including Mathematics, Physics, Engineering, Biology, and Medicine.

A Tribute to Professor Athanassios S. Fokas

On July 22 and 23, 2022, more than 20 scientists from many countries, who had collaborated with Prof. Fokas over the years, lectured at this Summer School-Conference. They presented important results related to these collaborations, in areas of Applied Mathematics, Theoretical Physics, Biology, and Neuroscience, that have already gained worldwide recognition.

More specifically, these authors reported results obtained with Prof. Fokas on the solution of key open problems concerning: integrable nonlinear evolution equations, the development of efficient algorithms for the solution of inverse problems arising in medical imaging, the asymptotic analysis of the Riemann zeta function, and a novel approach toward the solution of the Lindelöf hypothesis. Several speakers discussed various areas impacted by the so-called Fokas Transform method.

Professor Fokas' remarkable career started with a BSc in Aeronautics from Imperial College and continued with a PhD in Applied Mathematics from Caltech and an MD from the University of Miami. Since 2002, he has held the Chair of Nonlinear Mathematical Sciences at Cambridge University. Currently, he is the Director of the "Legendary Program in Mathematics" at the University of Cambridge and an Adjunct Professor of the Departments of Civil and Environmental Engineering, and Biomedical Engineering of the University of South California.

In 2000, he was awarded the United Kingdom's Naylor Prize in Applied Mathematics; in 2005, he was decorated with the Order of Phoenix by the President of the Hellenic Republic; and in 2006, he was awarded the Aristeion Prize of the Bodossaki Foundation in Greece. He is a member of the Academy of Athens (2004), a member of the European Academy of Science (2010), a Fellow of the American Institute for Medical and Biological Engineering (2019), and a member of the European Academy of Sciences and Arts (2021).

In 2023, he was elected member of the Academia Europaea and received the Blaise Pascal Medal in Mathematics from the European Academy of Sciences.

In closing, the Editors wish to dedicate this volume to Prof. Athanassios S. Fokas and his colleagues for their contributions to the 28th Summer School-Conference on “Dynamical Systems and Complexity” and wish Prof. Fokas many more “happy birthdays” and scientific achievements in his future career.

Patras, Greece

Tassos Bountis
Program Chair
28th Summer School-Conference on
“Dynamical Systems and Complexity”
Chania, Crete, July 18–26, 2022

Contents

Chaos

Diffusion Without Spreading of a Wave Packet in Nonlinear Random Models 3
Serge Aubry

Nonlinear Phenomena Shaping the Structure of Spiral Galaxies 37
P. A. Patsis

Phase Space Transport and Dynamical Matching in a Caldera-Type Hamiltonian System 47
Matthaios Katsanikas and Stephen Wiggins

The Building Blocks of Spiral Arms in Galaxies 57
Mirella Harsoula

Ordered and Chaotic Bohmian Trajectories 71
Athanasios C. Tzemos

A Brief Introduction to Quantum Chaos of Generic Systems 83
Marko Robnik

Dynamics and Statistics of Weak Chaos in a 4-D Symplectic Map 109
Tassos Bountis, Konstantinos Kaloudis, and Helen Christodoulidi

Fractals

Multifractal Analysis of SEM Images of Multiphase Materials: The Case of OPC Clinker 125
M. Chatzigeorgiou, V. Constantoudis, M. Katsiotis, and N. Boukos

Fractal Dimensional Analysis for Retinal Vascularization Images in Retinitis Pigmentosa: A Pilot Study 139
Francesca Minicucci, Fotios D. Oikonomou, and Angela A. De Sanctis

Extending the Bayesian Framework from Information to Action 153
Vasileios Basios, Yukio-Pegio Gunji, and Pier-Francesco Moretti

Complexity

Fokas on Medical Imaging: Analytic Reconstructions for Emission Tomography 169
Nicholas E. Protonotarios, Konstantinos Kalimeris, and George A. Kastis

Inverse EEG Problem, Minimization and Numerical Solutions 189
Georgia Parakevopoulou, Athanassios S. Fokas, Antonios Charalambopoulos, and Stavros Perantonis

Traveling Waves in Flowing Sand: The Dynamical Systems Approach 199
Ko van der Weele, Dimitrios Razis, and Giorgos Kanellopoulos

Identifying Discrete Breathers Using Convolutional Neural Networks 213
T. Dogkas, M. Eleftheriou, G. D. Barmparis, and G. P. Tsironis

Subthreshold Oscillations in Multiplex Leaky Integrate-and-Fire Networks with Nonlocal Interactions 221
K. Anesiadis, J. Hizanidis, and A. Provata

Networks' Modulation: How Different Structural Network Properties Affect the Global Synchronization of Coupled Kuramoto Oscillators 233
Juliette Courson, Thanos Manos, and Mathias Quoy

Neural Correlates of Human-Machine Trust in Autonomous Vehicles Context 245
Andrei Dragomir, Ioulietta Lazarou, Manuel S. Seet, Spiros Nikolopoulos, Ioannis Kompatsiaris, and Anastasios Bezerianos

Fokas and Mathematics

High-Order Localized Wave Solutions of the New (3+1)-Dimensional Kadomtsev-Petviashvili Equation 265
Yulei Cao, Athanassios S. Fokas, and Jingsong He

Progress in Initial-Boundary Value Problems for Nonlinear Evolution Equations and the Fokas Method 275
A. Alexandrou Himonas

Instabilities of Linear Evolution PDEs via the Fokas Method 289
A. Chatziafratis, L. Grafakos, S. Kamvissis, and I. G. Stratis

Fokas Diagonalization 301
D. A. Smith

A Novel Difference-Integral Equation Satisfied Asymptotically by the Riemann Zeta Function 319
Athanassios S. Fokas, Konstantinos Kalimeris, and J. Lenells

The Role of Periodicity in the Solution of Third Order Boundary Value Problems 333
B. Pelloni and D. A. Smith

The Fokas Method for the Well-posedness of Nonlinear Dispersive Equations in Domains with a Boundary 347
Dionyssios Mantzavinos

Athanassios S. Fokas: A Renaissance Scientist

Athanassios S. Fokas: A Renaissance Scientist 363
George Dassios

Author Index 373

Contributors

K. Anesiadis School of Applied Mathematical and Physical Sciences, National Technical University of Athens, Athens, Greece;
Institute of Nanoscience and Nanotechnology, National Center for Scientific Research "Demokritos", Athens, Greece

Serge Aubry Laboratoire Léon Brillouin, CEA Saclay, Gif-sur-Yvette, France

G. D. Barmparis Department of Physics, University of Crete, Heraklion, Greece

Vasileios Basios Service de Physique des Systèmes Complexes et Mécanique Statistique and Interdisciplinary Center for Nonlinear Phenomena and Complex Systems C.P.231 CeNoLi-ULB, Université Libre de Bruxelles (ULB), Brussels, Belgium

Anastasios Bezerianos Information Technologies Institute, Centre for Research and Technology Hellas (CERTH-ITI), Thessaloniki, Greece

N. Boukos Institute of Nanoscience and Nanotechnology, National Centre for Scientific Research Demokritos, Patr., Agia Paraskevi, Greece

Tassos Bountis Center for Integrable Systems, P.G. Demidov Yaroslavl State University, Yaroslavl, Russia

Yulei Cao School of Mathematics and Science, Nanyang Institute of Technology, Nanyang, Henan, P. R. China

Antonios Charalambopoulos School of Applied Mathematical and Physical Sciences, Department of Mathematics, National Technical University of Athens, Zografou Campus, Athens, Greece

A. Chatziafratis Department of Mathematics, National and Kapodistrian University of Athens, Athens, Greece;
Institute of Applied and Computational Mathematics, FORTH, Heraklion, Greece

M. Chatzigeorgiou Institute of Nanoscience and Nanotechnology, National Centre for Scientific Research Demokritos, Patr., Agia Paraskevi, Greece;
School of Chemical Engineering, National Technical University of Athens, Athens, Zografou, Greece

Helen Christodoulidi Department of Mathematics, Lincoln University, Lincoln, LN6 7TS, UK

V. Constantoudis Institute of Nanoscience and Nanotechnology, National Centre for Scientific Research Demokritos, Patr., Agia Paraskevi, Greece

Juliette Courson Laboratoire de Physique Théorique et Modélisation (LPTM), CNRS, UMR 8089, CY Cergy Paris Université, Cergy-Pontoise Cedex, France;
Equipes Traitement de l'Information et Systèmes (ETIS), CNRS, UMR 8051, ENSEA, CY Cergy Paris Université, Cergy-Pontoise Cedex, France;
Department of Computer Science, University of Warwick, Coventry, UK

George Dassios Department of Chemical Engineering, University of Patras, Patras, Greece

Angela A. De Sanctis Department of Management and Business Administration, University "G. d'Annunzio" of Chieti-Pescara, Pescara, Italy

T. Dogkas Department of Physics, University of Crete, Heraklion, Greece

Andrei Dragomir The N.1 Institute for Health, National University of Singapore, Singapore, Singapore

M. Eleftheriou Department of Physics, University of Crete, Heraklion, Greece

Athanassios S. Fokas Department of Applied Mathematics and Theoretical Physics, University of Cambridge, Cambridge, UK;
Viterbi School of Engineering, University of Southern California, Los Angeles, Los Angeles, CA, USA;
Mathematics Research Center, Academy of Athens, Athens, Greece

L. Grafakos Department of Mathematics, University of Missouri, Columbia, MO, USA

Yukio-Pegio Gunji Department of Intermedia Arts and Science, School of Fundamental Science and Technology, Waseda University, Tokyo, Japan

Mirella Harsoula Research Center for Astronomy and Applied Mathematics of the Academy of Athens, Athens, Greece

Jingsong He Institute for Advanced Study, Shenzhen University, Shenzhen, Guangdong, P. R. China

A. Alexandrou Himonas University of Notre Dame, Notre Dame, IN, USA

J. Hizanidis Department of Physics, University of Crete, Herakleio, Greece; Institute of Applied and Computational Mathematics, Foundation for Research and Technology – Hellas, Herakleio, Greece

Konstantinos Kalimeris Mathematics Research Center, Academy of Athens, Athens, Greece

Konstantinos Kaloudis Department of Statistics and Actuarial-Financial Mathematics, University of the Aegean, Samos, Greece

S. Kamvissis Institute of Applied and Computational Mathematics, FORTH, Heraklion, Greece;
Department of Pure and Applied Mathematics, University of Crete, Rethymno, Greece

Giorgos Kanellopoulos Department of Mathematics, University of Patras, Patras, Greece

George A. Kastis Mathematics Research Center, Academy of Athens, Athens, Greece;
Institute of Nuclear and Radiological Science and Technology, Energy and Safety, National Center for Scientific Research "Demokritos", Agia Paraskevi, Greece

Matthaios Katsanikas Research Center for Astronomy and Applied Mathematics of the Academy of Athens, Athens, Greece;
School of Mathematics, University of Bristol, Bristol, UK

M. Katsiotis Group Innovation and Technology, Athens, Greece

Ioannis Kompatsiaris Information Technologies Institute, Centre for Research and Technology Hellas (CERTH-ITI), Thessaloniki, Greece

Ioulietta Lazarou Information Technologies Institute, Centre for Research and Technology Hellas (CERTH-ITI), Thessaloniki, Greece

J. Lenells Department of Mathematics, KTH Royal Institute of Technology, Stockholm, Sweden

Thanos Manos Equipes Traitement de l'Information et Systèmes (ETIS), CNRS, UMR 8051, ENSEA, CY Cergy Paris Université, Cergy-Pontoise Cedex, France

Dionyssios Mantzavinos Department of Mathematics, University of Kansas, Lawrence, KS, USA

Francesca Minicucci Ophthalmology Clinics, Department of Medicine and Science of Ageing, University "G. d'Annunzio" of Chieti-Pescara and Department of Life, Health and Environmental Sciences, University of L'Aquila, L'Aquila, Italy

Pier-Francesco Moretti CNR, National Research Council, Rome, Italy

Spiros Nikolopoulos Information Technologies Institute, Centre for Research and Technology Hellas (CERTH-ITI), Thessaloniki, Greece

Fotios D. Oikonomou Department of Physics, University of Patras, Rio, Greece

Georgia Parakevopoulou School of Applied Mathematical and Physical Sciences, Department of Mathematics, National Technical University of Athens, Zografou Campus, Athens, Greece

P. A. Patsis Research Center for Astronomy and Applied Mathematics of the Academy of Athens, Athens, Greece

B. Pelloni Heriot-Watt University and Maxwell Institute for the Mathematical Sciences, Edinburgh, Scotland

Stavros Perantonis Institute of Informatics and Telecommunications, National Center for Scientific Research - "Demokritos", Patriarchou Gregoriou E' and 27 Neapoleos Str, Agia Paraskevi, Greece

Nicholas E. Protonotarios Department of Applied Mathematics and Theoretical Physics, University of Cambridge, Cambridge, UK;
Mathematics Research Center, Academy of Athens, Athens, Greece

A. Provata Institute of Nanoscience and Nanotechnology, National Center for Scientific Research "Demokritos", Athens, Greece

Mathias Quoy Equipes Traitement de l'Information et Systèmes (ETIS), CNRS, UMR 8051, ENSEA, CY Cergy Paris Université, Cergy-Pontoise Cedex, France;
IPAL CNRS Singapore, Singapore, Singapore

Dimitrios Razis Department of Mathematics, University of Patras, Patras, Greece;
Department of Civil Engineering, University of Thessaly, Volos, Greece;
Department of Physics, National and Kapodistrian University of Athens, Zografou Athens, Greece

Marko Robnik CAMTP - Center for Applied Mathematics and Theoretical Physics, Maribor, Slovenia

Manuel S. Seet The N.1 Institute for Health, National University of Singapore, Singapore, Singapore

D. A. Smith Division of Science, Yale-NUS College, Singapore, Singapore;
Department of Mathematics, National University of Singapore, Singapore, Singapore;
Yale-NUS College and National University of Singapore, Singapore, Island

I. G. Stratis Department of Mathematics, National and Kapodistrian University of Athens, Athens, Greece

G. P. Tsironis Department of Physics, University of Crete, Heraklion, Greece

Athanasios C. Tzemos Research Center for Astronomy and Applied Mathematics of the Academy of Athens, Athens, Greece

Ko van der Weele Department of Mathematics, University of Patras, Patras, Greece

Stephen Wiggins Department of Mathematics, United States Naval Academy, Annapolis, USA

Chaos

Diffusion Without Spreading of a Wave Packet in Nonlinear Random Models

Serge Aubry

Abstract We discuss the long time behavior of a finite energy wave packet in nonlinear Hamiltonians on infinite lattices at arbitrary dimension, exhibiting linear Anderson localization. Strong arguments both mathematical and numerical, suggest for infinite models that small amplitude wave packets may generate stationary quasiperiodic solutions (KAM tori) almost indistinguishable from linear wave packets. The probability of this event is non vanishing at small enough amplitude and goes to unity at amplitude zero. Most other wave packets (non KAM tori) are chaotic. We discuss the Arnold diffusion conjecture (recently proven) and propose a modified Boltzmann statistics for wave packets valid in generic models. The consequence is that the probability that a chaotic wave packet spreads to zero amplitude is zero. It must always remain focused around one or few chaotic spots which wander randomly over the whole system and generate subdiffusion. In this paper, we study a class of so-called Ding Dong models, where the nonlinearities are replaced by hard core potentials, which also generate subdiffusion. We prove rigorously for these models that spreading is impossible for any initial wave packet.

Keywords Diffusion · Spreading · KAM tori · Chaos · Arnold diffusion

1 Introduction

It is well-known that wave propagation becomes impossible in linear random systems with strong enough disorder because of Anderson localization [1]. Such a situation may also occur in other non random models which are for example incommensurate [2]. Our purpose is to study the same problem when nonlinearities are taken into account in models with a purely discrete linear spectrum that is where all the linear eigenmodes are square summable and spatially localized.

It has been suggested mostly on the basis of numerical simulations and rough arguments that, in such models, Anderson localization is destroyed. A fully chaotic

S. Aubry (✉)
Laboratoire Léon Brillouin, CEA Saclay, 91191 Gif-sur-Yvette, France
e-mail: serge.aubry91@gmail.com

T. Bountis et al. (eds.), *Chaos, Fractals and Complexity*, Springer Proceedings in Complexity, https://doi.org/10.1007/978-3-031-37404-3_1

dynamics is supposed to take place as a consequence of the extra nonlinearities which couple the localized linear modes to each other [3–6]. Nevertheless, it was also proved that large enough amplitude wave packets cannot spread in some models with norm conservation [7]. Otherwise, it was suggested that small enough amplitude wave packets [8, 9] should generate stationary quasiperiodic solutions (KAM tori) despite the number of degrees of freedom being infinite and moreover that this happens with probability going to 1, as the amplitude of the wave packet goes to zero. The initial wave packets which do not generate KAM tori, are chaotic as in the case of finite systems. In infinite systems, we argued that their spatial spreading will stop after some transient time [9], when the wave packet will have an amplitude small enough to reach the region where KAM tori are very dense.

However, we could not predict the behavior of the wave packet beyond this initial transient time. The aim of this paper is to complete this work. Note, however, that the end of spreading does not mean the end diffusion as we explain now.

1.1 Difference Between Spreading and Diffusion

Indeed, "spreading" and "diffusion" were generally considered by physicists as equivalent when considering the behavior of wave packets. We apologize that we also added to this confusion in our early works on these topics, before we realized that this was not only a semantic problem. In order to abide by this updated definition, the word "diffusion" should be replaced by "spreading" everywhere in all our early papers concerning this topic.

We say that a wave packet with amplitude $A(\mathbf{r}, t)$ is spreading when its maximum amplitude $\sup_{\mathbf{r}} |A(\mathbf{r}, t)|$ goes to zero as time goes to infinity. Thus, on spreading means that the wave packet amplitude does not go uniformly to zero as $t \to +\infty$. This property does not imply that the amplitude has a non vanishing limit, or even that it does have a limit. Actually, in our situation, the wave packet has no limit.

Diffusion originally concerns particles moving randomly. Their dynamics is intrinsically chaotic since at each time, if the particle is localized at some position $\mathbf{r}$, its position as the next time is not completely determined but chosen according to some probabilistic law. Then we can define the probability density $\mathcal{P}(\mathbf{r}, t)$ for the particle to be at position $\mathbf{r}$ at time t considering all possible realizations of the random walk starting for example from the origin at time zero. We say we have diffusion when $\sup_{\mathbf{r}} \mathcal{P}(\mathbf{r}, t)$ goes to zero as time goes to infinity.

Our problem here concerns the behavior of initially localized wave packets in nonlinear lattices at any finite dimension d, with sites labelled $\mathbf{i} \in \mathcal{Z}^d$ and a discrete linear spectrum. First we conclude that, in all cases, $M(t) = \sup_{\mathbf{i}} E_{\mathbf{i}}(t)$ does not go to zero at infinite time and thus wave packets are never spreading. Then two different behaviors may occur. Either the wave packet is a stationary regular solution which remains localized around its initial condition at all time. Or its dynamics looks chaotic while always remaining focused around one or few peaks which look randomly wandering through the lattice. This motion suggests a kind of random walk but it is not if one only consider a single realization.

Since the behavior of each realization is strictly determined by its initial condition (unlike for a random walk), for defining a probability density $\mathcal{P}(E_{\mathbf{i}}(t))$ for the energy distribution of a wave packet as a function of time (which have a physical meaning), we need to define a "good" measurable set of initial conditions for doing a statistics. When the phase space has infinitely many dimensions, no physical measure can be defined in it (unlike the standard Liouville measure for finite dimension hamiltonian system). The most common approach is to use the fact that the system is disordered that is its Hamiltonian is defined by an infinite set of uncorrelated random numbers chosen according to some probability law. Then considering a given initial conditions for a wave packet, it is possible to define the probability (in the space of disorder realizations) that it generates either a non chaotic stationary regular trajectory or a chaotic wave packet with a (sub) diffusive behavior. Then $\mathcal{P}(E_{\mathbf{i}}(t))$ can be defined as an average over the disorder realizations restricted to those which generates a chaotic wave packet. With this definition we claim $\lim_{t\to+\infty}\sup_i \mathcal{P}(E_{\mathbf{i}}(t) = 0$ so that we can say we have diffusion without spreading.

This kind of statistics has already been used in many numerical simulations up to now. The averaged results suggested subdiffusion of wave packets but could not claim that wave packets could not spread. Indeed this property should appear for each disorder realization and becomes invisible when averaging over many disorder realization. An equivalent approach (which has the advantage it could be also used in non random systems such as quasiperiodic systems with discrete linear spectrum) is to make the statistics from the infinite set of Hamiltonians generated from a single Hamiltonian realization and all its translated Hamiltonian in the lattice i.e. that is we make a statistics of the dynamical behavior for the same initial wave packet located at all sites of the same random realization.

Note however that using this statistics has the flaw that the wave packet has not the same energy for all realizations. Actually, for a given disorder realization, the detailed diffusive process of a wave packet depends on its initial energy as well as its initial location and shape. Thus we may choose modified probability measure according the problem to understand. For example Fröhlich, Spencer and Wayne [27] define the probability to have a KAM torus with respect to disorder not for an arbitrary initial wave packet but as the probability (with respect to the disorder realization) that an arbitrarily given wave packet (determined at the integrable limit by its set of actions using the angle representation) be continuable from this integrable limit and thus remains a regular stationary wave packet. It is also possible to consider different projected measures with only one disorder realization by choosing the initial conditions of the wave packet in a (measurable) submanifold with finite dimension in the phase space as we did numerically is [8] for only on site initial wave packets with variable amplitudes. This method also has a flaw which is that we only explore a negligible part of the phase space.

1.2 Discussion of the Boltzmann Ergodic Hypothesis

We go back to the original theory for statistical mechanics which was mostly developed by Boltzmann for Hamiltonian systems and do not consider other possible more recent approaches for statistical mechanics. Boltzmann assumed that most Hamiltonian dynamical systems with many degrees of freedom mostly generate chaotic trajectories so that their physical behavior cannot be described in a deterministic way. He proposed the well-known Boltzmann Ergodic Hypothesis (BEH) which assumes that most trajectories generated by a large but finite Hamiltonian system, are chaotic and uniformly dense in the phase subspace at constant energy (microcanonical ensemble) considered with the standard Liouville measure. The Boltzmann entropy is then $k_B \ln W$ where W is the accessible volume in the phase space. In other words, for most trajectories, after a long enough time, the probability that the system visits an arbitrary region of the phase space becomes simply proportional to its volume.

Most means that the trajectories which do not fulfill this property have zero measure inside this subspace and thus can be neglected in the statistics. Actually there are many special trajectories (e.g. periodic or otherwise) among chaotic trajectories which are not ergodic. These trajectories, however, are assumed to have zero measure and can be neglected in the global statistics. This hypothesis is fundamental for constructing the well known theory of statistical mechanics.

This hypothesis is by far not obvious. Indeed there are also special Hamiltonian systems (called integrable) the trajectories of which are not chaotic but quasiperiodic and consequently do not constitute ergodic systems. For example, commonly used harmonic systems are integrable. However, it was believed in such cases that infinitely small hamiltonian perturbations completely break integrability and restore full ergodicity. Since the Hamiltonian of real systems cannot be perfectly known, it is expected that generally arbitrarily small unknown perturbations should restore the validity of statistical mechanics. In any case, a weak coupling with a thermal bath (with a continuous spectrum) forces thermalization. Thus, whatever is the initial state of a real physical system even far from thermal equilibrium, it was believed it should relax to a thermal equilibrium at some temperature after some transient time, in other words it maximizes its entropy.

Fermi even proposed a rigorous proof for ergodicity thus proving the validity of BEH for it but it was erroneous [10]. Later, the Fermi Pasta Ulam Tsingou model raised an apparent paradox in chaos theory which is still under debate. On the contrary, the pioneering work of Kolmogorov [11] proved that the ergodic hypothesis was wrong at least for some finite Hamiltonian near an integrable limit. Later, Arnold and Moser extended these results and proved that very generally integrability is not completely destroyed under small but arbitrary perturbations and consequently that a finite Hamiltonian system, assumed to be perfectly isolated, may not thermalize spontaneously.

Our problem of relaxation of a wave packet looks similar: We consider an *infinite* system in its ground state in which we inject a *finite* energy as a localized wave packet initial condition. Consequently, if the system relaxes to thermal equilibrium and since the energy density is zero, its temperature at complete thermalization should be strictly zero. Thus the wave packet would spread to zero which is the general belief. Actually, the wave packet cannot spread completely but instead it may become diffusive.

The reason is that the KAM theory (KAM is the acronym of the names of the main pioneering contributors: Kolmogorov, Arnold and Moser), initially proven only for finite size systems, may remain valid in the vicinity of integrable Hamiltonian for infinite systems on lattices under two conditions which are: (1) The linearized spectrum is purely discrete and (2) the energy of the wave packet is finite.

2 Dynamics of Hamiltonians with a Finite Number of Degrees of Freedom

We briefly review the main results and some conjectures about KAM theory for perturbed integrable systems which highlight the role of resonances. For details, there are many textbooks and reviews in the literature about this topic. We briefly review the main known results.

2.1 KAM Tori in Finite Nearly Integrable Hamiltonian Systems

Nearly integrable Hamiltonian systems exhibit quasiperiodic as well as chaotic trajectories. According to Liouville, an integrable Hamiltonian involving $n + n$ degrees of freedom (i.e. described by n pairs of conjugate variables) is characterized by the existence of n independent time invariants. The Liouville-Arnold theorem states that under a few extra assumptions (e.g. the boundedness of the constant energy submanifolds so that no trajectories go to infinity), there exists a canonical change of variables which define a new set of conjugate variables $\{I_i, \theta_i\}$ where I_i are real numbers called actions and θ_i are angles defined modulo 2π. With these new variables, the Hamiltonian becomes independent of the angles and only a function $H_0(\{I_i\}$ of the actions. Then, the Hamilton equations $\dot{\theta}_i = \partial H_0/\partial I_i$, $\dot{I}_i = -\partial H/\partial \theta_i$ implies that the actions I_i are time invariant and the angles are rotating uniformly with frequencies $\omega_i(\{I_j\}) = \partial H_0/\partial I_i$. Then, most trajectories of such a system are quasiperiodic with n fundamental frequencies which depend generally on the actions.

However there is a dense subset of so–called resonant invariant tori, for which there exists a set $\{k_i\}$ of n integers so that

$$\sum_{i=1}^{i=n} k_i \omega_i = 0 \tag{1}$$

Perturbations of such integrable Hamiltonians which may be written as

$$H = H_0(\{I_i\}) + \epsilon h(\{I_i, \theta_i\}) \tag{2}$$

where ϵ is a small parameter and $h(\{I_i, \theta_i\})$ an arbitrary Hamiltonian, Then we know from Poincaré, that in general all resonant tori (1) and a neighborhood of them (resonance gap) are destroyed and replaced by trajectories which are often chaotic, or periodic, some of which are linearly stable (with nearby KAM tori).

However, it might be that for very special perturbations $h(\{I_i, \theta_i\})$, resonance gaps do not open for some resonant tori (when for example $h(\{I_i, \theta_i\})$ does not depend on the angles $\{\theta_i\}$ in some region of the phase space). We must conclude, therefore, that the existence of chaotic resonance gaps is a generic property, which means it might not be fulfilled in rare and special models. Actually the mathematical definition of the word "generic" which can be found in the literature (e.g. a property valid in Baire subsets), may not be physically acceptable in some cases. We do not debate here this question and explain what we mean by generic in the physical context we consider.

Despite the existence of infinitely many resonance gap, all trajectories of the perturbed Hamiltonian do not become chaotic. The frequencies of the non resonant (incommensurate) tori which survive the perturbation have to fulfill a diophantine condition [10], i.e. that is there exist two positive constants α and $\tau > n-1$ such that for some set of integer $\{k_i\} \in \mathcal{Z}^n$, we have

$$|\sum_{i=1}^{i=n} k_i \omega_i| \geq \frac{\alpha}{|k|^\tau} \tag{3}$$

where $|k| = \sum |k_i|$ It is easy to prove that the set incommensurate tori which fulfills this condition, has full measure in the phase space.

Under the smoothness conditions that the perturbed Hamiltonian is an analytic function of its variables (or at least $n-1$ differentiable with continuous derivatives), the KAM theorem states that each of these incommensurate tori can be continued up to some critical value of the perturbation $\epsilon_c(\{\omega_i\})$ providing the Jacobian matrix $\{\frac{\partial \omega_i}{\partial I_j}\} = \partial^2 H_0(\{I_i\})$ be invertible. Moreover KAM theory, states that the global measure in the phase space of these surviving KAM tori goes to full measure as the perturbation parameter ϵ goes to zero.

KAM theory can be used in many situations and especially in the vicinity of linearly stable periodic orbits (or just stable fixed point) in any non integrable Hamiltonian. It states as a corollary that the quasiperiodic solutions obtained within the linear approximation near this periodic orbit and which are sufficiently close to it, survive as exact quasiperiodic solutions on condition of fulfilling again non resonant conditions (3). Since it can be shown that linearly stable orbits initially appear in the resonance gaps generated by the perturbation to integrability, new KAM tori appear, which do not exist for the integrable Hamiltonian. They are called *secondary* KAM tori, while those existing at the integrable limit are called *primary*.

Otherwise we expect that near these linearly stable periodic orbits, tori which are resonant or too close to them are destroyed and new resonant gaps open, where new linearly stable periodic orbit appear... and so on. Consequently, we obtain a complex landscape in the phase space with many primary and, secondary KAM tori which occupy a non vanishing measure of the phase space [12]. The complementary part is mostly occupied by unstable chaotic trajectories. It has not yet been rigorously proven that this part of the phase space has a non vanishing measure, though there is much evidence that this is true.

2.2 *Arnold Diffusion Conjecture and Extension*

Thus, a consequence of KAM theory is that a finite Hamiltonian system cannot spontaneously thermalize according to the Boltzmann statistics because of the existence many non ergodic KAM trajectories which occupy a non vanishing measure in the phase subspace **E** at constant energy E. We could, therefore, split the constant energy subspace $\mathbf{E} = \mathbf{K} \cup \mathbf{C}$ into two disjoint complementary measurable parts. The set **K** consists of all KAM tori which are linearly stable quasiperiodic trajectories. The complementary part **C** consists of the rest which are mostly linearly unstable trajectories, and are generally chaotic and ergodic. Many other non–chaotic trajectories like periodic, quasiperiodic (such as Cantori), whiskered tori, homoclinic and heteroclinic trajectories also exist, which all together have zero measure (However, knowing more about these special solutions could help as a "scaffold" to understand better the fine structure of **C**).

The set of KAM tori **K** is infinitely disconnected, that is given arbitrarily two KAM tori and a continuous path connecting two points of them, in generic cases, this path necessarily crosses resonance gaps which contain points not in **K**. It is a fat Cantor set (fat because it has non vanishing measure). We also say that this subset is *porous*.

When the number of degrees of freedom n is strictly larger than 2, KAM tori which have dimension n cannot split in two parts, the energy submanifold **C** which has dimension $2n-1$ (there is no inside and outside regions for a given torus). This implies that there are continuous paths remaining in **C** connecting two arbitrary points of the set **C** of chaotic trajectories. Consequently **C** is a connected set when $n > 2$. This set is supposed to also have non vanishing measure in **E**, which is not equal to the full measure of **E**, but nevertheless is dense everywhere in **E**.

Thus, the conjecture of Arnold diffusion [13] claims that when this set **C** is connected,($n > 2$) that given an arbitrary finite ordered set of open balls in the phase space, there exists in the generic case, a trajectory which visit these open balls in the same order when the system is near enough from its integrable limit. Arnold gave a proof of this only in a special example, while a generic proof was given only recently [14].

Here, we propose a more general conjecture which extends the Arnold diffusion conjecture. We call it the Boltzmann-Arnold Ergodic Hypothesis because it is none other than the Boltzmann ergodic hypothesis restricted to the only ergodic subspace **C** accessible to chaotic trajectories.

Conjecture: Boltzmann-Arnold Ergodic Hypothesis *When the number* n *of degrees of freedom of the Hamiltonian is strictly larger than* 2*, then in generic situations, the subset* **C***is ergodic that is the average in time of a physical quantity over a given trajectory is equal (with probability* 1*) to the same quantity averaged over* **C** *with its induced Liouville measure.*

The induced Liouville measure on **C** is a different measure of the whole subspace **E** when the measure of all KAM tori has been removed. As well as the original Liouville measure, it is invariant under the Hamiltonian flow.

Ergodicity implies that most trajectories in **C** are dense everywhere in **C** (and consequently in the whole subspace **E** of the phase space at constant energy E). But note that this conjecture is not a direct consequence of the original Arnold conjecture. Indeed, it is possible that most the trajectories of a measure preserving dynamical system be dense everywhere in the phase space without being ergodic. There are few and rare examples where this kind of behavior may be found for example for billiard in polygons (which can be associated discontinuous twist maps) or for some symbolic dynamics [15]. However, we do not consider these very special behaviors as relevant for the dynamical systems we consider here because they are continuous and differentiable enough so that they can exhibit KAM tori.

In the case where no KAM tori would exist, the conjecture becomes identical to the original BEH.

This mathematical statement about the statistics for the distribution of states after a very long time, does not say anything about the time needed to achieve this statistics. This hypothesis needs a physical interpretation because the subset **C** contains an infinite number of tiny resonance gaps where, according to the Nekhoroshev theorem [16], the dynamics is almost the same as those of the neighbouring tori over very long times. Consequently diffusion through **C** becomes very slow in such regions, while it is much faster in regions with strong chaos. The escape time from such regions (and hence the dense filling of other regions) may become so long that their observation becomes numerically impossible. Otherwise, note these tiny gaps, where the trajectories could remain trapped for very long times, have negligible measure compared to the global measure of **C**.

2.3 Anti Integrable Limit

We propose an approach different of the earlier proofs where we explain why we believe this extended conjecture is right. Our arguments are not complete but in any case proving or disproving our conjectures would be an important help for understanding the Arnold diffusion conjecture.

We know from KAM theory that regular (quasiperiodic) trajectories may exist in Hamiltonian systems because they dawn near integrable limits from which a large subset of them can be continued (under some diophantine conditions). We also know from Poincaré that (non regular) chaotic trajectories may appear near resonances. We believe as a counterpart that all non regular (chaotic) trajectories may dawn at different (singular) limits we call anti-integrable. At such a limit, all trajectories should be no more deterministic but purely random (and thus chaotic) essentially determined by choosing randomly a sequence of numbers among some discrete set of numbers (as for example for a random walk). However, such trajectories can be continued (under some conditions) away from the anti-integrable limit thus remaining still chaotic but now obeying the deterministic dynamical equations of the system.

Up to now, the existence of chaotic trajectories in a deterministic Hamiltonian dynamics was proven long ago by pioneering works of Poincaré about the studies of homoclinic or heteroclinic orbits near unstable fixed points. We can also prove their existence from the concept of anti-integrability [19] for symplectic maps. However, it is generally not easy to identify such a limit in Hamiltonian with a continuous time. It is thus necessary to introduce dynamical systems with discrete time analogous to Hamiltonian systems. Such dynamical systems are known as symplectic maps and exhibit most features existing in continuous time Hamiltonian systems.

Symplectic maps were already introduced by Poincaré when studying Hamiltonian flow at $n+n$ dimensions. He defined a return map (called Poincaré map) near a periodic orbit which maps a submanifold $\mathbf{S}$ with dimension $2n-2$ into itself. This submanifold is defined as the transverse intersection $\mathbf{S} = \mathbf{M} \cap \mathbf{E}$ of two submanifolds with dimension $2n-1$, where $\mathbf{E}$ is the manifold with dimension $2n-1$ at constant energy (identical to those of the periodic orbit) and $\mathbf{M}$ is an arbitrary manifold with dimension $2n-1$ which intersects transversally $\mathbf{E}$ (not tangent) and such that the periodic orbit intersects $\mathbf{S}$ at some point P. By continuity a trajectory $(\mathbf{u}(t), \mathbf{p}(t))$ with arbitrary initial point close enough to P generates a trajectory close to the periodic cycle. Consequently, there exists a neighborhood $\mathbf{V} \subseteq \mathbf{S}$ of P so that the trajectory generated by $(\mathbf{u}(0), \mathbf{p}(0)) \in \mathbf{V}$ intersect again $\mathbf{S}$ at some strictly positive finite time $t_1 > t_0$ that is $(\mathbf{u}(t_1), \mathbf{p}(t_1)) \in \mathbf{S}$. Considering the smallest positive value of t_1, the Poincaré map $T : \mathbf{V} \to \mathbf{S}$ is then defined as $(\mathbf{u}(t_1), \mathbf{p}(t_1)) = T((\mathbf{u}(0), \mathbf{p}(0))$. This map is continuous by construction and moreover Poincaré have shown that it is symplectic in its domain of definition (subsequently he used symplectic representation for proving theorems the existence of chaotic trajectories. from the existence of homoclinic or heteroclinic points). A generating function (defined in the next) can be easily defined for such return map from the original Lagrangian action with continuous time which yields the trajectories as extrema.

However, the return symplectic map can describe over long time only the trajectories of the continuous Hamiltonian which never escape from $\mathbf{V}$ but sometime can't for many others. Thus in order to sidestep this problem, it is simpler to study directly symplectic map defined in a whole space $\mathbf{S}$ and with a generating function.

Some symplectic maps can be defined from (non unique) generating function $L : (\mathbf{u}, \mathbf{u}') \to L(\mathbf{u}, \mathbf{u}') \in \mathcal{R}$ where $\mathbf{u} \in \mathcal{R}^{n-1}$ and $\mathbf{u}' \in \mathcal{R}^{n-1}$ on condition of invertibility of the matrix of crossed derivatives $-\bar{\bar{\partial}}_{12} L(\mathbf{u}, \mathbf{u}')$ with respect to the components of $\mathbf{u}$ and of $\mathbf{u}'$) [18]. We also assume for convenience that this matrix is strictly positive with a bounded norm. This condition reduces to the twist map condition in the lowest dimensional case where $\mathbf{u}_i$ are scalar numbers may be considered as generalized twist map condition in higher dimension.

Then, a symplectic map is obtained from the set of equations fulfilled by the extrema of the formal sum

$$\mathcal{A} = \sum_i L(\mathbf{u}_i, \mathbf{u}_{i-1}) \tag{4}$$

which should define a map $(\mathbf{u}_i, \mathbf{u}_{i-1}) \to (\mathbf{u}_{i+1}, \mathbf{u}_i)$ equivalent to a symplectic map. $\partial\mathcal{A} = 0$, yields

$$\bar{\partial}_2 L(\mathbf{u}_{i+1}, \mathbf{u}_i) + \bar{\partial}_1 L(\mathbf{u}_i, \mathbf{u}_{i-1}) = 0 \tag{5}$$

where $\bar{\partial}_1 L(\mathbf{u}, \mathbf{u}')$ denotes the derivative vector of L with respect to the first set of variables $\mathbf{u}$ and $\bar{\partial}_2 L(\mathbf{u}, \mathbf{u}')$ with respect to the second set of variables $\mathbf{u}'$. $\mathbf{u}_{i+1}$ is uniquely determined from the knowledge of $\mathbf{u}_i, \mathbf{u}_{i-1}$, since we assumed above that $-\bar{\bar{\partial}}_{12} L(\mathbf{u}, \mathbf{u}')$ is always invertible. Note that changing $L(\mathbf{u}, \mathbf{u}')$ into $L(\mathbf{u}, \mathbf{u}') + F(\mathbf{u}) - F(\mathbf{u}')$ where F is an arbitrary function does not change (4) so that it generates the same map.

Then, setting $\mathbf{p}_i = \bar{\partial}_1 L(\mathbf{u}_i, \mathbf{u}_{i-1})$ as the conjugate variable of $\mathbf{u}_i$, Eq. (5) determines a map $\{\mathbf{p}_{i+1}, \mathbf{u}_{i+1}\}$ as a function of $\{\mathbf{p}_i, \mathbf{u}_i\}$ which is the initial symplectic map.

We may choose an action angle representation so that the components of $\mathbf{u}_i$ are angles. Then the model is invariant by global rotations of by multiples of π that is if $\{\mathbf{u}_i\}$ is a trajectory, $\{\mathbf{u}_i + 2\pi\mathbf{m}\}$ where $\mathbf{m}$ is any vector with integers components, is also a trajectory This condition implies that this generating function L has the periodicity property

$$L(\mathbf{u}, \mathbf{u}') = L(\mathbf{u} + 2\pi\mathbf{m}, \mathbf{u}' + 2\pi\mathbf{m})$$

It is convenient to redefine a new function $A(\mathbf{u}, \mathbf{u}' - \mathbf{u}) = L(\mathbf{u}, \mathbf{u}')$ so that $A(\mathbf{u}, \mathbf{y}) = A(\mathbf{u} + 2\pi\mathbf{m}, \mathbf{y})$ is periodic with respect to the first set of variables only. We then obtain a generalization of the well known twist map which maps a cylinder onto itself. Thus studying only the properties of discrete symplectic maps is a very nice method for obtaining properties of continuous time Hamiltonians (note that all the possible invariants of the initial Hamiltonian beside the energy can be removed by this method, each invariant removing a degree of freedom of the initial Hamiltonian).

Within this representation, two interesting limits appear: The first situation is when the generating function $A(\mathbf{u}, \mathbf{y})$ only depends on its second variable that is when $L(\mathbf{u}, \mathbf{u}') = W(\mathbf{u}' - \mathbf{u})$ where W is a convex function because we assumed $-\bar{\bar{\partial}}_{12} L$ strictly positive. Then the solution of Eq. (5) yields $\mathbf{u}_i - \mathbf{u}_{i-1} = \omega$ independent of the discrete time i where the rotation vector ω can be arbitrarily chosen. Then the conjugate momenta $\mathbf{p}_i$ are constant in time as well. This is an integrable limit near which KAM theory holds for non resonant tori were ω have to fulfil a Diophantine condition (3).

The other limit we call anti-integrable, is obtained when A depends only on its first variable when

$$L(\mathbf{u}, \mathbf{u}') = V(\mathbf{u}) \tag{6}$$

where V is 2π periodic with respect to each of the component of vector $\mathbf{u}$. Then Eq. (5) becomes just $\bar{\partial} V(\mathbf{u_i}) = 0$. The periodic function V have necessarily a finite number of extrema (including maxima, minima and others like saddle points) within a single periodic unit which are denoted $\mathbf{b}_\nu$ and of course infinitely many others which are the same modulo 2π. Then all the solutions of Eq. (5) have the form

$$\mathbf{u_i} = \mathbf{a}_i + 2\pi \mathbf{m_i} \tag{7}$$

where for each i, $\mathbf{a}_i$ is chosen arbitrarily among the set of vectors $\mathbf{b}_\nu$ as well as the integer components of vector $\mathbf{m_i}$. Note at this limit Eq. (5) does not determine a symplectic map, but it does as soon there is a small perturbation $A(\mathbf{u}, \mathbf{y}) = V(\mathbf{u}) + \epsilon B(\mathbf{u}, \mathbf{y})$ of the generating function which involves both variables. Then, it is easy to prove that each solution of (7) random or not, are continuable versus ϵ by the implicit function theorem up to a nonzero value of ϵ providing two conditions [19]. The first condition is that the extrema the periodic potential (6) are not singular that is the Jacobian matrix $\bar{\bar{\partial}}^2 V$ is invertible for $\mathbf{u_i} = \mathbf{b}_\nu$. The second condition is that for the considered solution (7), there exists a strictly positive number B so that

$$|\mathbf{m}_{i+1} + \mathbf{m}_{i-1} - 2\mathbf{m}_i| \leq B \quad \text{for all} \quad i \tag{8}$$

(Note that for a random set of $\mathbf{m}_i$ fulfilling (8)), the sequence $r_i = \mathbf{m}_{i+1} - \mathbf{m}_| \ \mathbf{m}_i$ may be viewed as a random walk of a particle in a n dimensional square lattice where each jump $\mathbf{r}_{i+1} - \mathbf{r}_i$ occurs randomly by integer steps in any direction but with a maximum length smaller or equal to B). A solution of eq. (5) is continuable when the Jacobian operator $\bar{\bar{\partial}} \mathcal{A}$ is invertible, which according to [20] is equivalent to say that the associated trajectory is uniformly hyperbolic. Consequently this set of chaotic trajectories has zero measure. Each of these chaotic trajectories fulfilling condition (8) at the integrable limit can be continued till a certain threshold where $\bar{\bar{\partial}} \mathcal{A}$ ceases to be invertible. Since these trajectories may be viewed as extrema of the generating function (which has infinitely many variables), the disappearance of an extrema should be associated to bifurcations.

Indeed, it was numerically observed in the standard map (where vectors $\mathbf{u}$ becomes scalars) that the continuation of these chaotic trajectories generally disappear through bifurcations, where they annihilate one another by pair. Pitchwork bifurcation may be also observed for hyperbolic trajectories which has a spatial symmetry. Of course our observation were done for large but finite systems. When the smaller value of B for which Eq. (8) is fulfilled for all i, is large, the bifurcations occur near the anti integrable limit, while they becomes more robust when B is smaller. These bifurcating trajectories are generally chaotic trajectories (with positive Lyapounov exponent) which are no more uniformly hyperbolic.

However, as numerically observed in the standard map, few of them where the sequence m_i and a_i are chosen non random in a special way where the smallest possible parameter in (8) is $B = 1$, can be continued without bifurcations from the anti-integrable limit till the opposite integrable limit. For example, this is observed for periodic orbits which are in minimum action configuration [21] and are obtained by choosing the set of integers m_i according to a well defined rule and a_i constant at the unique minimum of V [19]. The same rule applied to well-chosen quasiperiodic orbits yields Cantori (also called Aubry-Mather sets) which continue beyond their bifurcation as (primary) KAM tori till the integrable limit.

However, these special configurations do not undergo standard bifurcations, but a so called "breaking of analyticity". Beyond this point the trajectory is no more uniformly hyperbolic but becomes a KAM torus which require a more sophisticated proof (KAM theory) for continuation. We can also prove that there still exists (secondary) KAM tori arbitrarily close to the anti integrable limit. They appear because it can be proven that many periodic orbits must become linearly stable in some interval before reaching their bifurcation. Such a behavior requires to fulfil a simple parity rule concerning the choice of a_i [19] so that we can say that a large fraction of the periodic cycles obeys this rule. Though these trajectories become elliptic periodic cycles, they remain continuable with the constraint they remain periodic orbits but are no more uniformly hyperbolic trajectories because they are necessarily surrounded by secondary quasi periodic KAM tori (with non vanishing measure) in the linear stability interval near the bifurcation. These intervals can be chosen arbitrarily close to the integral limit by choosing the minimum constant B in (8) large. However, it does not prove that all these interval overlaps, that is for any map near the integrable limit, there are always secondary KAM tori so that the measure of the chaotic trajectory is never the full measure of the phase space. We conjecture however that this assertion is true.

Continuation from the anti integrable limit, generates only uniformly hyperbolic trajectories. Complications may appear when there are several possible anti integrable limits in models with more than one parameter which may generate entanglement of bifurcations between trajectories with different origin. However we suggest that there are simpler symplectic maps easier to study (such as perhaps the standard map), where all uniformly hyperbolic trajectories can be generated just by continuation from a unique anti integrable limit.

It is known that the set of uniformly hyperbolic trajectories **H** characterized by a finite gap in the spectrum of their Jacobian matrix [20], has necessarily zero measure in the phase space. However, we conjecture that the closure of this set (as well as the subset of uniformly hyperbolic periodic cycles) which includes zero gap non uniformly hyperbolic trajectories, is generically identical to the whole phase space of trajectories. The generic condition is that the set of KAM tori **K** is porous (that is when the Arnold diffusion conjecture should hold). Of course this conjecture does not hold at the integrable limit where uniformly hyperbolic trajectories do not exist anymore or when the system remains semi integrable in some domains (see the end of Sect. 5 for our definition of semi integrability). Thus in the generic case, we expect a dense scaffold of uniformly hyperbolic trajectories (extending the set of the trajectories which corresponds only to absolute minima of the generating function (4) we studied earlier for twist map on the cylinder). Then varying the model parameter from the anti integrable limit, the uniformly chaotic trajectories should disappear gradually through bifurcations at which they are no more uniformly hyperbolic but remains chaotic but now with finite measure in the phase space. Some special non chaotic trajectories at the anti integrable limit should undergo special bifurcations where they become KAM tori (primary or secondary).

More studies and proofs are needed to confirm (or disproof) these conjectures. In any case we believe that starting from anti -integrable limits may be a good approach for proving the earlier mentioned Boltzmann-Arnold diffusion conjecture at least in some class of simple symplectic maps.

3 Dynamics of Hamiltonians on Infinite Lattices

We consider now wave packets Hamiltonian on infinite square lattices at arbitrary finite dimension. If the system is spatially periodic, the linear spectrum is purely absolutely continuous with bands and delocalized eigen modes. When the system is random, the linear spectrum may become purely discrete but in higher dimension than 2, it may still contain an absolutely continuous part with mobility edges. When the linear spectrum contains an absolutely continuous part and when a wave packet is introduced in the nonlinear system, the dynamics of the wave packet cannot be quasiperiodic or chaotic because the Fourier spectrum of its dynamics which is dense, necessarily involves frequencies in the absolutely continuous part of the linear spectrum which radiates energy. However, the wave packet may generate a periodic solution if its frequency and all its harmonics do not overlap the absolutely continuous part of the linear spectrum. We may obtain stationary periodic solutions called Discrete Breathers (DB) [22–24]. When DBs exist, two well-known situations may occur for an arbitrary initial wave packet depending on its initial conditions. Either the initial wave packet spreads to zero or it converges to a Discrete Breather. In the later case, a part of its energy spread while the other part remains localized. Sometime we get a transiently mobile DB.

The situation which was poorly understood, is the topics of this paper. It occurs when the linear spectrum is purely discrete without any continuous part that is when the linear spectrum is not dissipative [25].

3.1 *KAM Tori in Infinite Hamiltonian Systems on Lattices with Linear Discrete Spectrum*

A common belief [26] is that as the number of degrees of freedom of a Hamiltonian increases, the relative measure of the KAM tori goes to zero. However this statement is too vague because it does specify how the infinite size limit is taken. Actually, this statement is likely true only in the thermodynamical limit, that is when the energy of the system diverges proportionally to its size. This is not the situation here since the initial wave packet energy remains finite.

In a pioneering work, Fröhlich, Spencer and Wayne (FSW) [27] proved rigorously that Anderson localization may persist in the presence of nonlinearities. Their rigorous proof holds only for some special class of models chosen so that the complex KAM machinery is easier to implement. They consider a random system on a square lattice at arbitrary dimension, with strong disorder where the eigenstates are mostly localized at single sites, and chose a quartic nonlinear perturbation, which couples only nearest neighbor oscillators. $i : j$ denote the bonds of the lattice between site i and its neighboring site j counted once since $i : j$ is equivalent to $j : i$.

The FSW Hamiltonian has the form

$$H_{FSW} = \sum_{i \in \mathcal{Z}^d} h_i(u_i, p_i) + \sum_{i:j} W_{i:j}(u_i, p_i, u_j, p_j) \tag{9}$$

where $h_i(u_i, p_i) = \frac{1}{2}p_i^2 + \frac{1}{2}\omega_i^2$ is the Hamiltonian of a linear eigenmode at site i of d dimension lattice arbitrary which has the random frequency ω_i. The nonlinear perturbation is the nonlinear coupling $W_{i:j}(u_i, p_i, u_j, p_j)$ only between neighboring oscillators $i : j$. It is an analytic function with respect to its variables which expands at order 4 at the lowest order.

FSW Theorem *Choosing arbitrarily a strongly localized linear solutions (i.e. decaying faster than exponential see* [27]*) of the linearized Hamiltonian which thus is quasi periodic in time, the probability that there exists an exact quasiperiodic (non resonant solution) of the perturbed Hamiltonian near this arbitrarily chosen linear solution (for the L_2 topology), is non vanishing when the amplitude of the initial solution becomes small enough. Moreover this probability goes to* 1 *as the amplitude goes to zero. The perturbed solution is also quasi periodic with perturbed frequencies.*

To be more precise, the linear solution is defined at the uncoupled limit by choosing arbitrarily but with fast decay, its action I_i for each oscillator i. Since there is an infinite number of oscillators, this solution at the integrable limit necessarily involves an infinite countable number of fundamental frequencies and harmonics a priori non resonant. Thus the terminology quasi periodic is not appropriate, we should say that almost periodic is the proper term as explained in the next. Note that in infinite systems the non resonance condition defined by Eq. 3 does not involve an infinite number of frequencies but only any arbitrarily finite subset of frequencies (otherwise this condition would be meaningless).

The probability of existence of these KAM solutions is defined with respect to the random distribution of frequencies ω_i, which are uncorrelated and distributed according to some probability law. FSW also conjecture that their theorem remains still valid in general for lattice Hamiltonians with short range interactions whose linear spectrum exhibits full Anderson localization, and also when the restrictions requiring fast decay for the initial wave packet are removed.

When the disorder becomes weaker, the maximum localization length of the eigenmodes increases so that the nonlinear interactions between these eigenmodes extend significantly much beyond the nearest neighbors. Then, although the interactions between them still decay exponentially as a function of the distance, model (9) is no longer a good model because the nonlinear interaction should extend beyond the nearest neighbours while remaining short ranged. However, we do not see any convincing reason to say that in that situation, this theorem does not remain true. We expect, however, that the small amplitude region where KAM tori becomes very dense smoothly shrink to zero as the longest localization length increases. The reason is the dynamical behavior should exhibit a kind of global continuity because, as mentioned above, KAM tori cannot exist anymore when the linear spectrum contains an absolutely continuous part.

3.2 *Numerical Observation of KAM Tori in Large Non Integrable Hamiltonian Systems*

The FSW theorem suggests that the existence of KAM tori as localized solutions in nonlinear infinite lattices with purely discrete spectrum is a ubiquitous phenomenon. In [8], we proposed an easy numerical method which does not require too much time consuming calculations. It is based on the theory of Almost Periodic Functions pioneered by Harald Bohr in 1924 [29].

Bohr Definition and Theorem *A function $f(t): \mathcal{R} \to C$ is said to be almost periodic when for any $\epsilon > 0$, there exists $L(\epsilon) > 0$ so that in any interval with length $L(\epsilon)$ there exists a pseudo period τ so that $|f(t+\tau) - f(t)| < \epsilon$ for all t.*

If $f(t)$ is bounded uniformly continuous and almost periodic, then $f(t)$ can be written as a generalized Fourier series $f(t) = \sum_n f_n e^{i\omega_n t}$ which is absolutely convergent and where $\{\omega_n\}$ is a countable set of frequencies.

Almost periodic functions involve an infinite number of arbitrary fundamental frequencies unlike quasiperiodic functions which involve only a finite number of fundamental frequencies (and their harmonics). In finite systems, KAM theory prove that there only quasi periodic KAM tori with a number of fundamental frequencies equal to the dimension of the torus. In that situation each of the coordinates of the solution is quasi periodic. In infinite system, the FSW theorem proves the existence of KAM tori we may still call quasiperiodic but actually must be almost periodic solutions. The two terminologies often tend to be confused. In any case, we do not see any very good reason to make a formal distinction. We still prefer to always use the terminology quasiperiodic meaning sometime almost periodic.

The above Bohr's theorem can be used to find KAM tori in large systems only by testing for Poincaré recurrences as we explain now.

The Poincaré recurrence theorem states that in a finite Hamiltonian system, a given trajectory return with probability 1 into any arbitrary neighborhood of its initial condition (of course after leaving this neighborhood in case of continuous time). Thus for most trajectories and an arbitrary neighborhood of their initial conditions, we can define an unbounded monotone increasing sequence of times t_n where $\lim_{n\to+\infty} t_n = +\infty$ and $\lim_{n\to-\infty} t_n = -\infty$ so that the considered trajectory returns at time t_n into this neighborhood. This property looks similar to those required by the Harald Bohr condition, except that the distribution of return times $t_{n+1} - t_n > 0$ is not necessarily bounded. However, considering trajectories so that for any initial neighborhood, there exist some constant T so that $0 < t_{n+1} - t_n < T$ is bounded for all n (we called this property weak periodicity [28]), is not enough to prove quasiperiodicity The reason is that the Bohr theorem uses the uniform topology not the weak topology as for Poincaré recurrence. Nevertheless we can prove that if the Poincaré recurrence time interval of a trajectories are bounded, this trajectory is minimal that is its closure (which is invariant under the Hamiltonian flow) does not contain any strictly smaller closed invariant subset.

Actually minimal trajectories which are linearly unstable (with positive Lyapounov exponents), are sensitive to the initial conditions (and imbedded in chaotic regions) cannot be observed numerically over long time (by integration from fixed initial conditions). Only the minimal trajectories which are linearly stable (with zero Lyapounov exponents) could be (approximately) observed numerically over reasonably long time. Since KAM tori (or possibly lower dimension linear stable quasiperiodic tori) are the only known minimal trajectories which are linearly stable, we may assume that if we observe Poincaré recurrence into any neighborhood of the initial condition a given trajectory so that the intervals of time between consecutive returns are bounded, le trajectory is likely a KAM torus.

Otherwise for testing numerically the possible existence of KAM tori, it is not necessary to test all coordinates of the trajectory as a function of time but a few of them since the Hamilton equations relate these functions of time one with each other. For example in 1D model knowing that the coordinates of two consecutive oscillators are almost periodic implies recursively that all coordinates are almost periodic. Finally in practice, testing only one coordinate is enough (and thus computer saving). If we find this coordinate is quasiperiodic, testing any of the other coordinates show they are also quasiperiodic. Otherwise it makes sense that the generic behavior of a hamiltonian dynamical system is that all coordinates behaves all together either quasiperiodic or chaotic (Nevertheless, non generic exceptions can be found in semi integrable models such as the Ding-Dong models Sect. 5.

In summary, our numerical method based of Bohr theorem is used in a very loose way for making it easy to use. It is thus rather efficient and fast to roughly discriminate between chaotic trajectories and KAM tori and get an intuitive idea of the phase space. But of course because of the unavoidable numerical errors, numerics cannot strictly discriminate either between weakly chaotic trajectories (near KAM tori) and these KAM tori. Otherwise, we may have trajectories which look chaotic for a relatively long time but which actually would generate a high order secondary KAM torus. It is thus hopeless to analyse the fine structure of the phase space at microscopic scale with this kind method. The GALI method [30] which has been recently proposed has the same flaws due these unavoidable numerical integration errors.

Using these argument, we searched in a Random DNLS model in 1D for wave packets localized initially at a single site [8]. We confirmed that for a given disorder realization, there are indeed many wave packets mostly at small amplitude which seems to fulfill the Bohr condition, i.e. they are recurrent many times over all the numerically observed evolution times.

We also observed "sticky" trajectories which appear Bohr recurrent over a time beyond which recurrences completely disappear. They can be interpreted as trajectories generated by initial conditions which are close to real KAM tori. The KAM region as expected from the Nekhoroshev theorem [16, 17] estimates the perturbation growth near quasiperiodic integrable solutions. This growth is algebraic and becomes slower and slower as the initial condition is closer and closer to an integrable limit that is close to some KAM tori.

Because of Arnold diffusion, such trajectory which is not a true KAM torus finally escapes from the resonance gap and rejoins the main chaotic region where it becomes fully chaotic. When this escape time becomes longer than the computing time, this effect is not numerically observable. On the other side, the width of the corresponding high order resonance gap also drastically goes to zero. Thus in practice we observe that the region of KAM tori at small amplitude looks connected while we expect the existence of infinitely many tiny resonance gaps.

We also observed Bohr recurrent trajectories for large amplitude wave packets which may be interpreted by the fact that initial conditions are close to linearly stable discrete breathers (which are known exist in that model at large enough amplitude). Studying special trajectories like periodic cycles, homoclinic or heteroclinic orbits, whiskered tori etc. which occupy a zero measure in the phase space in finite Hamiltonian, would require different methods, variational or else.

We also noticed that the trajectories which are not Bohr recurrent exhibit chaotic behavior. We checked that the trajectories that look like KAM tori, according to this Bohr criterion, have zero Lyapunov exponent, within numerical error, while the all others exhibit a chaotic trajectories with non zero Lyapunov exponent. Actually, the Bohr recurrence method is quite efficient for discriminating numerically between KAM trajectories and chaotic trajectories.

Recently, we became aware of recent works [30] which use a different method for discriminating between the chaotic trajectories and the others (KAM tori) in a different model (random discrete KG models). Actually they got basically the same conclusion as ours in the random DNLS [8] that is the probability to find regular trajectories (or KAM tori) goes to unity at small amplitude. Discriminating between localized and spreading chaos is another question discussed the next subsection.

4 Long Time Behavior of Wave Packets

When the initial wave packet generates a KAM torus, the long time behavior is known since its dynamics is stationary and quasiperiodic. There is no spreading and no diffusion at all. When it does not generate a chaotic trajectory, we consider our infinite systems as usual in physics as a finite system involving N sites where N becomes large.

The extended Arnold diffusion conjecture described above holds that most chaotic trajectories should visit densely the whole set **C** of chaotic trajectories which is itself dense in the whole phase subspace **C** at constant energy, despite having no full measure. We propose first a criterion which could be used numerically at least in systems that are not too large. We define the probability density of the wave packet in **C** as a function of its participation number.

4.1 A Criterion for the Long Time Behavior of a Wave Packet

Assume that a wave packet with energy E is spread over the whole system (whatever its dimensionality) which consists of a large number of sites N, while its energy density per site is about its average E/N and goes to zero as N grows to infinity. Near that limit, the energy of the wave packet is mostly obtained from the harmonic terms in the Hamiltonian, while the contribution to the energy of the nonlinear terms goes to zero. Consequently, the perturbation from integrability being very small, most wave packets in that regime should generate KAM tori. Consequently the measure of the trajectories which are both in **C** and have a small amplitude becomes negligible. On the contrary, wave packets which are not spread too much have non negligible probability to be in **C**.

In order to quantify this effect, we propose to use a quantitative criterion which measures how much a given wave packet has spread. The most commonly used criterion, i.e. the inverse participation number of the wave packet energy seems to be a convenient choice. To define it, we split the global Hamiltonian H of our system into a sum $H = \sum_i H_i$ of local hamiltonians H_i at the lattice sites which can be chosen positive and vanishing when the corresponding oscillator is at rest. We may choose for example, H_i as the Hamiltonian of the isolated nonlinear oscillator at site i plus half of its interacting potentials with its neighbors. Thus, we can define the inverse participation number of a wave packet as

$$S = \frac{\sum_i H_i^2}{(\sum_i H_i)^2} \tag{10}$$

In finite systems with size N, we only have $\frac{1}{N} \leq S \leq 1$ since the participation number cannot be larger than the size of the system. In the limit of infinite size N, this inequality becomes $0 \leq S \leq 1$.

Then, we define a probability density $P(S)$ for arbitrary configurations in **C** with the fixed energy E and a given participation number S. More precisely

$$P(S)dS = \frac{\mu(B([S, S+dS]))}{\mu(\mathbf{C})} \tag{11}$$

where $\mu(B([S, S + dS]))$ is the measure of the subset $\mathbf{B}([S, S + dS])$ of wave packets in the phase subspace $\mathbf{C}$ which have an inverse participation number in the small interval $[S, S + dS]$ while $\mu(\mathbf{C})$ is the total measure of $\mathbf{C}$.

If one assumes that there are no KAM tori that is $\mu(\mathbf{K}) = 0$ in the phase subspace $\mathbf{E}$ and consequently $\mu(\mathbf{C}) = \mu(\mathbf{E})$, then the BEH would hold in the whole space $\mathbf{E}$ instead of $\mathbf{C}$. We assume that the initial wave packet thermalizes. Instead of calculating $P(S)$ within the microcanonical ensemble, it is simpler to use Boltzmann statistics in the canonical ensemble where the system is not strictly isolated. Then we have to fix a different temperature $k_B T_N = 1/\beta_N$ for each system with size N in order that the total energy average of the whole system $\mathbf{E}$ be equal to the energy of the wave packet.

Since $e^{-\beta_N E_i}/Z(\beta_N) = \beta_N e^{-\beta_N E_i}$ is the Boltzmann probability energy density for each oscillator, we have $< E_i >= 1/\beta_N$ and $k_B T_N = E/N$. The total energy $\sum_i E_i$ of this system is not absolutely fixed, as it should be, but its expected fluctuation $< (\sum_i E_i - E)^2 >^{1/2}= E/\sqrt{N}$ goes to zero at large size N. We neglect this fluctuation. Then we can calculate the average of S defined by (10) which is $< S >= \int P_0(S) S dS = S_m = 2/N$ and $< (S - S_m)^2 >^{1/2}= \sqrt{20}/N$. This result shows that $P_0(S)$ exhibits a thin peak located very near zero, which becomes a Dirac peak $\delta(S)$ when $N = +\infty$.

The curve $P(S)$ becomes drastically different from $P_0(S)$ because most initial wave packets with a small participation number S generate KAM tori and have to be removed from the statistics. Thus, we expect that $P(S)$ exhibits a peak roughly at the participation number S_m which corresponds to the average cross over amplitude $A_m \approx E S_m$, below which the density of KAM tori sharply increases. (Note that because of model disorder, this cross-over depends on the location of the wave packet in the system. A_m is thus an average). Thus when the wave packet energy E increases, its peak at S_m moves toward smaller values proportionally to $1/E$. The shape of $P(S)$ should depend on the model and the amplitude of the disorder which determine its KAM tori. For example, we expect that A_m (as well as S_m for the same energy) becomes smaller for smaller disorder when the localization length increases, which does not favor the existence of KAM tori and decreases A_m.

We thus conjecture that $P(S)$ should look roughly like a single peak as shown in Fig. 1 with a maximum located at S_m not near zero but somewhere else in the interval $0 < S_m < 1$.

We think that this probability density $P(S)$ should be numerically calculable with a reasonable accuracy at least for small systems, which would yield more quantitative information about it. One method could be to make the statistics of the initial inverse participation number S over many random choice of wave packets at the same energy E discarding all initial wave packets which look KAM tori (actually in finite systems KAM do exist). Another method would be to integrate one of these initial chaotic wave packet over very long time and look at the distribution of their participation number over time. According to the Arnold–Boltzmann hypothesis, the two methods should be equivalent. However it may be more efficient to combine the two methods by repeating the time statistics method over many initial wave packets

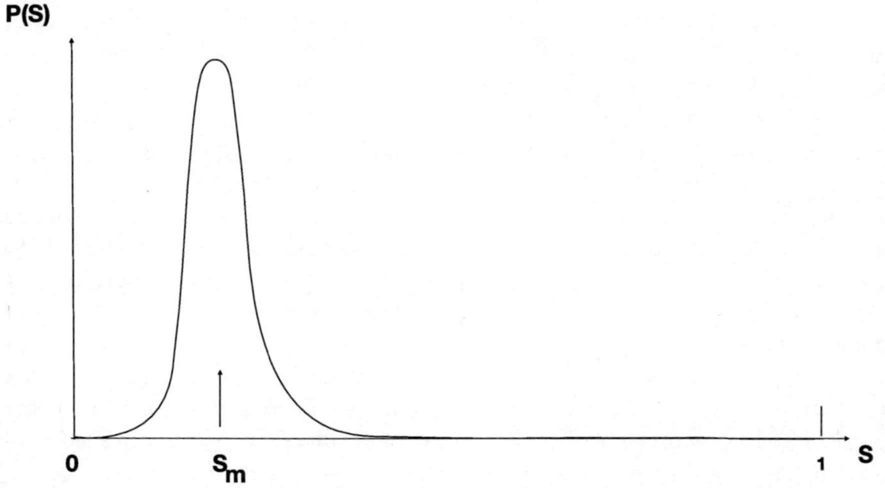

Fig. 1 Sketch of the expected probability density $P(S)$ of chaotic wave packets (in **C**) versus their inverse participation number S in the limit of large system. Its maximum is obtained for a participation number S_m. The shape of this curve depends on the model and its nonlinearities but for a given model S_m should be also roughly proportional to the inverse energy E of the wave packet. $P(S)$ can also be defined for finite systems but then it depends on the chosen disorder. However, its fluctuations should disappear in the limit of an infinite system

chosen randomly in **C**. Increasing the size as much as possible could give a more precise idea of $P(S)$ at infinite size...

Actually, up to now most numerical experiments were performed for initial wave packets systematically located only at a single site, or better few sites or a single linear mode. Their inverse participation numbers near unity were generally very different from the expected most probable value roughly around S_m at the pseudo thermal equilibrium in **C**. Consequently it should be necessary to wait for a certain transient time (which may be very long) to reach the core of **C** where the behavior of the wave packet becomes more typical. Otherwise, although the dynamics of the inverse participation numbers was investigated, no statistical study of its distribution was made to my knowledge...

4.2 Discussion of the Long Time Behavior of Wave Packets

The above studies and conjectures imply two essential conjectures about the long time behavior of finite energy wave packets (not rigorously proven but supported by strong arguments):

Conjecture 1 *In an infinite nonlinear lattice Hamiltonian system with a purely discrete linear spectrum (i.e. with Anderson localization), the probability that a wave* *packet with finite energy spreads (i.e. its maximum amplitude modulus goes to zero), is zero.*

In physical terms it just never spread. Consequently at any given time the wave packet is spatially localized in the sense that its inverse participation number is always non vanishing. However this condition does not imply that the wave packet always remain focused within a single peak. It may also break into a finite (only) number of smaller wave packets. In any case, since the wave packet does not spread, we expect it behaves almost the same as in (large enough) finite system over long time (till it reaches the boundary of this system). In any large finite system, the focused wave packet should be wandering through the lattice visiting the whole phase space if one assume the Boltzmann conjecture which says that most chaotic wave packet in **C** should go at some time arbitrarily close to any other wave packet in **C**. Since we have energy diffusion through any large enough finite system whatever is its size and we should get diffusion in the infinite system as well.

However this conclusion does not imply that the Boltzmann diffusion conjecture can be extended the system size is strictly infinite. The first reason first is that the Liouville measure is no more defined for infinite systems so that we have to find another appropriate definition for the word probability as we already noticed above. The second reason is it is really possible that the wave packet does not explore the whole phase space at constant energy (considered with the l_2 norm since we study wave-packets with finite energy thus square summable).

For illustrating our claim, we consider an intrinsically chaotic dynamical system (non Hamiltonian) the trajectories consist of random walks on a square lattice at arbitrary dimension. Though we believe that simple random walks are not good models for describing the wave packet diffusion they yield interesting informations. The phase space in which evolve the random walk just consists of all possible locations on the square lattice. This is the standard model for diffusion which yields the well-known diffusion laws from the statistics over all possible realizations of the random walk. The well-known Polya theorem (proven by combinatorial analysis) asserts that the probability $P(d)$ that a given random walk on a square lattice returns to the origin after some finite time is unity when the dimension of the lattice is $d = 1$ or 2. but at three dimensions and more, $P(d)$ is non vanishing but is no more unity.

A simple empirical physical argument allows to recover this result and more. It is well-known that after N steps, the random walk is mostly confined into a sphere with a radius proportional to $\sqrt{N}$. Since the volume of this sphere at d dimension is proportional to $N^{d/2}$ and since a random walk at time N visits only N sites (at most), at long time it is clear that when $d > 2$ there are too many sites to visit in the sphere so that considering only a single random walk realization, many sites of the lattice are never visited. However considering the statistics over many realizations, all sites have a nonvanishing probability strictly smaller than 1 to be visited (depending on their distance from the origin). Unlike for Arnold diffusion, a single chaotic trajectory may not visit the whole phase space when the dimension of the lattice is strictly larger than 2. However if we consider any finite connected part of the lattice (phase space) whatever its dimension, a single random walk fill densely the whole phase pace with probability one. Finally we conclude that Arnold diffusion conjecture is surely true in that model for any finite system and if the system is infinite but if its dimension is larger than 2, it is surely wrong.

Since we cannot say that the random walk is not diffusive in lattice with dimension larger than 2, we suggest a definition for having a diffusive wave packet which is acceptable in physics which avoid this problem. We require the condition of compact support for the wave packet in order to avoid spurious initial conditions where the finite energy wave packet is already broken into an infinite number of smaller wave packet spread over the whole system.

Definition of Diffusive *A wave packet is diffusive in a finite system if it generates a chaotic trajectory dense in the phase space (fulfilling the Arnold diffusion conjecture). A wave packet with compact support is diffusive in the infinite system if there exists* $L_0 > 0$ *so that it is diffusive in all finite connected subsystems defined by* $\|\mathbf{i}\| < L$ with $L > L_0$.

In other words we assume (as done in practice in numerical simulations) that the dynamics of the wave packet in the infinite system can be well approximate over arbitrary long time in finite subsystems providing they are large enough. Actually physical systems are always finite though they could be very large but we approximate them as infinite systems in theoretical models.

A simple observable numerical manifestation of the diffusivity of a wave packet is that the variation of the second momentum of the energy distribution as a function of time looks unbounded. Then we suggest the next conjecture valid for infinite systems:

Conjecture 2 *With the same conditions as for the above conjecture 1, wave packets with a given finite energy may generate either stationary KAM tori (in* $\mathbf{K}$*) or diffusive chaotic trajectories (in* $\mathbf{C}$*) both with finite probability (This probability is defined with respect to the disorder realization).*

Diffusion in our model is different from standard diffusion described by random walks. Actually we expect that the rate D of diffusion of the wave packet depends both on its inverse participation number S (which fluctuate as a function of time) and of its spatial location because of disorder. When S is larger than or of the order of S_m, the wave packet is fully chaotic because KAM tori are relatively rare, hence we should get rather faster diffusion. On the other hand, when S becomes smaller than or of the order of S_m, the KAM tori become denser, so that chaos becomes weaker and diffusion slower. It may also happen at smaller S that wave packets get trapped for a very long time in very weakly chaotic domains of tiny resonances gaps. Because of Nekhoroshev's theorem, the escape time from such quasi trapping region increases very fast by order of magnitudes as S becomes small and there is practically no diffusion during this trapping time.

However, though quasi trapping events last very long, they are also so rare so that the probability density $P(S)$ remains small for small S. Thus, the diffusion of the wave packet may be sometimes fast, slow or very slow depending not only on the variation of S but also on the local disorder. Such a behavior is reminiscent of a so called random walk in a random scenery. This fact may explain why standard energy diffusion is not observed, while slower diffusion with lower time exponents is found in many models both with disorder and nonlinearities.

Let us summarize now, the consequences of the above conjectures on the qualitative interpretation of the numerical observations on the wave packet behavior in lattice Hamiltonian with discrete linear spectrum:

1. For large energy wave packets, initially well focused and of large amplitude far above the KAM threshold which exists at small amplitude, the inverse participation number S of the wave packet is initially near unity while S_m is small. Thus its inverse participation should first decay to reach the region of maximum probability around S_m. This effect appears as a transient spatial spreading of the wave packet. After this transient time, the inverse participation number of the wave packet mostly fluctuates around its average value S_m (pseudo thermalization in **C**) while wave packet spreading stops on average. The wave packet remains chaotic but mostly at diffusive spots corresponding to the largest amplitude of the wave packet. We note that some aspects of certain recent numerical observations [31, 32] could be interpreted consistently within our predictive model.
2. When the initial wave packet still well focused has a smaller energy, S_m increases and becomes closer to 1. There is a certain probability that it generates a stationary KAM torus. When it does not, the transient spreading of the wave packet last a shorter time and chaos is weaker than in the large energy case.
3. When the initial energy of the wave packet is very small. In most cases, the wave packet is non chaotic and generates a quasiperiodic and stationary KAM trajectory similarly to the case without disorder. However, there is still a small probability that the wave packet be weakly chaotic near KAM tori. This is in the Nekloroshev regime which becomes numerically non observable. Thus there is practically no observable difference between the behavior of very small amplitude wave packets in the model with nonlinearities and those of the same wave packets in the same but linearized model. Indeed this feature was already mentioned in [3].
4. Another consequence of the fact that the complementary set **C** is dense everywhere which means that giving an arbitrary wave packet with energy E, either it belongs to **K** or it is possible to find arbitrarily close to it a wave packet with the same energy as in **C**. The consequence is that in principle spatially localized chaos cannot generically exist (in contradiction with numerical observations) [30].

Actually localized chaos is believed to be numerically observed [30]. We shall explain this observation in the next section where we study Ding Dong models which are not generic and where localized chaos is proven to exist.

Our general conclusions may not hold in some special models. Let us discuss first why our conclusions should be modified in case of non genericity in some special models. Our theory relies on the porosity of the set **K** of KAM tori at constant energy E described in subsection (2.2). However porosity may not always occur. It may be that in some region **R** of the phase subspace **E**, chaotic resonance gaps which generically open near resonant tori, do not do so and thus the KAM tori are locally dense and compact. Thus, in that region $\mathbf{R} \subseteq \mathbf{K}$, the nonlinear perturbation of the Hamiltonian preserves integrability (we call such an Hamiltonian semi-integrable) so that the set $\mathbf{C}$ of chaotic trajectories is not dense everywhere in $\mathbf{E}$. Moreover it may become disconnected because the connectivity argument also relies on porosity. Consequently, if the phase space $\mathbf{C}$ does break into many disconnected subsets $\mathbf{C}_n$, full ergodicity in $\mathbf{C}$ will no longer hold. However, ergodicity may still persist separately in each of the connected subset $\mathbf{C}_n$.

The Ding Dong (DD) model we discuss now is an example of a non generic model in the sense that it is semi- integrable as defined above. Its advantage is that exact results can be obtained on it and in particular we prove the first Conjecture 1 of this section. We also prove that $\mathbf{C}$ is disconnected unlike the generic case which implies the second conjecture does not hold for DD models. This is explained because of the absence of porosity of the regular (i.e. KAM) trajectories, since KAM barriers are not porous and hence the existence of localized chaos becomes possible. We also explain how a tiny perturbation of this model can restore its genericity.

5 Exact Results for Ding Dong Models

The dynamics of wave packets in special models belonging to this class, was numerically studied by Pikovsky [33]. It was observed that the behavior of such wave packets was also subdiffusive as well as in the other models quoted above earlier.

5.1 Definition

The hamiltonian form of the Ding Dong (DD) model is that of a modified FSW model where the nonlinear coupling is replaced by hard core potentials. It consists of an array of random anharmonic oscillators on a lattice of arbitrary dimension d with nearest neighbor hardcore coupling. We choose square lattices for simplicity but our results would hold for any networks even random, on the condition that the coordination number (number of bonds starting from a given site) be finite and bounded. Consider the Hamiltonian

$$\mathcal{H}_{DD} = \sum_i \left(\frac{1}{2m_i} p_i^2 + V_i(u_i)) + \sum_{i:j} W(u_j + b_{i:j} - u_i \right) \tag{12}$$

where the smooth anharmonic local potentials $V_i(x)$ are chosen randomly and expands $V_i(u_i) \approx \frac{1}{2} m_i \omega_i^2 + \cdots$. We assume the existence of a positive constant K so that

$$K x^2 \leq V_i(x) \quad \text{for all } i \quad \text{and} \quad x \tag{13}$$

so that the lowest order of the expansion never vanishes. The masses m_i and linear frequencies ω_i are uncorrelated random numbers, with smooth probability density bounded in intervals $0 < m_{min} \leq m_i \leq m_{max}$ and $0 < \omega_{min} \leq \omega_i \leq \omega_{max}$.

The interaction potential $W(x)$ is a hardcore potential

$$\begin{aligned} W(x) &= 0 \quad \text{for} \quad 0 \leq x \\ W(x) &= +\infty \quad \text{for } x < 0 \end{aligned} \tag{14}$$

$\{b_{i:j}\}$ is a given collection of positive numbers. These numbers are assumed to be randomly distributed in an interval $0 < b_{i:j} < b_{max}$, which may include 0 according to an arbitrary probability density but without divergence at zero. This condition has the advantage to allow the existence of chaotic trajectories in the model for very small energies as well as in the generic case where the coupling potentials are smooth and not semi-integrable.

Consequently, the algebraic distance between two neighboring oscillators $d_{i:j}(t) = u_j(t) - u_i(t) + b_{i:j}$ has to remain positive or zero for all trajectories. During some interval of time between two collisions, each oscillator i has a time constant energy $E_i = \frac{1}{2m_i} p_i^2 + V_i(u_i) \geq 0$ and oscillates periodically with a frequency $\Omega_i(E_i)$, which generally depends on its energy (except if we choose the potential to be harmonic). When a collision occurs within one bond $i : j$ among the $2d$ bonds connected to i (that is when $d_{i:j}(t)$ vanishes at some time t_c), a standard elastic collision occurs between the two neighboring oscillators i and j where its energy (and frequency) changes discontinuously according to the conservation laws of the bond energy $E_i + E_j$ and momenta $p_i + p_j$ at the collision time.

5.2 *Regular Trajectories and a Lemma*

Exact results about (non) wave packet spreading are rather easy to obtain on this class of models as a consequence of the following lemma which is straightforward to prove:

Lemma *Considering a bond $i : j$ assumed isolated from the rest of the system, the two oscillators i and j cannot collide if $E_i + E_j < B_{i:j}$ where $B_{i:j} = \min_x(V_i(x) + V_j(b_{i:j} - x))$*

Because of the inequality (13), we have

$$\min_x K(x^2 + (b_{i:j} - x)^2) = \frac{K}{2} b_{i:j}^2 < \min_x (V_i(x) + V_j(b_{i:j} - x)) = B_{i:j} \tag{15}$$

so that the collection of random numbers $B_{i:j}$ is bounded from below by uncorrelated number $\frac{K}{2} b_{i:j}^2$

As a remark this lemma is also a necessary condition for having a pair of non colliding oscillators with the extra condition that the frequencies $\Omega_i(E_i)$ and $\Omega_j(E_j)$

of these two oscillators be incommensurate, that is $n_i\Omega_i(E_i) + n_j\Omega_j(E_j) \neq 0$ for any choice of integers n_i and n_j. Otherwise, there may exist non colliding solutions at larger energy for example when the two oscillators i and j oscillates at the same frequency and in phase. In any case, we only need to use the lemma in the next as a sufficient condition.

Considering now the infinite system, we attach a strictly positive number $B_{i:j}$ at each bond $i : j$. We thus have a collection of random numbers only correlated between nearest neighbor bonds $i : j$ and $i : k$ or $k : i$ sharing a common lattice site. We could remove this correlation by choosing $m_i\omega_i^2$ constant for all i but this is not necessary. As a corollary of the above lemma, we obtain

Corollary *If at a given time, the oscillator energies E_i of a trajectory generated by the DD Hamiltonian (12), fulfill the inequalities*

$$E_i + E_j \leq B_{i:j} \quad \text{for all bond} \quad i : j \tag{16}$$

then no collision occurs between any couple of neighboring oscillators anywhere and at any time past and future so that this condition holds at all time. Moreover, condition (16) is also generally necessary for having no collision.

However, when the oscillator frequencies $\Omega_i(E_i)$ and $\Omega_j(E_j)$ are commensurate (i.e. their ratio is a rational number), condition (16) is no longer necessary since the bound $B_{i:j}$ can be chosen larger by choosing appropriately their relative phase. Such situations can be neglected, however, since commensurability has probability zero to occur.

When inequalities (16) is fulfilled, each oscillator is oscillating periodically with its own frequency $\Omega_i(E_i)$ independently one from the other, and thus the global solution is almost periodic. We consider them as KAM tori which are also almost periodic solutions. Thus, considering in the phase space $\{p_i, u_i\}$, conditions $E_i + E_j < B_{i:j}$ for all $i : j$ and condition $\sum_i E_i = E$ determines a bounded closed set $\mathbf{R} \subseteq \mathbf{K} \subset \mathbf{E}$ which contains $\{0\}$.

Conversely, if only one (or more) of the inequalities (16) are not fulfilled at some arbitrary time, the considered trajectory necessarily exhibits at least one collision and then it can be readily proven that it must exhibit an infinite number of collisions in the past and in the future. Numerical simulations suggest that most colliding solutions are not almost periodic but are chaotic. Consequently, the two sets $\mathbf{R} = \mathbf{K}$ are equivalent neglecting only a zero measure subset. The complementary set in $\mathbf{E}$ of colliding trajectories where most trajectories are chaotic, define the set $\mathbf{C}$.

5.3 Some Exact Results

Using the above lemma and inequalities (15), we readily obtain:

Theorem 1 *No wave packet in a DD model (12) can spread*

Proof Let us assume that we find a wave packet with finite energy which would spread. Then for any $\epsilon > 0$, there should exist a time $t(\epsilon)$ so that $0 < E_i(t) < \epsilon$ for all i and all time $t > t(\epsilon)$.

A bond $i : j$ where inequalities (16) is fulfilled, cannot transfer any energy during some time. It is called blocked. When ϵ is chosen small enough, the local energies E_i become small enough so that there are many bonds (but not all of them) which become blocked for all time $t > t(\epsilon)$. When ϵ decreases the density $D(\epsilon)$ of blocked bonds increases and reaches 1 as ϵ goes to zero. Simultaneously, the density $1 - D(\epsilon)$ of unblocked bonds decreases and reaches a percolation threshold for a non vanishing $\epsilon < \epsilon_p$ where they only form disconnected and finite clusters. The consequence is that for $t > t(\epsilon_p)$, the energies confined in each of the finite clusters on unblocked bonds should remain constant and thus cannot go to zero. Thus it is impossible at this stage that for $t > t(\epsilon_p)$, the wave packet continue to spread. Consequently complete spreading is impossible for any wave packet with finite energy.

For defining a percolation threshold, we assumed that the random numbers $B_{i:j}$ in (16) are uncorrelated which is not perfectly true. To be absolutely rigorous, we could consider a smaller set of blocked bonds defined by the weaker condition $E_i + E_j < B^{\star}_{i:j} = \frac{K}{2} b^2_{i:j}$ (which implies (16)) but where $B^{\star}_{i:j}$ are uncorrelated numbers. The set of unblocked bonds defined by this condition becomes larger, thus it percolates more easily, which implies that the original set of unblocked bonds also exhibit their percolation transition for a larger critical value of ϵ.

We define a critical energy E_c, small enough and chosen such that all the bonds where $B_{i:j} < E_c$, do not percolate that is there are only finite clusters C_n of unblocked bonds disconnected on from each other by the blocked bonds where $B_{i:j} > E_c$. Then if we choose an initial wave packet with energy $E < E_c$ and also initially excited oscillators only inside one of the finite clusters of unblocked bonds, it is impossible that during its time evolution, this wave packet could transfer any energy outside this cluster since it is surrounded only by blocked bonds where $E_j + E_l < E < B_{k:l}$.

The dynamical behavior of the wave packet could be either quasi periodic or chaotic. It suffices for that to focus enough energy only on the two sites i and j of a single bond inside the cluster so that $E = E_i + E_j > B_{i:j}$. Consequently, the wave packet will necessarily involve collisions and be chaotic in general.

One can do the same trick when distributing the initial energy E over several bonds inside the same cluster but then it is necessary to choose appropriate finite clusters C_n where there are several (connected) neighboring bonds where $B_{i:j}$ are small enough. This is always possible to find since the lower bound of $B_{i:j}$ is zero.

Whatever the initial energy $E > E_c$ of the wave packet, it is possible to split it into a finite sum $E = \sum_k E_k$ where $0 < E_k < E_c$ and to distribute each component of the energy E_k on different clusters as explained above. Consequently, we have the theorem:

Theorem 2 *For any choice of the initial energy E, there exist wave packets at this energy which generate either quasiperiodic trajectories, or localized chaos (which may consist of one or several chaotic spots). This result implies that the set* $\mathbf{C}$ *of chaotic trajectories is not fully connected.*

However, it may still be true that when the initial energy E of the wave packet is large enough, a connected subset $\mathbf{C}' \subset \mathbf{C}$ of wave packets which generates delocalized chaos should exist and be predominant for large initial energy. We have no proof of this statement but it is necessary to assume its validity to understand why subdiffusion has been observed at least in some DD models [33]. Otherwise, when $E < E_c$ the above arguments prove that spreading wave packets cannot exist.

It is possible to calculate the probability density $P(S)$ of the inverse participation number S in $\mathbf{C}$, but since it is not connected, $\mathbf{C}$, it does not yield useful information about the long term behavior because an initial wave packet cannot visit the whole space of chaotic wave packets at the same energy. Calculating $P_n(S)$ for each connected component $\mathbf{C}_n$ of $\mathbf{C}$ would be more useful.

Other exact statements can be easily proven using the above lemma:

Theorem 3 *Let us consider a DD model (12) at some non vanishing temperature $k_B T = 1/\beta$, and a random configuration chosen according to the Boltzmann statistics, then the probability that this configuration be quasi periodic is zero.*

This theorem implies that the existence of regular trajectories does not play any role for the thermal equilibrium. It confirms that most trajectories are colliding solutions (likely chaotic) so that the BEH is true and that we have spontaneous Boltzmann thermalization. Another theorem can be proven for confirming that the dynamical behavior of a finite energy and chaotic wave packet cannot be described according to Boltzmann statistics in the full phase space, but within a statistics restricted to each of the zero measure ergodic subset $\mathbf{C}_n$ we proposed above:

Theorem 4 *Let us consider a sequence of finite size N DD models at non vanishing temperature $k_B T_N = 1/\beta_N$, chosen in order that the average total energy E (finite and non vanishing) of each system is the same and equal to E independently of N. Then the probability Q_N according to the Boltzmann statistics that a configuration generates a quasi periodic trajectory, goes to unity when N goes to infinity.*

5.4 Non Genericity of DD Models

Finally, let us explain in this subsection why our results concerning the existence of localized chaotic wave packets in DD models, seems to disagree with the generic results described in the previous section. Actually arbitrarily small perturbations of the Hamiltonian may suppress semi- integrability and replace its generic behavior with a kind of physical continuity. For example, we may replace in the Hamiltonian (12), the singular hardcore potential $W(x)$ (14) by a smooth potential $W_\nu(x) = (1 + \tanh x)^\nu$ with $\nu > 0$ large so that $\lim_{\nu \to +\infty} W_\nu(x) = W(x)$. KAM theory holds in the modified model which is no longer semi-integrable.

Considering wave packets in the DD model with energy E, located at a blocked bond $i:j$ characterized $B_{i:j} > E = E_i + E_j$, we obtain only non colliding quasiperiodic solutions at energy E in the DD model. They are replaced in the modified model by quasi periodic KAM tori looking almost the same but infinitely many very tiny chaotic resonance gaps open near each resonant tori. As a result chaotic wave packets at energy E may occupy this impenetrable DD barrier. They may penetrate inside the barrier through this tiny resonance gaps thus becoming locally very weakly chaotic (Nekhoroshev diffusion). The consequence is that DD barriers can no longer stop energy transfer and consequently wave packet diffusion. However, although between these pseudo barriers, the dynamical behavior of wave packets is almost unchanged, the DD barriers persist as pseudo barriers the crossing time of which may become very long. In the limit $\nu \to +\infty$, that is when recovering the DD model, the crossing time of such pseudo barrier should diverge so that physical continuity with the DD model is restored.

6 Concluding Remarks

Our predictions have been developed essentially in order to be consistent with the known mathematical theorems and well believed conjectures without any numerical help. Thus we have only qualitative predictions to offer without quantitative information. Some of our conclusions agree with the numerical observations which were done up to now, while for others, there is no agreement. We summarize our main predictions versus numerical observations and encourage some of our colleagues to reexamine their interpretations.

1. An initial wave packet may generate a stationary quasi periodic solution with some non vanishing probability which increases and tends to unity at small amplitudes. Such trajectories are non chaotic and look the same as in the linear case. Despite their existence as exact solutions, they were not believed as important for the wave packet diffusion problem. Our early numerics deeply supported their existence [8] in the random DNLS model and more recent papers also observed such regular trajectories in different models and with the GALI method [30]. We emphasize that their existence is indeed essential for understanding the behavior of chaotic subdiffusive wave packets.
2. For the non stationary (chaotic) wave packets, we expect that subdiffusion characterized by the divergence of the second moment of the energy distribution as a function of time occurs as a consequence of the Boltzmann diffusion conjecture. Moreover, we have arguments which suggest that this wave packet diffusion may be similar to a random walk in a random scenery. This fact could explain a diffusion exponent different and smaller than for standard diffusion. Indeed, subdiffusion has been numerically observed in many models quoted in this paper.
3. We also predict that the wave packet cannot spread to zero which is equivalent to say the inverse participation number does not go to zero. We expect it fluctuates around an average value S_m (see Fig. 1). Though this statement may still be considered as a conjecture in general, it is rigorously proven to be true in the Ding Dong models where numerical observations done for some of them, exhibit

wave packet subdiffusion similar to those observed in the other nonlinear random models [33].

4. On the contrary, it has been claimed on the base of numerical observations [31], that the inverse participation number goes to zero with an exponent σ related to the subdiffusion exponent α But this is impossible if the wave packet does not spread.
5. It has been recently claimed on the basis of numerical observations [32] that chaos persists in the wave packet after a long time [32]. We expect that this result is a consequence of non spreading since the average value S_m of the inverse participation number does not vanish. Otherwise this observation is in contradiction with the previous statement that the inverse participation number would go to zero, since the nonlinear terms should become numerically negligible after a long enough time. As a consequence we should see the wave packet behave as if it were linear with very weak Nekhoroshev chaos practically non observable and no chaotic spots.
6. Note also that the measure of the initial chaotic wave packets (if any) asymptotic to a quasiperiodic KAM solution at infinite time is zero (the proof of this statement is a straightforward consequence of the Liouville theorem which states that a Hamiltonian flow conserves the volume in the phase space). Consequently, the probability that the dynamics of an initially chaotic wave packet becomes regular after a long time and non chaotic is zero conversely to expectations.
7. Finally we predict also the existence of a transient regime during which an initially well focused wave packet should reach a kind of thermal equilibrium according to the Boltzmann Arnold statistics described above in this paper. During this transient time, the wave packet exhibits both diffusion and spreading that is both the second moment and the participation number grow. After this transient time only the second moment continue to grow, while the inverse participation number stops to decay and fluctuates around its well-defined average value S_m corresponding to the maximum of curve Fig. 1. Up to now this transient regime has not been numerically identified, but this might have been due to a bias following the early interpretations of numerical observations. Otherwise, this relaxation time may become very long when the average inverse average participation number S_m is small and when the wave packet is initially well focused at a single site.

The cause of the absence of spreading of wave packets in the class of models considered here is essentially an entropic effect, that is the accessible volume in the phase space that corresponds to spreading wave chaotic packets becomes negligible compared to the accessible volume for the focused wave packets. This is different from the absence of spreading which may occur in models which have an extra time invariant (or more) beside the total energy and where the absence of spreading is due to topological constraints. An example are DNLS models (random or not) which conserve both energy and total norm. In such models any trajectories must stay both in the $(2n-1)$ dimensional manifold $\mathbf{E}$ corresponding to its constant energy E and the $(2n-1)$ dimensional manifold $\mathbf{N}$ corresponding to its constant initial norm N. When the ratio of the energy E/N is too large it turns out that the manifold $\mathbf{E} \cap \mathbf{N}$

does not contain {0} as an accumulation point [7] so that spreading becomes just impossible when the system size n goes to infinity. Otherwise, this subspace may contain a subset of KAM tori $\mathbf{K} \subset \mathbf{E} \cap \mathbf{N}$ as well as chaotic trajectories with Arnold diffusion in the complementary set $\mathbf{C} \subset \mathbf{E} \cap \mathbf{N}$.

Similar behavior to the one expected in the class of models considered here should also occur in other non random systems for example quasiperiodic systems [2]. Our universal conclusion is that spreading of wave packet cannot occur when the linear spectrum of the system is purely discrete and when nonlinearities are taken into account. Randomness is not necessary. In some sense, Anderson localization is not suppressed by non linearities in contradiction with early papers.

This work should suggest new questions about the thermal behavior of the same models at very low temperature when the average amplitude of the thermal fluctuations becomes smaller than the crossover amplitude, where KAM tori appear. Then the system should exhibit only few slowly diffusive chaotic spots (or chaotic breathers) surrounded by an ocean of regular non diffusive quasi periodic fluctuations. Such a situation should be associated with a dramatic drop of the thermal conductivity so that complete thermalization may become impossible within reasonable times. Numerical experiments of fast quenching in some nonlinear systems have revealed the spontaneous formation of (time periodic) discrete breathers slowing down the thermalization, although in these examples the linear spectrum was not discrete but completely absolutely continuous [34]. The same kind of numerical experiments in systems with discrete spectrum (for example random) should dramatically enhance the effect. This may indeed be the situation in real glasses, where many questions remain unanswered [35], even though the linear phonon spectrum in such bond disorder systems should not be purely discrete, but exhibit an absolutely continuous part corresponding to low frequency and long wave length acoustic phonons.

I acknowledge G. Tsironis and T. Bountis for interesting discussions and questions. I also thank again T. Bountis for inviting me to this meeting and with Ch. Skokos for communicating their very interesting recent papers right after my talk. I am also indebted to R. MacKay for reading in detail the manuscript and his many interesting comments helping improvements.

References

1. Anderson, P.W.: Absence of diffusion in certain random lattices. Phys. Rev. **109**, 1492–1505 (1958)
2. Aubry, S., André, G.: Analyticity breaking and Anderson localization in incommensurate lattices Ann. Israel Phys. Soc. **3**(133), 18 (1980)
3. Pikovsky, A.S., Shepelyansky, D.L.: Destruction of Anderson localization by a weak nonlinearity. Phys. Rev. Lett. **100**, 094101 (2008)
4. Flach, S., Krimer, D.O., Skokos, C.: Universal spreading of wave packets in disordered nonlinear systems. Phys. Rev. Lett. **102**, 024101 (2009)
5. Skokos, C., Krimer, D.O., Komineas, S., Flach, S.: Delocalization of wave packets in disordered nonlinear chains. Phys. Rev. E **79**, 056211 (2009)

6. Laptyeva, T.V., Bodyfelt, J.D., Krimer, D.O., Skokos, C., Flach, S.: The crossover from strong to weak chaos for nonlinear waves in disordered systems. EPL (Europhys. Lett.) **91**(3), 30001 (2010)
7. Kopidakis, G., Komineas, S., Flach, S., Aubry, S.: Absence of wave packet diffusion in disordered nonlinear systems. Phys. Rev. Lett. **100** (2008)
8. Johansson, M., Kopidakis, G., Aubry, S.: KAM tori in 1D random discrete nonlinear Schroinger model? EPL (Europhys. Lett.) **91**(5), 50001 (2010)
9. Aubry, S.: KAM Tori and absence of diffusion of a wave-packet in the 1D random DNLS model. Int. J. Bifur. Chaos **21**(08), 2125–2145 (2011)
10. Pöschel, J.: A lecture of the classical KAM theorem. In: Proceedings of Symposia in Pure Mathematics, vol. 69, pp. 707–732 (2001)
11. Kolmogorov, A.N.: On the conservation of conditionally periodic motions for a small change in Hamiltonian function. Dokl. Akad. Nauk SSSR **98** (1954)
12. Biasco, L., Chierchia, L.: Explicit estimates on the measure of primary KAM tori. Annali di Matematica **197**, 261–281 (2018)
13. Arnold, V.I.: Instability of dynamical systems with many degrees of freedom. Dokl. Akad. Nauk SSSR **156**(1), 9–12 (1964)
14. Cheng, C.-Q., Xue, J.: Arnold diffusion in nearly integrable Hamiltonian systems of arbitrary degrees of freedom (2019). arXiv:1503.04153v5
15. https://mathoverflow.net/questions/74279/example-of-a-measure-preserving-system-with-dense-orbits-that-is-not-ergodic
16. Nekhoroshev, N.N.: An exponential estimate of the time of stability of nearly integrable Hamiltonian system. Uspehi Mat. Nauk **32**(6(198)), 5–66, 287 (1977)
17. Bambusi, D., Langella, B.: A simple proof for a C^∞ Nekhoroshev theorem (2020). ArXiv:2002.06985v1
18. Meiss, J.D.: Symplectic maps, variational principles, and transport. Rev. Mod. Phys. **64**, 795 (1992)
19. Aubry, S.: The concept of anti-integrability: definition, theorems and application to the standard map. In: McGehee, R., Meyer, K.R. (eds.) Twist Mappings and Their Applications, pp. 7–54 (1992)
20. Aubry, S., MacKay, R.S., Baesens, C.: Equivalence of uniform hyperbolicity for symplectic twist maps and phonon gap for Frenkel-Kontorova models. Phys. D: Nonlinear Phenom. **56**, 123–134 (1992)
21. Aubry, S., Le Daëron, P.-Y.: The discrete Frenkel-Kontorova model and its extensions: I. Exact results for the ground-states. Phys. D: Nonlinear Phenom. **8**(3), 381–422 (1983)
22. MacKay, R.S., Aubry, S.: Proof of existence of breathers for time-reversible or Hamiltonian networks of weakly coupled oscillators. Nonlinearity **7**, 1623 (1994)
23. Aubry, S.: Breathers in nonlinear lattices: existence, linear stability and quantization. Phys. D: Nonlinear Phenom. **103**, 201–250 (1997)
24. Aubry, S.: Discrete breathers: localization and transfer of energy in discrete Hamiltonian nonlinear systems. Phys. D: Nonlinear Phenom. **216**, 1–30 (2006)
25. Aubry, S., Schilling, R.: Anomalous thermostat and intraband discrete breathers. Phys. D **238**, 2045–2061 (2009)
26. Froeschlé, C., Scheidecker, J.P.: Stochasticity of dynamical systems with increasing number of degrees of freedom. Phys. Rev. A **12**, 2137 (1975)
27. Fröhlich, J., Spencer, T., Wayne, C.E.: Localization in disordered, nonlinear dynamical systems. J. Stat. Phys. **42**, 247 (1986)
28. Aubry, S.: Weakly periodic structures and example. J. Phys. Colloq. **50**, C3-97–C3-106 (1989). https://doi.org/10.1051/jphyscol:1989315
29. See for example Brom, J.: The theory of almost periodic functions in constructive mathematics. Pac. J. Math. **70**, 67–81 (1977)
30. Senyange, B., Skokos, C.: Identifying localized and spreading chaos in nonlinear disordered lattices by the Generalized Alignment Index (GALI) method. Phys. D **432**, 133–154 (2022)

31. C.G. Antonopoulos, Ch. Skokos, T. Bountis , S.Flach *Analyzing chaos in higher order disordered quartic-sextic Klein-Gordon lattices using q -statistics* Chaos, Solitons and Fractals 104 (2017) 129-134
32. Skokos, C., Gerlach, E., Flach, S.: Frequency map analysis of spatiotemporal chaos in the nonlinear disordered Klein-Gordon lattice. Int. J. Bifur. Chaos **32**(5), 2250074 (2022)
33. Pikovsky, A.: Scaling of energy spreading in a disordered Ding-Dong lattice. J. Stat. Mech.: Theory Exp. **2020** (2020)
34. Tsironis, G.P., Aubry, S.: Slow relaxation phenomena induced by breathers in nonlinear lattices. Phys. Rev. Lett. **77**, 5225 (1996)
35. See for example Yu, C.C., Carruzzo, H.M.: Two-level systems and the tunneling model: a critical view. https://doi.org/10.48550/arXiv.2101.02787

Nonlinear Phenomena Shaping the Structure of Spiral Galaxies

P. A. Patsis

Abstract The structures observed in disk galaxies can be explained by the presence of nonlinear phenomena associated with dynamical mechanisms acting in their stellar and gaseous components. Successful models can reproduce the observed morphologies and their evolution in time. Here, I summarize, from a personal point of view, the basic results of nonlinear, orbital galactic dynamics, which explain the presence of bars and spiral arms in the disks. I also mention the main ideas that have been discussed in the field during the last sixty years and I refer to some open issues and alternative possibilities for structure formation in spiral galaxies.

Keywords Galactic dynamics · Hamiltonian systems · Orbital theory · Gas-response models

1 Introduction

Spiral galaxies are complex dynamical systems. Their global morphology is the result of dynamical processes taking place mainly in their stellar component (disk and bulge), in the gas that lies in the equatorial plane of the galaxy and in the dark matter halo that surrounds the disk. The stellar and the gaseous components interact among themselves, as well as with the dark matter halo. The evolution of each one of these components has to take into account the presence of the others and their dynamical evolution. In order to understand the dynamics of the structures, which are observed in this type of galaxies, we have to understand the global dynamics of a complex system.

The structures that appear in disk galaxies, are the bars, the spiral arms and the rings (nuclear, inner and outer). Galaxies with prominent and well-defined spiral arms are called "grand design". The presence of the spiral arms may be accompanied by

P. A. Patsis (✉)
Research Center for Astronomy and Applied Mathematics of the Academy of Athens, Soranou Efessiou 4, 11527 Athens, Greece
e-mail: patsis@academyofathens.gr

T. Bountis et al. (eds.), *Chaos, Fractals and Complexity*, Springer Proceedings in Complexity, https://doi.org/10.1007/978-3-031-37404-3_2

Fig. 1 The grand-design spiral galaxy NGC 5248 dominates in the lower left corner of the figure. In the upper part of the image are discernible a disk galaxy with a ring and a disk galaxy with an edge-on orientation. (Observation in B filter with the 2.2 m ESO/MPA telescope, La Silla, Chile, by Patsis, Heraudeau & Grosbøl, 2000)

the presence of a bar and so we speak about normal (non-barred) and barred spiral galaxies. A typical grand design example (NGC 5248) is given in Fig. 1.

A major contribution to the field came in the 1990s, with the development of near-infrared detectors. Observations in near-infrared wavelengths allowed the imaging of the old stellar population of the disk, which traces much better the mass distribution than observations in the optical. The conspicuous differences in the morphologies of a galaxy in near-infrared and optical images, gives valuable information to be used as input in theoretical modeling. Stellar models have to be compared with data from near-infrared observations, while gaseous models with morphologies encountered mainly in the optical.

Plausible assumptions that reduce the degree of complexity of a galactic system are necessary in order to be able to construct models that reproduce the dynamical behavior of galactic disks, remaining, to a large degree, realistic. There are two main ways of studying the dynamics of galaxies. Either by means of N-body simulations, or by means of orbital models. N-body models are self-consistent, combine the evolution of the stellar and gaseous components and offer the possibility to include a live dark matter halo (see [1] for a review). Although such models are the best way to describe the time evolution of galactic systems, it is difficult to study with them the details of the dynamical phenomena that are in action as the system evolves. For that purpose have been used orbital models, simple in their initial set up, in most cases in the form of autonomous Hamiltonian systems that refer to the stellar dynamics of the galaxy (for a complete introduction in the subject see [2]). The potentials used are either well behaving analytic functions that match general properties of galactic disks (see e.g. Chap. 2 in [3]) or, in some cases, potentials that have been estimated directly from near-infrared images of specific galaxies (e.g. [4, 5]).

A key element for understanding galactic disk dynamics is to find out the location of the resonances between the epicyclic frequency, $\kappa(R)$, and the angular velocity of the stars, $\Omega(R)$, in the rotating with pattern speed Ω_p frame of reference (R is the radial distance of a test particle, in cylindrical coordinates). Especially the resonances $\kappa/(\Omega(R) - \Omega_p) = \pm 2/1$ (Inner and Outer Lindblad resonances respectively), the 4/1 resonance (defined in a similar way as the 2/1 one), as well as corotation, a resonance for which $\Omega(R) = \Omega_p$, play a crucial role for understanding the dynamics of barred and spiral galaxies. These resonances are defined on the equatorial plane of the galaxy. However, in the same way, we can specify vertical resonances, between the vertical frequency of the stars, $\nu(R)$ and $\Omega(R) - \Omega_p$ (see e.g. [6]).

2 Order and Chaos

2.1 *Two Dimensional (2D) Models*

As their name indicates, disk galaxies are flat objects. Thus, two-dimensional (2D) modeling has been extensively used as a good approximation for their study. The initial idea was to associate ordered motion in the vicinity of stable periodic orbits with the reinforcement of morphological features. Stable periodic orbits trap around them regular orbits, which remain close to the periodic ones forever. In this way they enhance the local density and thus they enhance structures that have a certain similarity with the topology of the periodic orbits (see e.g. Chap. 2 in [2]). This is a straightforward scenario, that gave the following results:

- The most well appreciated result of nonlinear orbital theory in galactic disk dynamics concerns the orbital content of galactic bars. The bars of barred-spiral galaxies are supported by orbits trapped around stable periodic orbits of the family “x1”, the orbits of which have elliptical-like shapes [7]. Beyond the inner 4/1 resonance, towards corotation, the existing families of periodic orbits in rotating barred potentials are mainly unstable and practically are found within a chaotic zone. This zone prevents the bars reaching corotation [8]. In this case Order forms a structure and Chaos hinders its extent beyond a certain distance, approaching the region characterized by $\Omega(R) \approx \Omega_p$, in which we find the Lagrangian equilibrium points [9].
- In the absence of a bar, a bisymmetric spiral pattern is also supported by a backbone of elliptical x1 orbits, which however precess in a characteristic way, so that their apocenters are aligned with the loci of the spiral arms. In this way the stars stay

longer time in the apocenters regions and enhance locally the surface density of the disk, forming the arms. This is the idea of the classic density wave theory [10] expressed by means of periodic orbits (see Fig. 3 in [11]).

The essential parameters for assessing this hypothesis, are the pattern speed and the amplitude of the spiral perturbation, i.e. of the spiral arms. The pattern speed determines the location of the resonances and consequently the local morphology of the model, while the amplitude of the perturbation defines the degree of nonlinearity, in other words the importance of chaotic phenomena. The response morphology can be directly compared with images of galaxies, while the presence of chaotic phenomena affects, among others, kinematic features, such as the profile of the dispersion of velocities in the disks. Such profiles provide constraints for the appearance of chaotic phenomena. Both quantities (pattern speed and amplitude of the spirals) are very difficult to be estimated from observations. Thus, modeling is needed, so that the right values can be deduced by comparing the situation predicted by the models with the observational data.

In normal (non-barred) spiral models it has been realized that an open spiral structure has major problems crossing the 4/1 resonance region. Due to the rhomboidal shape of the orbits in this region and their relative orientation, the 4/1 resonance becomes a main obstacle for the continuation of the spiral structure towards corotation [12, 13]. This time it is not the presence of Chaos, but the misalignment of the building blocks (i.e. of the periodic orbits) that imposed the damping of the density wave. A set of self-consistency tests have shown that this is the case for the *symmetric* part of grand-design galaxies of Hubble types Sb to Sc. Such spiral patterns rotate slowly, so that the end of their symmetric parts corresponds to the 4/1 resonance, while the estimated amplitudes are characterized by perturbing forces of the order of 5–10% with respect to the axisymmetric background. On the other hand, the tightly wound arms of Sa galaxies, could be modeled with spirals with 1% perturbation in the forces and could reach corotation [14]. However, in both cases, i.e. in models with big and small pitch angles, order dominates and this is reflected to the observed velocity dispersions in real galaxies (see e.g. [15, 16]). The relation between pitch angle, amplitude of the perturbation and pattern speed is also recently investigated in [17].

Gaseous response models have confirmed the above results. In addition, the inclusion of asymmetries in the imposed potentials, for example in the form of $m = 1$ components, made the models able to reproduce at the right place even secondary features appearing in images of open spiral galaxies, such as asymmetric bifurcations of the arms (see e.g. Fig. 4 in [18]). The inner *symmetric* part of the grand-design has been always identified with the location of the 4/1 resonance, while off-phase, with respect to the imposed spiral perturbation, extensions, could be found between 4/1 and corotation in the responses (see also the results of three dimensional models in [19]). In all the above cases, the most sensitive parameter in order to obtain a morphological similarity of the model with the modeled galaxy was the pattern speed (Ω_p). This should be that slow, as to put corotation beyond the end of the inner symmetric part of the spiral arms. Later, models that have

considered kinematic data as well, have also confirmed this result, by pushing corotation at, or beyond, the end of the overall observed spiral structure [20].
A characteristic, gaseous (by means of Smoothed Particle Hydrodynamics, SPH) response model for open normal spirals (pitch angle 25°) is given in Fig. 2. The model includes, besides a main $m = 2$ logarithmic spiral component, also an $m = 1$ term, with the same pattern speed and shape as the $m = 2$ one. The ratio of the amplitudes $A_{m=1}/A_{m=2} = 0.15$, while the relative force perturbation at the end of the symmetric part of the spiral pattern is of the order of 10% of the axisymmetric force. The long arrows in Fig. 2 point to the locations of the end of the symmetric part of the spiral pattern, at the 4/1 resonance region, while the short, thicker, arrows at larger distances from the center, point to weak extensions of the arms, to which we refer in the next paragraph.

- Another mechanism for supporting the spiral structure of galaxies started being discussed in the middle 2000s, applied to barred-spiral systems. Mainly two groups, elaborated the idea that the spiral structure observed beyond the ends of the bars in this type of galaxies is guided primarily by the unstable manifolds emanating from the unstable Lagrangian points L1, L2 at the corotation region [21, 22]. The idea has been presented earlier [23], without a detailed description of the dynamical mechanism. The later was known in studies of the three-body problem, however without relating it with the support of an emerging structure in that case [24, 25]. According to this mechanism, stars following the paths dictated by the manifolds are on chaotic orbits, as their Lyapunov numbers indicate, so the formed spirals, have been called "chaotic" spirals. Evidence that the orbital content of these spirals is associated with the so-called "hot orbital" population [26] is given in [27]. Such chaotic orbits in autonomous Hamiltonian systems have Jacobi constants, E_J, larger than those of L1 and L2 and for some time they may exhibit a 4/1-resonance orbital behaviour inside corotation. They enhance the spiral arms of the barred-spiral morphology as they cross corotation through the bottlenecks formed by the isocontours of the effective potential at various E_J's. They are of the same type of orbits as those building the envelope of the bar in the case of NGC 4314 studied in [28]. Further orbits of this type have been presented in [29], in models for NGC 1300.
  Since in the chaotic seas we can find unstable periodic orbits around the equilibrium points (Lyapunov orbits) as well as unstable periodic orbits belonging to the 4/1, 6/1 etc. families, it is natural to conclude that all the families of unstable periodic orbits near and beyond corotation contribute to the same phenomenon [30]. Manifolds of unstable 4/1 periodic orbits associated with the reinforcement of chaotic spirals have been presented in [31].
  Besides the spirals, the same mechanism has been proposed for explaining several types of rings observed around the bars ([22, 32] and subsequent papers by the same authors). Also in this case, orbits classified by chaos indicators as "chaotic", reinforce a well defined morphological structure.
  Models in which the two mechanisms for supporting two different spiral patterns coexist, as the one presented in Fig. 2 (the arrows point to two different sets of spiral arms), lead to rare, but known grand design morphologies, as in the cases

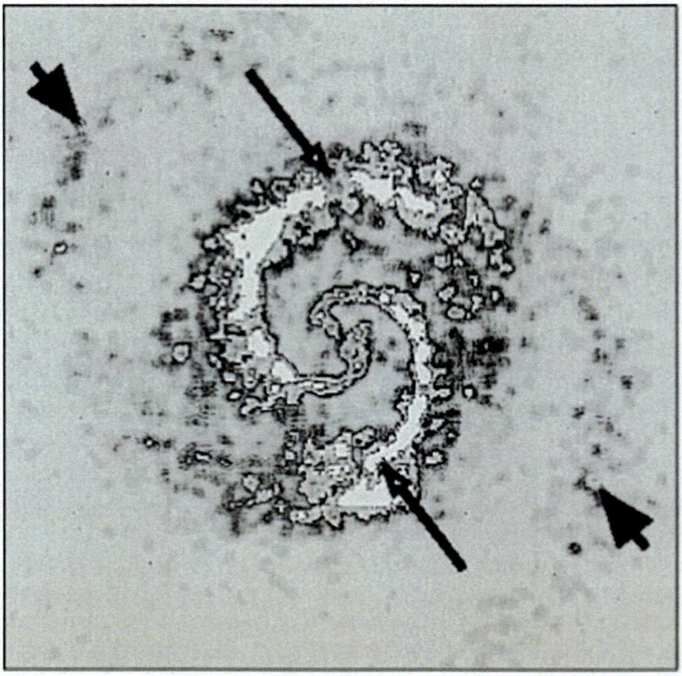

Fig. 2 An SPH response model to a logarithmic spiral potential with pitch angle 25° that includes a main $m = 2$ and a secondary $m = 1$ component. The long arrows point to the end of the inner, symmetric spiral arms, at the 4/1 resonance, while the short arrows at larger distances to a weak continuation of the spiral arms beyond corotation, which are formed following the mechanism of "chaotic" spirals (see text)

of NGC 1566 or NGC 5248 [31, 33]. In these cases the inner spiral structure is supported by regular orbits trapped around precessing x1 periodic orbits, while the outer spiral structure by chaotic orbits that cross the region of the L1 and L2 points and continue beyond corotation. These spiral arms are those indicated with the short arrows in Fig. 2. The main difference among the regular and the chaotic spirals is in the flow of material in their regions. In arms supported by regular orbits, the flow is accross the arms, while in chaotic spirals, the flow is in general along them [27, 33].

2.2 *Three Dimensional (3D) Models*

Disk galaxies can be considered in a first approximation as two dimensional objects. However, the inner part of the bars extends well above the equatorial plane of the galaxy, reaching heights up to 2 kpc in some cases, forming a characteristic boxy,

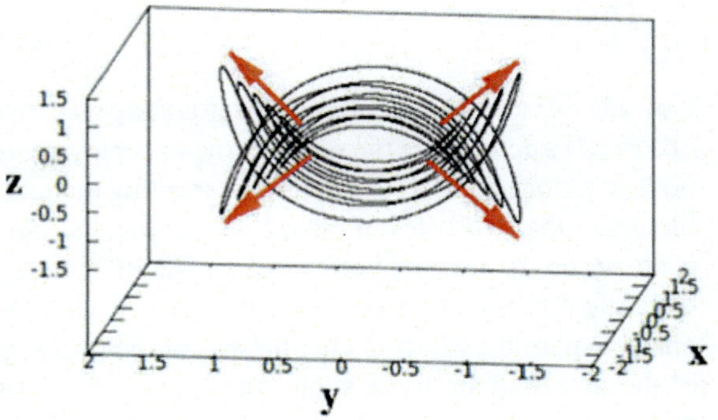

Fig. 3 A set of successive stable x1v1 periodic orbits in a Ferrers bar model in a nearly side-on projection. The wings of the X are formed along the $|z|$-maxima of the orbits, i.e. along the directions indicated with the arrows

peanut-shape morphology when viewed edge-on. In some cases the side-on profiles of galactic bars outline the pronounced morphology of an "X" shape [34]. These structures, and their relation with bars, have been identified in early N-body simulations [35].

The study of the orbital content of the peanuts, has been based on the analysis of the stability of families of periodic orbits in 3D autonomous Hamiltonians [36] as a parameter of the system, usually E_J, varies [37–40]. The main dynamical mechanism proposed to be in action in this case, is again that of regular orbits trapped around stable periodic orbits. A family that offers the appropriate orbital content to make this scenario feasible, is the 3D bifurcation of x1 at the vertical 2/1 resonance, called x1v1 [38].

As indicated in [40] (their Fig. 11), the "X" structure is not formed as the density is enhanced along the shape of the orbits of x1v1, but it appears along the maxima heights from the equatorial plane, of successive orbits of this family. An example of such a backbone of stable x1v1 orbits supporting a X/peanut structure in a Ferrers bar model [37, 38] is given in Fig. 3.

The study of 3D autonomous Hamiltonian systems is a field, where various nonlinear phenomena appear and affect their evolution, such as inverse bifurcations, collisions of bifurcations and complex instability [36]. Especially complex instability, a kind of orbital instability that appears when the four eigenvalues of the 4×4 monodromy matrix M^1, are complex and off the unit circle (see e.g. [38]), appears in orbits that may participate in supporting the peanut for considerable time intervals [40]. This is one more case in which chaotic orbits contribute to the reinforcement of structures by behaving for times significant for the dynamical scales of the system we study, as regular. Essentially, we have to do with the phenomenon of stickiness [41], which is ubiquitous in galactic stellar dynamics and upgrades the role of chaotic orbits in supporting structures.

[1] In autonomous Hamiltonian systems, the monodromy matrix relates the final deviation of a neighbouring orbit from the periodic one $\bar{\xi}$, with an initially introduced deviation $\bar{\xi_0}$, in a space of section, i.e. $\bar{\xi} = M\bar{\xi_0}$.

3 Discussion

Nowadays there is a general consensus among researchers working in the field that the observed structures are the result of the interplay between Order and Chaos. Usually, both situations coexist in structure-supporting models. Regular orbits are the building blocks of the structures in most of the cases, but not in all of them. The alternative is sticky orbits near the borders of an island of stability in the phase space of a 2D dynamical system, or orbits remaining sticky near the unstable asymptotic curves of unstable periodic orbits in chaotic seas of the phase space. Examples of sticky orbits of the first kind are those supporting outer boxy envelopes of barred galaxies [28] or inner boxy structures in the central regions of the bars [42, 43]. Orbits sticky to unstable asymptotic curves, are those supporting the spiral arms outside corotation. An extreme example of an ansae-type bar supported mainly by chaotic orbits is given in [29]. In a case presented in that paper, the shape of the bar is defined by the shape of the isocontours of the effective potential, allowing particles in chaotic motion to visit all regions inside the area outlined by the ansae-type isocontour (see their Figs. 2 and 3).

It is not always easy to distinguish which mechanism is behind an observed morphological feature in a real galaxy. For example, the outer boxiness of a bar may be due to regular orbits trapped around boxy 4/1 resonance orbits, or due to chaotic orbits sticky to tiny stability islands in the 4/1 resonance region. When we see a pair of spiral arms emerging from the ends of a bar, we may conclude that they are due to chaotic orbits associated with Lyapunov orbits around L1 and L2, provided that the ends of the bar are close enough to the Lagrangian points. However, the ratio of the corotation radius, R_c, to the length of the semi-major axis of a bar, R_b, is in general $1 < R_c/R_b < 1.4$ [44], while in some cases it can be assumed even larger, reproducing successfully barred-spiral morphologies (see e.g. models in [31, 33]. For bars that end away from the Lagrangian points, other mechanisms have to be invoked for explaining the spirals.

In that respect, imaging in the optical and in the near-infrared, as well as detailed kinematic data, are always needed to be compared with the predictions of the models in order to qualify the best scenario behind the emergence of a specific morphological feature.

References

1. Athanassoula, E.: Bars and secular evolution in disk galaxies: theoretical input. In: Falcon-Barroso, J., Knapen, J.H. (eds.) Secular Evolution of Galaxies, pp. 305–352. Cambridge University Press, Cambridge, UK (2013)
2. Contopoulos, G.: Order and Chaos in Dynamical Astronomy (Astronomy and Astrophysics Library). Springer, Berlin Heidelberg (2002)
3. Binney, J., Tremaine, S.: Galactic Dynamics. Princeton University Press, Princeton (2008)
4. Quillen, A.C., Frogel, J.A., Gonzalez, R.A.: The gravitational potential of the bar in NGC 4314. ApJ **437**, 162–172 (1994). https://doi.org/10.1086/174984

5. Kalapotharakos, C., Patsis, P.A., Grosbøl, P.: NGC 1300 dynamics–I. The gravitational potential as a tool for detailed stellar dynamics. Mon. Not. R. Astron. Soc. **403**, 83–95 (2010). https://doi.org/10.1111/j.1365-2966.2009.16127.x
6. Patsis, P.A., Grosbøl, P.: Thick spirals: dynamics and orbital behavior. Astron. Astrophys. **315**, 371–383 (1996)
7. Contopoulos, G., Grosbøl, P.: Orbits in barred galaxies. Astron. Astrophys. Rev. **1**, 261–289 (1989). https://doi.org/10.1007/BF00873080
8. Contopoulos, G.: The effects of resonances near corotation in barred galaxies. Astron. Astrophys. **102**, 265–278 (1981)
9. Contopoulos, G.: Periodic orbits near the particle resonance in galaxies. Astron. Astrophys. **64**, 323–332 (1978)
10. Lin, C.C., Shu, F.H.: On the spiral structure of disk galaxies. Astrophys. J. **140**, 646–655 (1964). https://doi.org/10.1086/147955
11. Kalnajs, A.J.: Spiral structure viewed as a density wave. Proc. ASA **2**, 174–177 (1973). https://doi.org/10.1017/S1323358000013461
12. Contopoulos, G., Grosbøl, P.: Stellar dynamics of spiral galaxies: nonlinear effects at the 4/1 resonance. Astron. Astrophys. **155**, 11–23 (1986)
13. Contopoulos, G., Grosbøl, P.: Stellar dynamics of spiral galaxies: self-consistent models. Astron. Astrophys. **197**, 83–90 (1988)
14. Patsis, P.A., Contopoulos, G., Grosbøl, P.: Self-consistent spiral galactic models. Astron. Astrophys. **243**, 373–380 (1991)
15. Thomasson, M., Donner, K.J., Elmegreen, Bruce G.: Simulations of the effect of spiral arms on the cloud-ensemble velocity dispersion. Astron. Astrophys. **250**, 316–323 (1991)
16. Zasov, A.V., Khoperskov, A.V., Tyurina, N.V.: Stellar velocity dispersion and mass estimation for galactic disks. Astron. Lett. **30**, 593–602 (2004)
17. Harsoula, M., Zouloumi, K., Efthymiopoulos, C., Contopoulos, G.: Precessing ellipses as the building blocks of spiral arms. Astron. Astrophys. **655**, A55 (2021). https://doi.org/10.1051/0004-6361/202140984
18. Patsis, P.A., Grosbøl, P., Hiotelis, N.: Interarm features in gaseous models of spiral galaxies. Astron. Astrophys. **323**, 762–774 (1997)
19. Chaves-Velasquez, L., Patsis, P.A., Puerari, I., et al.: Dynamics of thick, open spirals in PERLAS potentials. Astrophys. J. **871**, 79 (2019) https://doi.org/10.3847/1538-4357/aaf6a6
20. Kranz, T., Slyz, A., Rix, H.-W.: Probing for dark matter within spiral galaxy disks. Astrophys. J. **562**, 164–178 (2001). https://doi.org/10.1086/323468
21. Voglis, N., Stavropoulos, I., Kalapotharakos, C.: Chaotic motion and spiral structure in self-consistent models of rotating galaxies. Mon. Not. R. Astron. Soc. **372**, 901–922 (2006). https://doi.org/10.1111/j.1365-2966.2006.10914.x
22. Romero-Gomez, M., Masdemont, J.J., Athanassoula, E., et al.: The origin of rR1 ring structures in barred galaxies. Astron. Astrophys. **453**, 39–45 (2006). https://doi.org/10.1051/0004-6361:20054653
23. Danby, J.M.A.: The formation of arms in barred spirals. Astron. J. **70**, 501–512 (1965). https://doi.org/10.1086/109773
24. Koon, W.S., Lo, M.W., Marsden, J.E., Ross, S.D.: Heteroclinic connections between periodic orbits and resonance transitions in celestial mechanics. Chaos **10**, 427–469 (2000). https://doi.org/10.1063/1.166509
25. Gomez, G., Koon, W.S., Lo, M.W., et al.: Connecting orbits and invariant manifolds in the spatial restricted three-body problem. Nonlinearity **17**, 1571–1606 (2004). https://doi.org/10.1088/0951-7715/17/5/002
26. Kaufmann, D.E., Contopoulos, G.: Self-consistent models of barred spiral galaxies. Astron. Astrophys. **309**, 381–402 (1996)
27. Patsis, P.A.: The stellar dynamics of spiral arms in barred spiral galaxies. Mon. Not. R. Astron. Soc.: Lett. **369**, L56–L60 (2006). https://doi.org/10.1111/j.1745-3933.2006.00174.x
28. Patsis, P.A., Athanassoula, E., Quillen, A.C.: Orbits in the bar of NGC 4314. Astrophys. J. **483**, 731–744 (1997). https://doi.org/10.1086/304287

29. Patsis, P.A., Kalapotharakos, C., Grosbøl, P.: NGC1300 dynamics—III. Orbital analysis. Mon. Not. R. Astron. Soc. **408**, 22–39 (2010). https://doi.org/10.1111/j.1365-2966.2010.17062.x
30. Tsoutsis, P., Efthymiopoulos, C., Voglis, N.: The coalescence of invariant manifolds and the spiral structure of barred galaxies. Mon. Not. R. Astron. Soc. **387**, 1264–1280 (2008). https://doi.org/10.1111/j.1365-2966.2008.13331.x
31. Tsigaridi, L., Patsis, P.A.: The backbones of stellar structures in barred-spiral models—the concerted action of various dynamical mechanisms on galactic discs. Mon. Not. R. Astron. Soc. **434**, 2922–2939 (2013). https://doi.org/10.1093/mnras/stt1207
32. Athanassoula, E., Romero-Gómez, M., Masdemont, J.J.: Rings and spirals in barred galaxies—I. Building blocks. Mon. Not. R. Astron. Soc. **394**, 67–81 (2009). https://doi.org/10.1111/j.1365-2966.2008.14273.x
33. Patsis, P.A., Tsigaridi, L.: The flow in the spiral arms of slowly rotating bar-spiral models. Astrophys. Space Sci. **362**, 129–145 (2017). https://doi.org/10.1007/s10509-017-3109-9
34. Laurikainen, E., Salo, H.: Observed properties of boxy/peanut/barlens bulges. In: Laurikainen, E., Peletier, R.F., Gadotti, D.A. (eds.) Galactic Bulges, Astrophysics and Space Science Library, vol. 418, pp. 77–106 (2016). https://doi.org/10.1007/978-3-319-19378-6_4
35. Combes, F., Debbasch, F., Friedli, D., et al.: Box and peanut shapes generated by stellar bars. Astron. Astrophys. **233**, 82–95 (1990)
36. Contopoulos, G., Magnenat, P.: Simple three-dimensional periodic orbits in a galactic-type potential. Celest. Mech. **37**, 387–414 (1985). https://doi.org/10.1007/BF01261627
37. Pfenniger, D.: The 3D dynamics of barred galaxies. Astron. Astrophys. **134**, 373–386 (1984)
38. Skokos, Ch., Patsis, P. A., Athanassoula, E.: Orbital dynamics of three-dimensional bars—I. The backbone of three-dimensional bars. A fiducial case. Mon. Not. R. Astron. Soc. **333**, 847–860 (2002). https://doi.org/10.1046/j.1365-8711.2002.05468.x
39. Patsis, P.A., Skokos, C., Athanassoula, E.: Orbital dynamics of three-dimensional bars—III. Boxy/peanut edge-on profiles. Mon. Not. R. Astron. Soc. **337**, 578–596 (2002). https://doi.org/10.1046/j.1365-8711.2002.05943.x
40. Patsis, P.A., Katsanikas, M.: The phase space of boxy-peanut and X-shaped bulges in galaxies—I. Properties of non-periodic orbits. Mon. Not. R. Astron. Soc. **445**, 3525–3545 (2014). https://doi.org/10.1093/mnras/stu1988
41. Contopoulos, G., Harsoula, M.: Stickiness in chaos. Int. J. Bifurc. Chaos **18**, 2929–2949 (2008). https://doi.org/10.1142/S0218127408022172
42. Patsis, P.A., Katsanikas, M.: The phase space of boxy-peanut and X-shaped bulges in galaxies—II. The relation between face-on and edge-on boxiness. Mon. Not. R. Astron. Soc. **445**, 3546–3556 (2014). https://doi.org/10.1093/mnras/stu1970
43. Chaves-Velasquez, L., Patsis, P.A., Puerari, I., et al.: Boxy orbital structures in rotating bar models. Astrophys. J. **850**, 145, 17 (2017). https://doi.org/10.3847/1538-4357/aa961a
44. Athanassoula, E.: The existence and shapes of dust lanes in galactic bars. Mon. Not. R. Astron. Soc. **259**, 345–364 (1992). https://doi.org/10.1093/mnras/259.2.345

Phase Space Transport and Dynamical Matching in a Caldera-Type Hamiltonian System

Matthaios Katsanikas and Stephen Wiggins

Abstract The goal of this paper is to review the phase space mechanism by which a Caldera-type potential energy surface (PES) exhibits the dynamical matching phenomenon. Using the method of Lagrangian descriptors, we can easily establish that the non-existence of dynamical matching is a consequence of heteroclinic connections between the unstable manifolds of the unstable periodic orbits (UPOs) of the upper index-1 saddles (entrance channels to the Caldera) and the stable manifolds of the family of UPOs of the central minimum of the Caldera, resulting in the temporary trapping of trajectories. Moreover, dynamical matching will occur when there is no heteroclinic connection, which allows trajectories to enter and exit the Caldera without interacting with the shallow region of the central minimum. Knowledge of this phase space mechanism is relevant because it allows us to effectively predict the existence, and non-existence, of dynamical matching. In this work we explore a stretched Caldera potential by means of Lagrangian descriptors, allowing us to accurately compute the critical value for the stretching parameter for which dynamical matching behavior occurs in the system.

Keywords Phase space transport · Hamiltonian systems · Lagrangian descriptors · Periodic orbits · Invariant manifolds · Symmetry · Caldera potential · Poincaré sections

M. Katsanikas (✉)
Research Center for Astronomy and Applied Mathematics of the Academy of Athens, Soranou Efesiou 4, 11527 Athens, Greece
e-mail: mkatsan@academyofathens.gr

M. Katsanikas · S. Wiggins
School of Mathematics, University of Bristol, Fry Building, Woodland Road, BS8 1UG Bristol, UK
e-mail: s.wiggins@bristol.ac.uk

S. Wiggins
Department of Mathematics, United States Naval Academy, Chauvenet Hall, 572C Holloway Road, 21402-5002 Annapolis, USA

T. Bountis et al. (eds.), *Chaos, Fractals and Complexity*, Springer Proceedings in Complexity, https://doi.org/10.1007/978-3-031-37404-3_3

1 Introduction

Dynamical matching is an interesting mechanism originally proposed in [6, 7] that arises in Caldera-type potential energy surfaces (PES). These potentials are relevant in chemistry since they provide good approximations for the description of many organic chemical reactions, such as those that occur in the vinylcyclopropane-cyclopentene rearrangement [3, 14], the stereomutation of cyclopropane [10], the degenerate rearrangement of bicyclo[3.1.0]hex-2-ene [11, 12] or that of 5-methylenebicyclo[2.1.0]pentane [24].

A study of the nature of trajectories that cross a two dimensional caldera potential was given in [8]. The caldera potential energy surface studied in that paper possessed a symmetry (to be described shortly), and the effect of asymmetry, or "stretching" of the potential, on trajectories was also considered. In [18] an analysis of the phase space structures that determined the different behaviors of trajectories was given for the symmetric caldera potential. In particular, we investigated the mechanisms of trapping trajectories and of dynamical matching in the symmetrical caldera potential energy surface. The trajectories that have initial conditions on the dividing surfaces of the unstable periodic orbits of the lower saddles are guided from the invariant manifolds of the periodic orbits until they are trapped from the invariant manifolds of the unstable periodic orbits that exist in the central region of the caldera. The trajectories that have initial conditions at the central region of the caldera have two options. The first option is to lie on or are inside the Kolmogorov-Arnold-Moser (KAM) tori that surround the stable periodic orbits of the central area. The second option is to be trapped by the invariant manifolds of the unstable periodic orbits of the central region until they are transported from the unstable invariant manifolds to the exit from the caldera through the four different regions of saddles. The trajectories that have initial conditions on the dividing surfaces of the unstable periodic orbits of the upper saddles are not trapped but, on the contrary, we have the phenomenon of dynamical matching [18]. Dynamical matching is the behaviour of trajectories having initial conditions on the periodic orbit dividing surfaces of the upper saddles, that go straight across the caldera and exit via the opposite lower saddle. We showed that this occurs when there is no interaction of the invariant manifolds of the unstable periodic orbits of the upper saddles with the central region of the caldera [18]. This implies that there is no trapping of these trajectories in the symmetric caldera potential energy surface.

In this paper we investigate the possibility of a mechanism for trapping of trajectories that have initial conditions on the periodic orbit dividing surfaces of the upper saddles for the stretched version of the caldera potential in a manner that does not exist in the symmetric caldera potential energy surface. By "stretching" of the potential we mean that we scale the coordinate x (see (1)) by a parameter λ $0 < \lambda \leq 1$ and $\lambda \to 1$. The classical case of the potential is obtained for $\lambda = 1$. In this situation the saddles move away from the central minimum in the x-direction as the stretching parameter λ becomes smaller.

We begin our investigation of the stretched potential by first considering if there is a critical value of λ that controls the existence of dynamical matching? The fact that the parameter λ plays a role in the dynamical matching phenomenon was evident in the trajectory studies in [8]. However, no explanation of this behavior was given in terms of phase space structure and transport. In this paper we provide such an explanation.

We present our model in Sect. 2. Furthermore, we give a concise introduction to the Lagrangian Descriptors method used in this paper (see Sect. 3). Finally, we present our results and the conclusions in Sects. 4 and 5.

2 Hamiltonian Model

We give a brief description of the caldera potential energy surface and Hamiltonian as described in [8]. The caldera potential has a stable equilibrium point at the center, referred to as the central minimum. This potential has an axis of symmetry, the y-axis. We have also the existence of potential walls around the central minimum. On these potential walls we encounter four 1-index saddles (two for lower values of energy, referred to as the lower saddles, and two for higher values of energy, referred to as the upper saddles). In this paper we consider the stretched version of the caldera potential:

$$V(x, y) = c_1(y^2 + (\lambda x)^2) + c_2 y - c_3((\lambda x)^4 + y^4 - 6(\lambda x)^2 y^2) \tag{1}$$

The potential parameters are $c_1 = 5$, $c_2 = 3$, $c_3 = -3/10$ and $0 < \lambda \leq 1$. For $\lambda = 1$ we have the symmetric caldera potential [8, 18]. The contours of this potential and the stationary points are depicted for different values of λ in Fig. 1. The positions of the saddles move away from the center of the caldera as the parameter lambda decreases.

The Hamiltonian of the system is:

$$H(x, y, p_x, p_y) = \frac{p_x^2}{2m} + \frac{p_y^2}{2m} + V(x, y) \tag{2}$$

with potential $V(x, y)$ (see the Eq. (1)) and $m = 1$. The numerical value of the Hamiltonian we will call it as energy E.

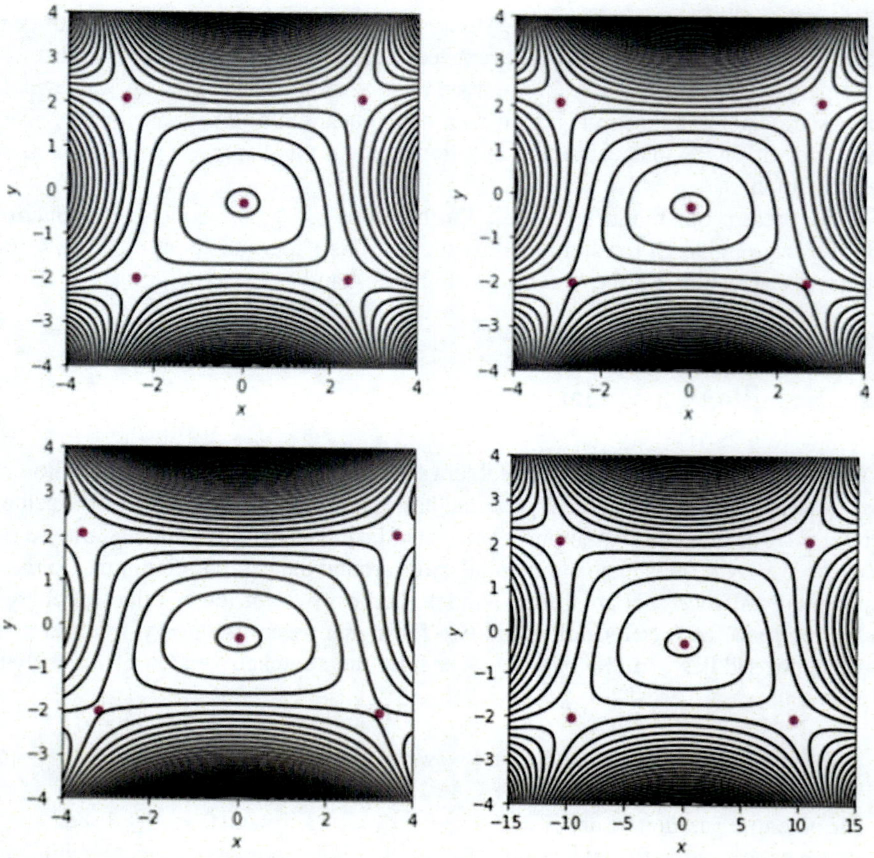

Fig. 1 The stationary points (depicted by purple points) and contours of the potential for $\lambda = 0.8$ (upper left panel), $\lambda = 0.72$ (upper right panel), $\lambda = 0.6$ (lower left panel) and $\lambda = 0.2$ (lower right panel)

3 Lagrangian Descriptors in Caldera-Type Hamiltonian Systems

The Lagrangian Descriptors (LDs) is a diagnostic tool that can reveal the phase space structures. The first time that this technique was used was in the paper [21], were was aiming to study transport and mixing in geophysical flows. In the last years this technique has been broadly used not only in fluid mechanics [4, 5, 13, 20, 22, 23] but also in the area of chemical reaction dynamics [1, 2, 9, 17]. The LDs method works in the following way: In order to reveal the phase space structures in a given slice using the method of LDs the steps that we need to follow are very simple. First we choose the slice and we define in this slice a grid of initial conditions. Then we integrate this initial conditions forward and backward in time for a given integration time τ and while we are integrating we accumulate along these trajectories a positive quantity defined from the vector field that determines the dynamical system that we are studying. If you integrate trajectories forwards in time, the LD function is going to detect the stable manifolds while the backwards in time integration will detect the unstable manifolds. The scalar output obtained from the method will highlight the location of the invariant stable and unstable manifolds intersecting this slice, which are detected at points where the values of LDs display an abrupt change.

There are several definitions for the LDs. In this work we are using the p-norm definition of the method that relies on variable time integration, where $p \in (0, 1]$ (the reader can find more information about the application of variable time LDs to Caldera potentials in [15]). We have fixed the value of p to be $p = 1/2$. This definition of the LDs is preferable here due to the nature of the Caldera's potential energy surface (PES), that is an open potential. That can lead to an increasingly fast pace escape of the trajectories.

Let's consider the following dynamical system with time dependence: consider the following dynamical system with time dependence:

$$\frac{d\mathbf{x}}{dt} = \mathbf{v}(\mathbf{x}, t), \quad \mathbf{x} \in \mathbb{R}^n, \ t \in \mathbb{R} \tag{3}$$

where $\mathbf{v}(\mathbf{x}, t) \in C^r (r \geq 1)$ in $\mathbf{x}$ and it is continuous in time. Given an initial condition x_0 at time t_0, take a fixed integration time $\tau > 0$ and $p \in (0, 1]$. The method of LDs in the p-norm definition is as follows:

$$M_p(\mathbf{x}_0, t_0, \tau) = \sum_{k=1}^{n} \left[\int_{t_0-\tau}^{t_0+\tau} |v_k(\mathbf{x}(t; \mathbf{x}_0), t)|^p \, dt \right] = M_p^{(b)}(\mathbf{x}_0, t_0, \tau) + M_p^{(f)}(\mathbf{x}_0, t_0, \tau), \tag{4}$$

where $M_p^{(b)}$ and $M_p^{(f)}$ its backward and forward integration parts:

$$\begin{aligned} M_p^{(b)}(\mathbf{x}_0, t_0, \tau) &= \sum_{k=1}^{n} \left[\int_{t_0-\tau}^{t_0} |v_k(\mathbf{x}(t; \mathbf{x}_0), t)|^p \, dt \right], \\ M_p^{(f)}(\mathbf{x}_0, t_0, \tau) &= \sum_{k=1}^{n} \left[\int_{t_0}^{t_0+\tau} |v_k(\mathbf{x}(t; \mathbf{x}_0), t)|^p \, dt \right], \end{aligned} \tag{5}$$

The formulation of the p-norm definition that we apply to this model is the following:

$$M_p(\mathbf{x}_0, t_0, \tau) = \sum_{k=1}^{n} \left[\int_{t_0-\tau_{x_0}^-}^{t_0+\tau_{x_0}^+} |v_k(\mathbf{x}(t; \mathbf{x}_0), t)|^p \, dt \right] \tag{6}$$

where

$$\tau_{x_0}^{\pm} = min\{\tau_0, |t^{\pm}|_{|x(t^{\pm};x_0 \notin R)}\}$$

Note that the integration time τ_0 is fixed and that t^+ is the time that the trajectory exits the interaction region R in forward time whereas t^- is for backward time.

4 Numerical Results

In this section we compute Lagrangian descriptors with $\tau = 4$ in order to study the phase space structures close to the UPOs associated with the upper index-1 saddles (for a fixed value of energy above the energy of the upper index-1 saddles, $E = 29$ see [15, 16]). For this purpose we use the Poincaré surfaces of section $y = 1.88409$ with $p_y > 0$, which was also used in [18, 19]. This analysis is carried out for different values of λ. Our goal is to understand how LDs are capable of detecting the dynamical matching mechanism.

The phenomenon of dynamical matching refers to the lack of a mechanism that would enable transport of trajectories from the region of the upper saddles to the central area of the Caldera. As we know, trajectories with initial conditions on the invariant manifolds of unstable periodic orbits move away from the periodic orbit (unstable manifold) or approach the periodic orbit (stable manifold). A mechanism that could be responsible for the transport of trajectories from the region of the upper saddles to the central area of the Caldera, would be heteroclinic intersections of the unstable invariant manifolds of the unstable periodic orbits of the upper saddles with the stable manifolds of the unstable periodic orbits that exist in the central area. We will show that the non-existence or existence of this mechanism determines if we have dynamical matching or not. For this reason, we compute the invariant manifolds for different values of λ starting from $\lambda = 1$ to zero in order to find the values of λ that correspond to dynamical matching and trapping.

1. **Dynamical matching:** The gap in Fig. 2 (for $\lambda = 0.8$) indicates that we have no interaction (heteroclinic intersections) of the unstable invariant manifold of the periodic orbits associated with the upper saddle with the central area and this means that we have no mechanism of transport of trajectories from one region to the other. Consequently, we have in this case the phenomenon of dynamical matching, the trajectories that have initial conditions on the dividing surfaces of the periodic orbits of upper saddles go straight across the Caldera and they exit via the lower opposite saddle as we know from previous papers [18, 19]. An example of this is given in Fig. 2 for $\lambda = 0.8$. As we can see in this figure [15] we choose an initial conditions (circle) inside the region of the unstable invariant manifold of the unstable periodic orbits of upper saddles. If we integrate backward the initial condition that corresponds to the circle the resulting trajectory exits via the region of the upper saddle. If we integrate it forward the resulting trajectory goes straight across the caldera and exits via the lower opposite saddle. This means

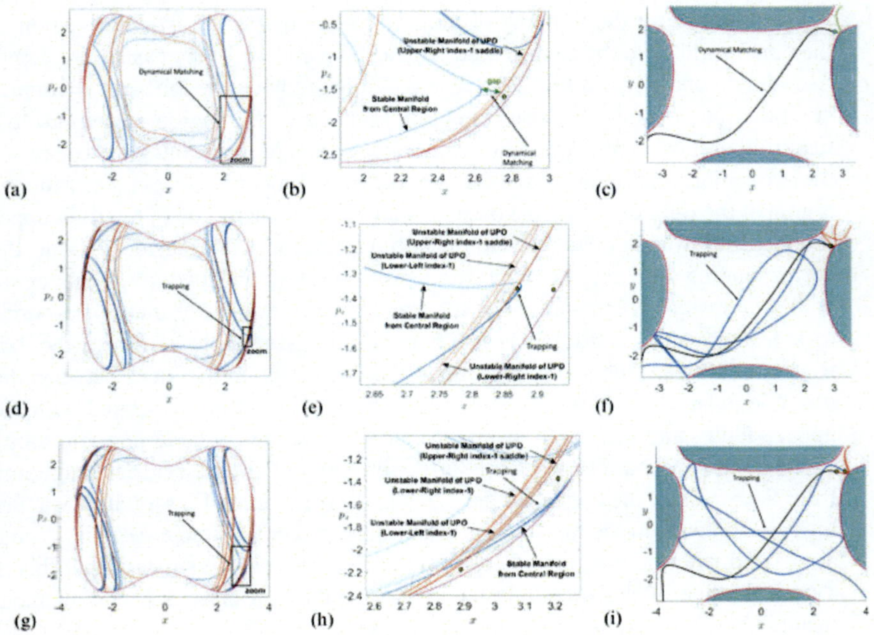

Fig. 2 The phase space close to the unstable periodic orbits associated with the upper saddles (first column) and the enlargement of the region of the phase space that is indicated by a rectangle in the figures of the first column (figures in the second column) using Lagrangian Descriptors (with $\tau = 4$). The figures in the third column depict the trajectories in the configuration space that correspond to a circle and diamond in the figures in the second column. In the first row, the green line indicate the part of the trajectory at backward integration that corresponds to the circle. In the second and third row, the red line indicate the part of the trajectories at backward integration that correspond to both of them, circle and diamond. In addition, black and blue line indicate the part of the trajectories at forward integration that correspond to the circle and diamond respectively (in all rows). **a**, **b**, **c** are for $\lambda = 0.8$, **d**, **e**, **f** are for $\lambda = 0.778$ and **g**, **h**, **i** are for $\lambda = 0.7$

that the trajectory comes from the region of the upper saddle and it exhibits the phenomenon of dynamical matching. This gap decreases in size as we decrease the stretching parameter λ until we reach a critical value of λ.

2. **The critical value:** In Fig. 2 we observe for $\lambda = 0.778$ (middle row of figures) the unstable manifolds of the periodic orbits of upper saddles start to interact with the stable manifolds of the unstable periodic orbits of the central area, resulting in heteroclinic connections and forming lobes between them. These lobes are very narrow and cannot be distinguished initially as we can see in Fig. 2. In order to observe these lobes we magnify the region of the upper saddles, for example the region of the upper right saddle in Fig. 2. When we magnify these regions, we see the heteroclinic connections and the lobes between the unstable invariant manifolds of the unstable periodic orbits of upper saddles and the stable manifolds of the unstable periodic orbits that exist in the central area. These lobes

are responsible for the trapping of the trajectories that come from the region of the upper saddles to the central area. This can be checked very easily. We depict two initial conditions in Fig. 2 for $\lambda = 0.778$, one inside the lobe (the diamond) and other one outside the lobe (the circle) but inside the region of the unstable manifold of the unstable periodic orbit of upper saddle. If we integrate backward the two initial conditions, we see that the corresponding trajectories come from the region of the right upper saddle because they exit via the region of the right upper saddle. But if we integrate forward the initial condition, that is inside the lobe, the corresponding trajectory is trapped and after a long time exits through the region of the opposite lower saddle. On the contrary, the trajectory that corresponds to the other initial condition is not trapped and go straight across to the exit from the caldera. This means that the initial conditions in the lobes between the unstable invariant manifolds of the unstable periodic orbits associated with the upper saddles and the stable invariant manifolds of the unstable periodic orbits of the central area are responsible for the trapping of the trajectories that come from the region of the upper saddles. This is the first value of λ for which we find interaction between the unstable invariant manifolds of unstable periodic orbits, associated with the upper saddles, with the central area. This means that this is a critical value of the stretching parameter for the non-existence of dynamical matching, as we have observed in a previous paper [19]).

3. **Trapping:** Now if we decrease the value of λ, starting from the critical value, we have again interaction of the unstable invariant manifolds of unstable periodic orbits of upper saddles with the central area. We have again lobes between the unstable invariant manifolds of the unstable periodic orbits with the stable invariant manifolds of the unstable periodic orbits that exist in the central region as we can see for example for $\lambda = 0.7$ in Fig. 2. This means that we have again trapping for values of λ lower than the critical value.

5 Conclusions

In this paper we have shown that heteroclinic connections are the phase space mechanism that controls dynamical matching. While we have demonstrated this behavior for a two DoF caldera model, the notion of a heteroclinic trajectory is valid for dynamical systems with an arbitrary number of dimensions. Hence, it would be interesting to explore the formation of this phase space structure as a mechanism for dynamical matching in systems with three or more DoF.

Acknowledgements We acknowledge the financial support provided by the EPSRC Grant No. EP/P021123/1.

References

1. Agaoglou, M., Aguilar-Sanjuan, B., García-Garrido, V.J., García-Meseguer, R., González-Montoya, F., Katsanikas, M., Krajňák, V., Naik, S., Wiggins, S.: Chemical Reactions: A Journey into Phase Space. Bristol, UK, Zenodo (2019)
2. Agaoglou, M., Aguilar-Sanjuan, B., García-Garrido, V.J., González-Montoya, F., Katsanikas, M., Krajňák, V., Naik, S., Wiggins, S.: Lagrangian Descriptors: Discovery and Quantification of Phase Space Structure and Transport (2020). zenodo: https://doi.org/10.5281/zenodo.3958985
3. Baldwin, J.E.: Thermal rearrangements of vinylcyclopropanes to cyclopentenes. Chem. Rev. **103**(4), 1197–1212 (2003)
4. Balibrea-Iniesta, F., Xie, J., García-Garrido, V.J., Bertino, L., Mancho, A.M., Wiggins, S.: Lagrangian transport across the upper arctic waters in the Canada basin. Q. J. R. Meteorol. Soc. **145**(718), 76–91 (2019)
5. Balibrea-Iniesta, F., Lopesino, C., Wiggins, S., Mancho, A.M.: Lagrangian descriptors for stochastic differential equations: a tool for revealing the phase portrait of stochastic dynamical systems. Int. J. Bifurc. Chaos **26**(13), 1630036 (2016)
6. Carpenter, B.K.: Trajectories through an intermediate at a fourfold branch point: implications for the stereochemistry of biradical reactions. J. Am. Chem. Soc. **107**(20), 5730–5732 (1985)
7. Carpenter, B.K.: Dynamic matching: the cause of inversion of configuration in the [1, 3] sigmatropic migration? J. Am. Chem. Soc. **117**(23), 6336–6344 (1995)
8. Collins, P., Kramer, Z.C., Carpenter, B., Ezra, G.S., Wiggins, S.: Nonstatistical dynamics on the caldera. J. Chem. Phys. **141**(034111), 034111 (2014)
9. Crossley, R., Agaoglou, M., Katsanikas, M., Wiggins, S.: From Poincaré maps to Lagrangian descriptors: the case of the valley ridge inflection point potential. Regul. Chaotic Dyn. **26**(2), 147–164 (2021)
10. Doubleday, C., Bolton, K., Hase, W.L.: Direct dynamics study of the stereomutation of cyclopropane. J. Am. Chem. Soc. **119**(22), 5251–5252 (1997)
11. Doubleday, C., Nendel, M., Houk, K.N., Thweatt, D., Page, M.: Direct dynamics quasiclassical trajectory study of the stereochemistry of the vinylcyclopropane—cyclopentene rearrangement. J. Am. Chem. Soc. **121**(19), 4720–4721 (1999)
12. Doubleday, C., Suhrada, C.P., Houk, K.N.: Dynamics of the degenerate rearrangement of bicyclo[3.1.0]hex-2-ene. J. Am. Chem. Soc. **128**(1), 90–94 (2006)
13. García-Garrido, V.J., Curbelo, J., Mancho, A.M., Wiggins, S., Mechoso, C.R.: The application of Lagrangian descriptors to 3D vector fields. Reguar Chaotic Dyn. **23**(5), 551–568 (2018)
14. Goldschmidt, Z., Crammer, B.: Vinylcyclopropane rearrangements. Chem. Soc. Rev. **17**, 229–267 (1988)
15. Katsanikas, M., García-Garrido, V.J., Wiggins, S.: Detection of dynamical matching in a caldera Hamiltonian system using Lagrangian descriptors. Int. J. Bifurc. Chaos **30**, 2030026 (2020)
16. Katsanikas, M., García-Garrido, V.J., Wiggins, S.: The dynamical matching mechanism in phase space for caldera-type potential energy surfaces. Chem. Phys. Lett. **743**, 137199 (2020)
17. Katsanikas, M., García-Garrido, V.J., Agaoglou, M., Wiggins, S.: Phase space analysis of the dynamics on a potential energy surface with an entrance channel and two potential wells. Phys. Rev. E **102**(1), 012215 (2020)
18. Katsanikas, M., Wiggins, S.: Phase space structure and transport in a caldera potential energy surface. Int. J. Bifurc. Chaos **28**(13), 1830042 (2018)
19. Katsanikas, M., Wiggins, S.: Phase space analysis of the nonexistence of dynamical matching in a stretched caldera potential energy surface. Int. J. Bifurc. Chaos **29**(04), 1950057 (2019)
20. Lopesino, C., Balibrea-Iniesta, F., Wiggins, S., Mancho, A.M.: Lagrangian descriptors for two dimensional, area preserving, autonomous and nonautonomous maps. Commun. Nonlinear Sci. Numer. Simul. **27**(1), 40–51 (2015)
21. Madrid, J.A.J., Mancho, A.M.: Distinguished trajectories in time dependent vector fields. Chaos **19**, 013111 (2009)

22. Mancho, A.M., Wiggins, S., Curbelo, J., Mendoza, C.: Lagrangian descriptors: a method for revealing phase space structures of general time dependent dynamical systems. Commun. Nonlinear Sci. Numer. Simul. **18**(12), 3530–3557 (2013)
23. Mendoza, C., Mancho, A.M.: The hidden geometry of ocean flows. Phys. Rev. Lett. **105**, 038501 (2010)
24. Reyes, M.B., Lobkovsky, E.B., Carpenter, B.K.: Interplay of orbital symmetry and nonstatistical dynamics in the thermal rearrangements of bicyclo[n.1.0]polyenes. J. Am. Chem. Soc. **124**, 641–651 (2002)

The Building Blocks of Spiral Arms in Galaxies

Mirella Harsoula

Abstract This is a review paper on the prevailing theories that explain the orbital content of galactic potentials that can support the spiral arms. In the case of grand design spiral galaxies there exist families of stable periodic orbits with an approximate elliptical shape that create waves having a spiral shape. These waves are similar to those observed in real spiral galaxies. On the other hand, in the case of barred spiral galaxies, the spiral structure is supported by chaotic orbits which stay, due to stickiness phenomena, close and along the unstable asymptotic manifolds of the Lagrangian points L_1 and L_2 for long enough time compared to the Hubble time. This is called manifold theory and it has been tested for the case of one pattern speed as well as of two different pattern speeds, for the bar and the spiral structure.

Keywords Galaxies · Chaos · Spiral arms · Periodic orbits

1 Introduction

A great percentage of spiral galaxies possess a bar. Approximately two-thirds of all spiral galaxies are thought to be barred spiral galaxies. On the other hand, grand-design spiral galaxies are also observed (Fig. 1 on the left), possessing no bar at their center and having symmetrical spiral arms. The well known "density wave" theory can explain the spiral structure in this case. However, there are still questions about the longevity of these spiral waves (see [1, 9, 13] for reviews).

The amplitude of the spiral perturbation of the galaxy must not exceed a value of $10-20\%$ in order to use the density wave theory. In this case, nonlinear corrections are necessary [6, 27, 40] to the linear Lindblad-Lin-Shu theory [21–24]. Within the framework of the density wave theory one can prove the existence of approximately elliptical stable periodic orbits. These orbits have a main axis that change its

M. Harsoula (✉)
Research Center for Astronomy and Applied Mathematics of the Academy of Athens, Soranou Efesiou 4, 11527 Athens, Greece
e-mail: mharsoul@academyofathens.gr

T. Bountis et al. (eds.), *Chaos, Fractals and Complexity*, Springer Proceedings in Complexity, https://doi.org/10.1007/978-3-031-37404-3_4

Fig. 1 The grand design spiral galaxy NGC 4321, on the left (copyright Barry Wilson/LTA/Ruben Barbosa) and the barred spiral galaxy NGC 1365 on the right (*Credit* ESO/IDA/Danish 1.5 m/ R. Gendler, J-E. Ovaldsen, C. Thöne, and C. Feron.)

orientation with the energy and in such a way they can form a spiral density wave. Kalnajs [20] gave such examples of density waves made out of precessing ellipses and he even computed the response potential due to these ellipses. On the other hand, Contopoulos (1970, 1975) gave all the analytical theory of these orbits using resonant perturbation theory. He used action-angle variables for constructing the corresponding phase space and studied also the number and the stability of these orbits. These elliptical orbits supporting spiral density waves have been studied numerically in analytical models in many papers [5, 8, 14, 28, 29, 33, 36].

We present the results of a recent work [19] of ours using a galactic potential with an axisymmetric component and a logarithmic spiral potential simulating a Milky-Way like model. We study the spiral density waves derived from the main stable periodic orbits in the region between the Inner Lindblad Resonance (ILR) and the 4:1 resonance. We also find the dependence on the amplitude of the spiral perturbation.

On the other hand, in the case of barred spiral galaxies (Fig. 1, on the right), the main theory that prevails is the manifold theory [35, 41]. According to this theory the spiral arms are supported by chaotic orbits that are connected, through stickiness phenomena, with the unstable asymptotic curves emanating from the bar's Lagrangian points L_1 and L_2 (see also [11]). We use here the "apocentric manifold' version of the manifold theory. These chaotic orbits in general create a flow along the spiral arms [30]. Whereas, in the case of grand design galaxies the flow of the regular orbits is intersecting the spiral arms. In [31] it is emphasized that this difference between the flow of the orbits can serve as an observational tool in order to distinguish regular from chaotic orbits supporting spiral arms.

In [15] we study the manifold theory in the case where the spiral arms have a different pattern speed than the bar, in barred spiral galaxies. In this case, the apocentric manifolds are related to the unstable asymptotic curves emanating from Lagrangian equilibrium orbits, (instead of equilibrium points L_1 and L_2, in the case of one pattern speed). These new apocentric manifolds are time dependent in the rotating frame of the bar.

The paper is structured as follows: in Sect. 2 we present the galactic model of a grand design spiral galaxy and we study the dependence of spiral density waves on the amplitude of the spiral perturbation. In Sect. 3 we give a Milky Way like model of a barred spiral galaxy and we study the construction of a spiral apocentric manifold supporting the spiral structures for one pattern speed, as well as for two pattern speeds. Finally in Sect. 4 we draw our conclusions.

2 Spiral Arms in Grand Design Galaxies

2.1 The Model

The galactic potential that we use consists of an axisymmetric and a spiral potential:

$$V = V_{ax} + V_{sp}. \tag{1}$$

The axisymmetric potential V_{ax} includes a disc, a bulge and a halo:

$$V_{ax} = V_d + V_b + V_h. \tag{2}$$

The disc potential V_d is a Miyamoto-Nagai model [26] given by the relation:

$$V_d = \frac{-GM_d}{\sqrt{r^2 + (a_d + \sqrt{z^2 + b_d^2})^2}}, \tag{3}$$

where $M_d = 8.56 \times 10^{10}\ M_\odot$ is the total mass of the disc, $a_d = 5.3$ kpc and $b_d = 0.25$ kpc. In our case we take $z = 0$ and $r = \sqrt{x^2 + y^2}$. The component of the bulge is represented by a Plummer potential V_b given by the relation:

$$V_b = \frac{-GM_b}{\sqrt{r^2 + b^2}}, \tag{4}$$

where $M_b = 5 \times 10^{10}\ M_\odot$ is the total mass of the bulge, $r = \sqrt{x^2 + y^2}$ and $b = 1.5$ kpc.

The halo potential is a γ-model [12] with parameters as in [32]:

$$V_h = \frac{-GM_{h(r)}}{r} - \frac{-GM_{h,0}}{\gamma r_h}\left[-\frac{\gamma}{1 + (r/r_h)^\gamma} + \ln(1 + \frac{r}{r_h})^\gamma\right]_r^{r_{h,max}}, \tag{5}$$

where $r_{h,max} = 100$ Kpc, $\gamma = 1.02$, and $M_{h,0} = 10.7 \times 10^{10} M\odot$, and $M_{h(r)}$ is given by the function:

$$M_{h(r)} = \frac{M_{h,0}(r/r_h)^{\gamma+1}}{1 + (r/r_h)^{\gamma}}. \tag{6}$$

The spiral potential is represented by a logarithmic spiral model $V_{sp}(r, \phi, z)$ introduced by Cox and Gomez [10]. We have on the disc plane:

$$V_{sp} = 4\pi G h_z \rho_0 \, G(r) \, \exp\left(-\left(\frac{r - r_0}{R_s}\right)\right) \frac{C}{KB} \cos\left[2\left(\varphi - \frac{\ln(r/r_0)}{\tan(\alpha)}\right)\right], \tag{7}$$

where

$$K = \frac{2}{r|\sin(\alpha)|}, \quad B = \frac{1 + Kh_z + 0.3(Kh_z)^2}{1 + 0.3Kh_z} \tag{8}$$

and $C = 8/(3\pi)$, $h_z = 0.18$ kpc, $r_0 = 8$ kpc, $R_s = 7$ kpc, $\alpha = -13°$ is the pitch angle of the spiral arms. This value is proposed as a mean global pitch angle in the Milky Way's spiral arms [39]. The function $G(r)$ is given by the relation $G(r) = b - c \arctan(R_{s0} - r)$, with $R_{s0} = 6$ kpc, $b = 0.474$, $c = 0.335$. The density of the spiral arms is $\rho_0 = 5 \times 10^7$, or 30×10^7 $M\odot/\text{kpc}^3$ in the two different models under study, respectively. These two values of the density correspond to a weak and a strong spiral perturbation respectively (see for example [2]).

2.2 *Spiral Density Waves from Precessing Ellipses*

The Hamiltonian, in the rotating frame of reference (with a fixed pattern speed Ω_{sp}), is given in polar coordinates by the relation:

$$H = \frac{p_r^2}{2} + \frac{p_\varphi^2}{2r^2} - \Omega_{sp} p_\varphi + V_{ax}(r) + V_{sp}(r, \varphi), \tag{9}$$

where, p_r is the radial velocity and p_φ is the angular momentum per unit mass.

Now, by using the Hamiltonian (9) we find the stable periodic orbits that have elliptical shapes and support the spiral density wave. These orbits are the continuation of the circular orbits of the axisymmetric part of the potential and are found in the region between Inner Lindblad Resonance (ILR) and corotation in the model of Sect. 2.1. For further details see [19]. We fix the value of the pattern speed $\Omega_{sp} = 15\,\text{km} \cdot \text{s}^{-1} \cdot \text{kpc}^{-1}$ and the pitch angle $a = -13°$, which are realistic values for grand design galaxies. In order to easily locate these stable periodic orbits we use action-angle variables of epicyclic theory. The pair (φ, p_φ) of Eq. (9) are action-angle variables but the pair (r, p_r) are not, so we define some new "radial" action-angle variables (ξ, P_ξ) (as defined in [19]). We construct Poincaré surfaces of sections

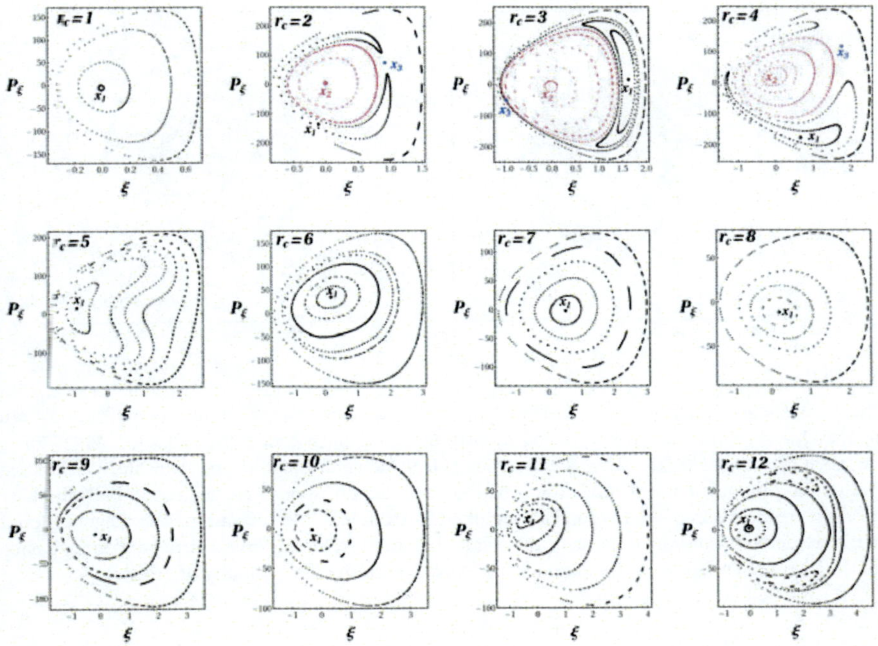

Fig. 2 The Poincaré phase space of the galactic model for action-angle variables (ξ, $P\xi$) for $\rho_0 = 5 \times 10^7 M \odot /\text{kpc}^3$ (Eq. (7)), and for radii $r_c = 1, 2, \ldots, 12$ kpc. We locate three different periodic orbits, i.e., the stable x_1 family (black dots), the stable x_2 family (magenta dots) and the unstable x_3 family (blue dots). The periodic orbits that support the spiral density wave are the periodic orbits of the x_1 family that correspond to radii between approximately 5 kpc (second ILR) and 11 kpc (4:1 resonance). Figures from [19]

(ξ, P_ξ) for a fixed Jacobi constant $E_j \equiv H$. For a fixed value of $\varphi = \pi/2$ we find a Poincaré sequence of surfaces of sections of our model.

We now focus on the form of the periodic orbits that support the spiral density wave in the region between Inner Lindblad Resonance (ILR) and corotation in the model of Sect. 2.1, as well as the shape of the phase space around these orbits. For further details see [19]. We choose two different values for the spiral amplitude (parameter ρ_0 in Eq. (7)), with pattern speed $\Omega_{sp} = 15\,\text{km.sec}^{-1}.\text{kpc}^{-1}$ and pitch angle $a = -13°$, which are realistic values for grand design galaxies.

Figure 2 shows the Poincaré surfaces of section (ξ, P_ξ) for $\rho_0 = 5 \times 10^7 M \odot/\text{kpc}^3$ in Eq. (7). It shows the phase portraits for twelve different values of the radius r_c namely $r_c = 1, 2, \ldots, 12$ kpc, corresponding to a region from the center of the galaxy and up to a radius just outside the 4:1 resonance.

Contopoulos [7] introduced the nomenclature of the families of periodic orbits in a spiral galaxy and named x_1, x_2 (stable periodic orbits) and x_3 (unstable periodic orbit) the three basic families. The family of orbits that supports the spiral density wave are the stable periodic orbits of the x_1 family.

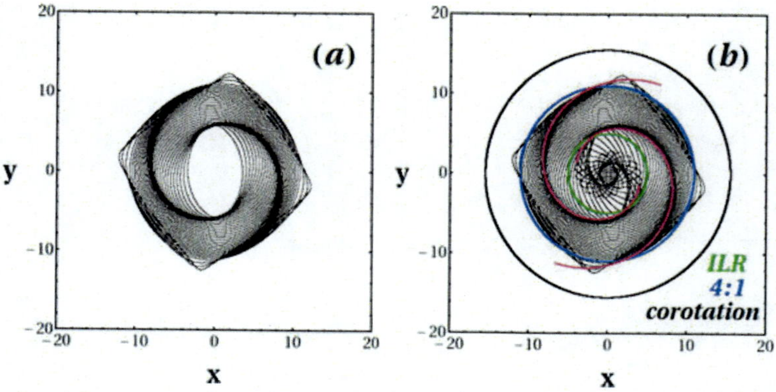

Fig. 3 **a** The x_1 family of elliptical orbits support a spiral density wave, of the model (9) for pattern speed $\Omega_{sp} = 15\,\text{km} \cdot \text{s}^{-1} \cdot \text{kpc}^{-1}$ and density of the spiral potential $\rho_0 = 5 \times 10^7 M\odot/\text{kpc}^3$, between the second ILR and the 4:1 resonance. **b** Same as in **a** but we also plot the x_1 elliptical periodic orbits inside the second ILR. The green circle corresponds to the second ILR, the blue circle corresponds to the 4:1 resonance and the black circle to the corotation. The magenta spiral arms are derived from the minima of the spiral potential (7). We observe a very good coincidence of the imposed spirals with the spiral density wave created by the x_1 family of orbits

In Fig. 2, the black dots (inside the black islands of stability) correspond to the stable x_1 family of periodic orbits. They exist for all the radii r_c from the center of the galaxy and up to the 4:1 resonance. The magenta dots (inside the magenta islands of stability) correspond to the stable x_2 family of periodic orbits and the blue dots correspond to the unstable periodic orbit x_3. The x_2 and x_3 family of orbits exist only in the region between the first and the second ILR. Their corresponding orbits do not support the spiral density wave. The x_1 family remains stable at all radii up to the 4:1 resonance (which corresponds to $r_c \approx 11$).

The orbits of the x_1 family support a well defined spiral density wave (Fig. 3a) between the second ILR ($\approx$ 5 kpc) and the 4:1 resonance, where the corresponding orbits become more rectangular. In Fig. 3b the x_1 family of orbits is plotted also inside the second ILR. Between the first ILR ($\approx$ 1.5 kpc) and the second ILR ($\approx$ 5 kpc) the orbits of the x_1 family form a less dense and well defined spiral density wave. On the other hand, the orbits of x_1 family are circular inside the first ILR and do not support a density wave at all. The green circle in Fig. 3b corresponds to the Inner Lindblad resonance, the blue circle to the 4:1 resonance and the black circle to the radius of corotation. The spiral arms (in magenta) correspond to the minima of the spiral potential of Eq. (7) and coincide very well with the spiral density wave made out of the stable elliptical orbits of the x_1 family.

In Fig. 4 the same phase portrait as in Fig. 2 is plotted in the case of a much greater amplitude of the spiral perturbation, i.e. $\rho_0 = 30 \times 10^7 M\odot/\text{kpc}^3$. We observe that chaos is introduced outside the ILR and the x_1 family becomes unstable. Therefore, there are no more islands of stability around this orbit and there is no material that can support the spiral density wave. In fact, this value of ρ_0 is unrealistic for real

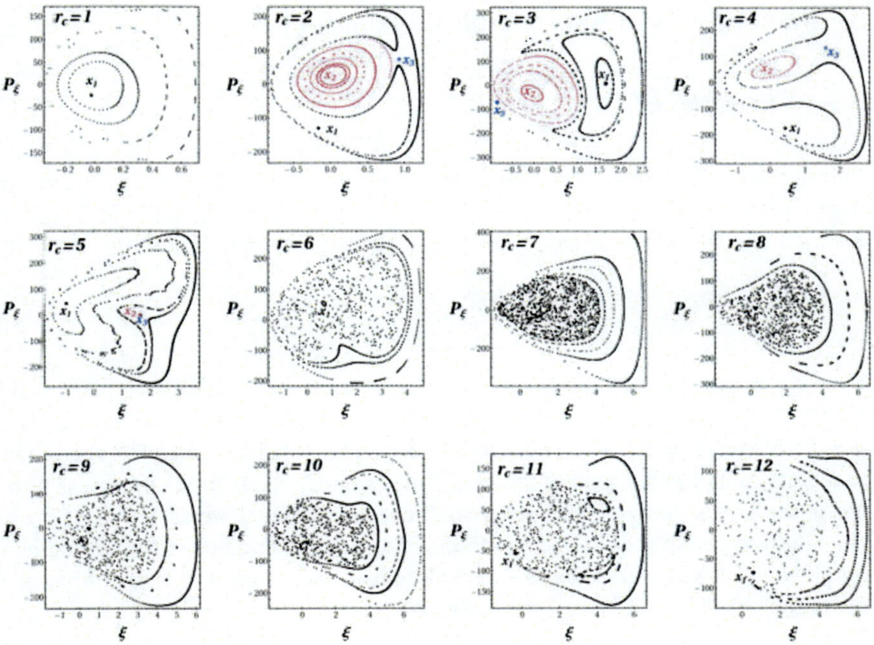

Fig. 4 Same as in Fig. 2, but for $\rho_0 = 30 \times 10^7 M\odot/\mathrm{kpc}^3$. Figures from [19]

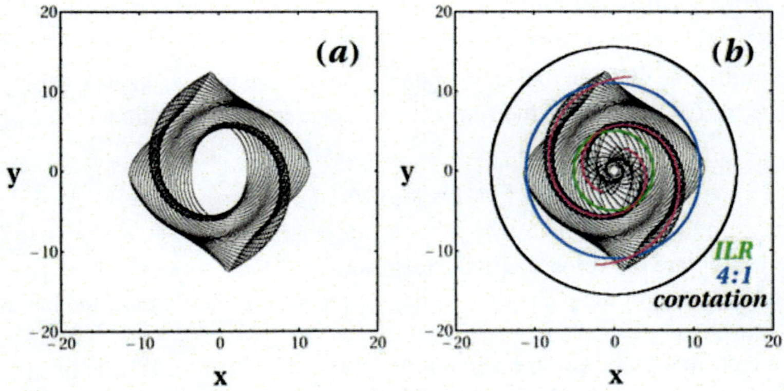

Fig. 5 Same as in Fig. 3, but for $\rho_0 = 30 \times 10^7 M\odot/\mathrm{kpc}^3$. Figures from [19]

grand design spiral galaxies, but we still plot the precessing ellipses of the unstable x_1 family in Fig. 5 in order to compare it with the realistic spiral density wave of Fig. 3. Some interesting correlations between the free parameters of the galactic model ρ_0, Ω_{sp} and the pitch angle α are found in [19].

3 Spiral Arms in Barred Spiral Galaxies

3.1 The Model

In order to study a Milky way like barred spiral galactic model we use the axisymmetric potential of Eqs. (3), (4) and (5) as well as the spiral potential of Eq. (7) of Sect. 2.1 and we introduce the following bar potential model, as in [25]:

$$V_{bar} = \frac{GM_b}{2a} \ln\left(\frac{x - a + T_-}{x + a + T_+}\right) \tag{10}$$

with $T_{\pm} = \sqrt{[(a \pm x)^2 + y^2 + (b + \sqrt{c^2 + z^2})^2]}$, $M_b = 6.25 \times 10^{10} M\odot$, $a = 5.25$ kpc, $b = 2.1$ kpc and $c = 1.6$ kpc. The values of a and b correspond to the major and minor axis of the bar, while, c corresponds to the thickness of the bar in the z-axis [4, 16, 34]. For a two dimensional model we set $z = 0$. If we set the pattern speed of the bar $\Omega_{bar} = 45$ km/s/kpc, then the radius of the corotation is set close to 5.4 kpc and the bar's length close to 4 kpc, as the radius of corotation is approximately $1.2 - 1.3$ times the bar's length.

3.2 Manifold Spirals

The Hamiltonian of the barred spiral galactic model that we use, in the rotating frame of reference (of the bar and the spiral arms), is given by the relation:

$$H = \frac{1}{2}(p_x^2 + p_y^2) - \Omega_p(xp_y - yp_x) + V(x, y) \tag{11}$$

where p_x, p_y are the velocities in the rest frame and $V(x, y) = V_{ax} + V_{bar} + V_{sp}$ is the total potential of the galaxy consisting of the axisymmetric part, the bar and the spiral arms.

The manifold theory, in the version of "apocentric manifolds" has been studied in many papers [15, 18, 37, 38, 41]. We give a brief description of the apocentric manifolds here: we find the equilibrium Lagrangian point L_1 (and its symmetric point L_2) of the Hamiltonian (11), in the rotating frame of reference, which correspond to a solution $(x, y, p_x, p_y) = (x_{L_1}, y_{L_1}, p_{x_{L_1}}, p_{y_{L_1}})$ of the equilibrium equations $\partial H/\partial x = \partial H/\partial y = \partial H/\partial p_x = \partial H/\partial p_y = 0$. We then plot the unstable asymptotic manifolds of these points (L_1 and L_2) on the phase space and take initial conditions along these manifolds. We integrate the orbits with these initial conditions for a long enough time. The apocenters of all these orbits, which are chaotic, form structures that support the spiral arms on the configuration space with many recurrences back and forth before escaping from the system. The same is true for all the unstable periodic orbits in the region of corotation as well as for the sticky

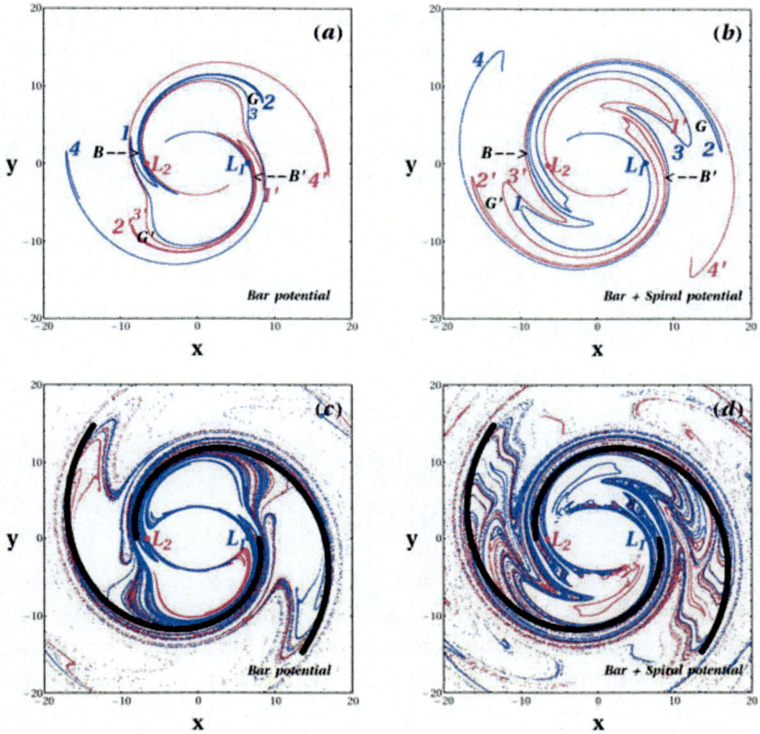

Fig. 6 **a** The apocentric manifolds of the L_1 and L_2 Lagrangian points of the galactic model (11) possessing only a bar and no spiral arms, in the rotating frame of reference. With blue we plot the manifolds from L_1 and with magenta the manifolds from L_2. We observe a R_1-type of ring around the bar **b** Same as in **a** but with an additional spiral potential term which rotate with the same pattern speed as the bar. Here we observe 'lobes', 'bridges' and 'gaps'. **c** and **d**: same as in **a** and **b** but here the apocentric manifolds are integrated for a much longer time. The black spiral arms are derived from the minima of the spiral potential of Eq. (7)

chaotic orbits along these manifolds. Therefore, the conclusion is than these manifolds form the paths along which all the chaotic orbits are forced to move, due to stickiness phenomena, before escaping from the system. These apocentic manifolds of chaotic orbits support the spiral structure for long enough time.

In Fig. 6 we plot the apocentric manifolds of the unstable Lagrangian points L_1 (blue) and L_2 (magenta), in the rotating frame of the Hamiltonian (11), in two different cases: (i) a potential having an axisymmetric part and a bar (Figs. 6a, c), and (ii) a potential having an axisymmetric part, a bar and spiral arms (Figs. 6b, d). The spiral arms, in this case, rotate with the same pattern speed as the bar. In the bar only case (Figs. 6a, c) we observe that the apocentric manifolds induce a R_1-type ring-like structure (see [3] for a review). This kind of ring structure has a main axis which is perpendicular to the main axis of the bar. On the other hand, in the case were a spiral potential is added (Figs. 6b, d) we observe 'lobes', 'bridges' and 'gaps' instead of ring structures. See [15] for a detailed description of this figure. In Figs. 6c, d, the apocentric manifolds are integrated for much longer time. We observe that in the bar-only case spiral arms are developed, while the R_1-type of ring is enhanced. On the other hand, in the case were the spiral potential is added we observe that the diffusion of these sticky chaotic orbits along the manifolds is very slow and the spiral arms make many recurrences back and forth supporting the spiral structure for long enough time compared with the Hubble time.

If we now consider that the spiral arms rotate with a different pattern speed than the bar, we conclude that the apocentric manifolds, are now time dependent (see [15]).

The Hamiltonian of the system, in the rotating system of the bar and in polar coordinates (ρ, ϕ), is now time dependent:

$$H = \frac{1}{2}\left(p_\rho^2 + \frac{p_\phi^2}{2\rho^2}\right) - \Omega_{bar} p_\phi + V_{ax}(\rho) + V_{bar}(\rho, \phi) + (\Omega_{sp} - \Omega_{bar}) I_2 + V_{sp}(\rho, \phi - \phi_2) \tag{12}$$

where $\phi_2 = (\Omega_{sp} - \Omega_{bar})t$, the pattern speed of the bar is Ω_{bar} and the pattern speed of the spiral arms is Ω_{spr}. In a time dependent Hamiltonian, there exist no longer equilibrium points, but equilibrium orbits, named GL_1 (and GL_2). The new potential is time dependent with period $T = \pi/(\Omega_{bar} - \Omega_{sp})$. In order to construct the time dependent apocentric manifolds we use a stroboscopic map at times $t_{sec} = \kappa * T$, where $\kappa = 0, 1/4, 1/2, 3/4$ and plot only the apocenters of the orbits for these times. The computation of the time dependent unstable apocentric manifolds of the orbits GL_1 and GL_2 is described analytically in [15]. These time dependent apocentric manifolds are exactly the same every $t_{sec} = mT$, $m = 1, 2, 3, \ldots$.

Figure 7 shows the main result: the apocentric manifolds of the GL_1 (blue points) and GL_2 (magenta points) equilibrium orbits are shown at four different times t_{sec}, namely $t_{sec} = 0$, $T/4$, $T/2$ and $3T/4$. The black dotted curves superposed to the manifolds correspond to the maxima of the surface density (or the minima of the spiral potential). These figures repeat periodically after the time $t_{sec} = T$. These spiral arms (black) rotate clockwise with respect to the rotating frame of the bar with angular velocity equal to $(\Omega_{bar} - \Omega_{sp})$. These time dependent apocentric manifolds coincide rather well with the minima of the spiral potential apart the time $t_{sec} = T/2$, where the spiral maxima are displaced by an angle $\pi/2$ with respect to the bar's horizontal axis (Fig. 7c).

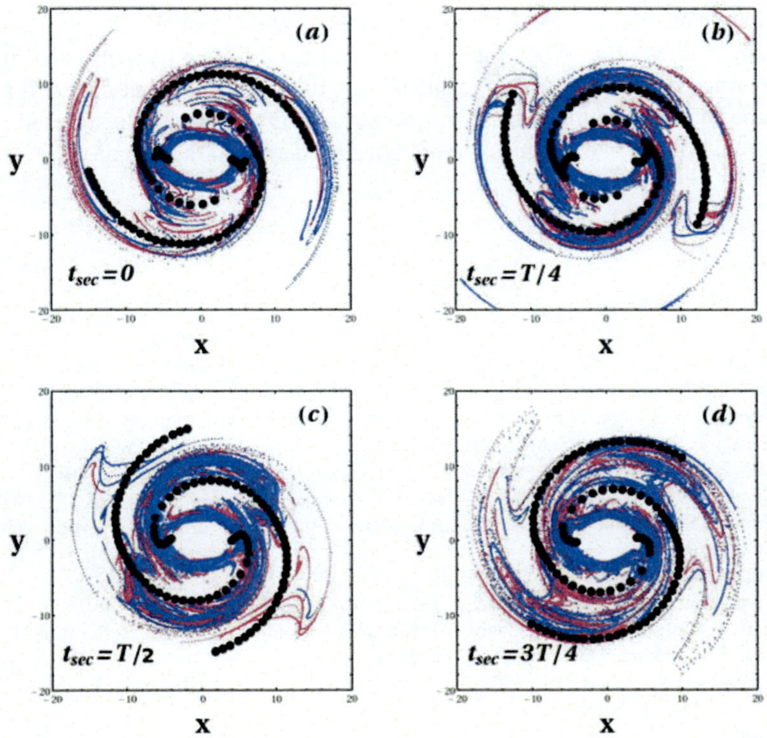

Fig. 7 The time dependent apocentric manifolds of the GL_1 equilibrium orbit (blue) and GL_2 (magenta) are plotted at four different times t_{sec} as indicated in each panel. Superimposed are plotted the local minima of the spiral potential of Eq. (7) at the same times $t = t_{sec}$

4 Conclusion

In the present paper we review the prevailing theories of the kind of orbits that support the spiral arms in two different cases, i.e. in the case of grand design spiral galaxies and in the case of barred spiral galaxies. In the case of grand design galaxies the prevailing theory is the density wave theory. We use a galactic model with an axisymmetric component and a spiral potential and we study the different types of periodic orbits that exist in this model. We conclude that the orbits that support the spiral density wave are the approximately elliptical x_1 orbits that extend from the Inner Lindblad Resonance (ILR) up to the 4:1 resonance where the orbits become more rectangular and can no longer support the spiral arms. The x_1 family of orbits becomes unstable after the ILR, when the spiral perturbation ρ_0 in Eq. (7) becomes $\rho_0 \geq 30 \times 10^7\, M\odot/\mathrm{kpc}^3$. This is a limiting value above which the precessing ellipses can no longer support the spiral density waves.

On the other hand, in the case of barred spiral galaxies, there exist no longer regular orbits in the region of corotation. In this case, chaotic orbits along the unstable asymptotic manifolds of the Lagrangian points L_1 and L_2 support the spiral structure for long enough time. When the bar rotates with a different pattern speed from the spiral arms then the apocentric manifolds are time dependent but they still support the time dependent spiral arms. The apocentric manifolds produce features like rings, lobes and bridges like the ones observed in real barred spiral galaxies.

References

1. Bertin, G., Lin, C.C., Lowe, S.A., Thurstans, R.P.: ApJ **338**, 78 (1989)
2. Block, D.L., Buta, R., Knapen, J.H., Elmegreen, D.M., Elmegreen, B.G., Puerari, I.: ApJ **128**, 183 (2004)
3. Buta, R.: In: Falcon-Barroso, J., Knapen, J.H. (eds.) 'Secular Evolution of Galaxies', XXIII Canary Islands Winter School of Astrophysics. Cambridge University Press (2013)
4. Cao, L., Mao, S., Nataf, D., Rattenbury, N.J., Gould, A.: MNRAS **434**, 595 (2013)
5. Chaves-Velasquez, L., Patsis, P.A., Puerari, I., Moreno, E., Pichardo, B.: ApJ **871**, 79 (2019)
6. Contopoulos, G.: In: Proceedings of I.A U. Symposium No 38 (Dordrecht: D. Reidel Publishing Co) (1970)
7. Contopoulos, G.: ApJ **201**, 566 (1975)
8. Contopoulos, G., Grosbøl, P.: A & A **155**, 11 (1986)
9. Contopoulos, G.: Order and Chaos in Dynamical Astronomy. Springer, Berlin (2002)
10. Cox, D.P., Gómez, G.C.: ApJ Sup. **142**, 261 (2002)
11. Danby, J.M.A.: AJ **70**, 501 (1965)
12. Dehnen, W.: MNRAS **265**, 250 (1993)
13. Donner, K.J., Thomasson M.A.: A & A **290**, 475 (1994)
14. Efthymiopoulos, Ch.: Eur. Phys. J. Spec. Top. **186**, 91 (2010)
15. Efthymiopoulos, C., Harsoula, M., Contopoulos, G.: A & A **636**, A44 (2020)
16. Gerhard, O.: In: Da Costa, G.S., Sadler, E.M., Jerjen, H. (eds.) The Dynamics, Structure and History of Galaxies: A Workshop in Honour of Professor Ken Freeman. ASP Conference Series, vol. 73 (2002)
17. Gerhard, O.: Pattern speeds in the milky way. Mem: Soc. Astron. It. Sup. **18**, 185 (2011)
18. Harsoula, M., Efthymiopoulos, C., Contopoulos, G.: Analytical forms of chaotic spiral arms. MNRAS **459**, 3419 (2016)
19. Harsoula, M., Zouloumi, K., Efthymiopoulos, C., Contopoulos, G.: Precessing ellipses as the building blocks of spiral arms. A & A **655**, A55 (2021)
20. Kalnajs, A.J.: Proc. ASA **2**, 174 (1973)
21. Lin, C., Shu, F.: ApJ **140**, 646 (1964)
22. Lin, C., Shu, F.: PNAS **55**, 229 (1966)
23. Lindblad, B.: ApJ **92**, 1 (1940)
24. Lindblad, B.: Stockholm Obs. Ann. **21**, 8 (1961)
25. Long, K., Murali, C.: ApJ **397**, 44L (1992)
26. Miyamoto, M., Nagai, R.: Publ. Astron. Soc. Japan **27**, 533 (1975)
27. Norman, C.A.: MNRAS **182**, 457 (1978)
28. Patsis, P.A., Contopoulos, G., Grosbøl, P.: A & A **243**, 373 (1991)
29. Patsis, P.A., Grosbøl, P.: A & A **315**, 371 (1996)
30. Patsis, P.A.: In: Contopoulos, G., Patsis, P.A. (eds.) Chaos in Astronomy, Conference 2007, Astrophysics and Space Science Proceedings, pp. 33–44. Springer Berlin, Heidelberg (2009)
31. Patsis, P.A., Tsigaridi, L.: Astosh. Spac. Sci. **362**, 129 (2017)
32. Pettitt, A.R., Dobbs, C.L., Acreman, D.M., Price, D.J.: MNRAS **444**, 919 (2014)
33. Pichardo, B., Martos, M., Moreno, E., Espresate, J.: ApJ **582**, 230 (2003)
34. Rattenbury, N.J., Mao, S., Sumi, T., Smith, M.C.: MNRAS **378**, 1064 (2007)

35. Romero-Gomez, M., Masdemont, J.J., Athanassoula, E., Garcia-Gomez, C.: A & A **453**, 39 (2006)
36. Tsigaridi, L., Patsis, P.A.: MNRAS **434**, 2922 (2013)
37. Tsoutsis, P., Efthymiopoulos, C., Voglis, N.: MNRAS **387**, 1264 (2008)
38. Tsoutsis, P., Kalapotharakos, C., Efthymiopoulos, C., Contopoulos, G.: A & A **495**, 743 (2009)
39. Vallée, J.P.: MNRAS **450**, 4277 (2015)
40. Vandervoort, P.O.: ApJ **166**, 37 (1971)
41. Voglis, N., Tsoutsis, P., Efthymiopoulos, C.: MNRAS **373**, 280 (2006)

Ordered and Chaotic Bohmian Trajectories

Athanasios C. Tzemos

Abstract We make a quick review of some highlights of our studies in 2d Bohmian order and chaos which are: (a) the development of a generic theoretical mechanism responsible for the emergence of chaos in arbitrary 2d Bohmian systems (b) the relation between chaos and Bohm's quantum potential and (c) the relation between chaos and entanglement in Bohmian qubit systems and its impact on the establishment of Born's rule by arbitrary initial distributions of Bohmian particles.

Keywords Bohmian Mechanics · Chaos · Entanglement · Born's rule

1 Introduction

Bohmian Quantum Mechanics (BQM) [1–6] is a non relativistic pilot wave quantum theory in which the quantum particles follow certain deterministic trajectories in spacetime according to the so called Bohmian equations of motion:

$$m_i\frac{dx_i}{dt}=\hbar Im\left(\frac{\nabla_i\Psi}{\Psi}\right). \tag{1}$$

The wavefunction Ψ is the solution of the Shrödinger equation (SE): $-\frac{\hbar^2}{2m}\nabla^2\Psi+V\Psi=i\frac{\partial\Psi}{\partial t}$. Thus Ψ acts as a background pilot wave which dictates the evolution of Bohmian particles. BQM predicts the same experimental results as standard Quantum Mechanics.

If we insert the polar decomposition of the wavefunction $\Psi=R\exp(iS/\hbar)$ in the SE and split the real from the imaginary part we find the equations $\frac{\partial R}{\partial t}=-\frac{1}{2m}[R\nabla^2S+2\nabla R\nabla S]$ and $\frac{\partial S}{\partial t}=-\left[\frac{|\nabla S|^2}{2m}+V+Q\right]$. The first equation acts as a continuity equation for the probability density ρ and the Bohmian velocity field

A. C. Tzemos (✉)
Research Center for Astronomy and Applied Mathematics of the Academy of Athens, Soranou Efesiou 4, Athens GR-11527, Greece
e-mail: atzemos@academyofathens.gr

T. Bountis et al. (eds.), *Chaos, Fractals and Complexity*, Springer Proceedings in Complexity, https://doi.org/10.1007/978-3-031-37404-3_5

$v = \frac{\nabla S}{m}$ (provided that $\rho = R^2$), while the second equation differs from the classical Hamilton-Jacobi equation by the term

$$Q = -\frac{\hbar^2}{2m}\frac{\nabla^2 R}{R}, \tag{2}$$

which is the so called 'quantum potential' [7–10] and depends on the curvature of the wavefuntion. The introduction of quantum potential brings the Bohmian evolution in a Hamiltonian form.

Bohmian equations are nonautonomous and, in general, highly nonlinear. Therefore they allow us to study chaos in quantum systems in a straightforward way by applying all the techniques of classical dynamical system theory.[1] Thus, in general, ordered and chaotic Bohmian trajectories coexist in a given Bohmian system. Bohmian chaos has attracted a lot of interest in the last decades (see our review papers [13, 14] and references therein). In fact, it is an active research field of the Research Center for Astronomy and Applied Mathematics (RCAAM) of the Academy of Athens in the last two decades. Our group made several important contributions in Bohmian Dynamics and in the next sections we are going to quickly review some of our most significant results in 2d Bohmian quantum systems.

The structure of the paper is the following: in Sect. 2 we present the nodal point-X-point mechanism for the generation of chaos in 2-d Bohmian systems and its relation with the quantum potential. In Sect. 3 we talk about chaos, entanglement and Born's rule (BR) in BQM by use of Bohmian qubits. Finally in Sect. 4 we make our summary and talk about our future research plans.

2 Nodal Point-X-Point Complex Mechanism

From the very first works in Bohmian Dynamics it was well understood that chaos is closely related with the moving nodal points of the wavefunction where Ψ becomes zero (i.e. $\Psi_{Real} = \Psi_{im} = 0$) . In fact, the Bohmian equations become singular at the nodal points and very close to them the quantum particles evolve very fast in a spiral way forming the so called Bohmian vortices.

However, in [15, 16] it was shown that in the frame of reference of a moving nodal point N there exists an unstable fixed point X, the 'X-point'. Together they form the so called 'nodal point-Xpoint complex' (NPXPC), a characteristic geometrical form of the Bohmian flow in the close neighbourhood of N. The larger the velocity of N the smaller the distance between N and X. The NPXPCs evolve in the configuration space and whenever a Bohmian particle comes close to them it gets scattered by the X-point. The cumulative effect of many such scattering events is the saturation of

[1] While in standard Quantum Mechanics quantum chaos refers to the characterization of the universal properties of quantum systems that reflect the regular or chaotic behaviour of their classical analogues [11, 12], in BQM chaos/order refers to the high/low sensitivity of the quantum trajectories on the initial conditions, as in the classical case.

the maximum Lyapunov exponent at a positive value, i.e. the emergence of chaos in Bohmian trajectories. Trajectories that do not encounter the NPXCS are ordered. The NPXPC mechanism has been extensively tested in the case of the unperturbed 2d harmonic oscillator with incommensurable frequencies which is the standard system in the study of Bohmian chaos (see for example [17–19]).

We proceed now with the wavefunction [20]

$$\Psi(x, y, t) = a\Psi_{0,0} + b\Psi_{1,0} + c\Psi_{1,1}, \tag{3}$$

where $\Psi_{n_1,n_2}(\mathbf{x}, t) = \Psi_{n_1,n_2}(\mathbf{x})e^{-iE_it/\hbar}$ and $\Psi_{n_1,n_2}(\mathbf{x})$ are eigenstates of the 2d harmonic oscillator of the form

$$\Psi_{n_1,n_2}(\mathbf{x}) = \prod_{k=1}^{2} \frac{\left(\frac{m_k\omega_k}{\hbar\pi}\right)^{\frac{1}{4}} \exp\left(\frac{-m_k\omega_k x_k^2}{2\hbar}\right)}{\sqrt{2^{n_k} n_k!}} H_{n_k}\left(\sqrt{\frac{m_k\omega_k}{\hbar}}x_k\right), \tag{4}$$

Moreover, n_1, n_2 are the quantum numbers, ω_1, ω_2 the frequencies and E_1, E_2 the corresponding energies. Hereafter we set $m_i = \hbar = 1$ and write x_1, x_2 as x, y. Thus for every doublet (n_1, n_2) we have $E = \sum_{i=1}^{2}(n_i + \frac{1}{2})\omega_i$, This wavefunction has a single nodal point $N = \left(x_N = -\frac{a\sin(t\omega_x + t\omega_y)}{\sqrt{2\omega_x}b\sin(t\omega_y)},\ y_N = -\frac{b\sin(t\omega_x)}{\sqrt{2\omega_y}c\sin(t\omega_x + t\omega_y)}\right)$.

We work with $a = b = 1, c = \sqrt{2}/2$ and $\omega_x = 1, \omega_y = \sqrt{2}/2$.[2] Its trajectory is shown in Fig. 1a, where we observe the characteristic arcs which lead to infinity with very high velocities (when the denominators of x_N, y_N go to 0). In Fig. 1b we present two trajectories which are captured by a moving nodal point. The smaller their distance from N the longer the existence of the Bohmian vortices. When N accelerates the trajectories cannot follow it anymore and wander around the configuration space until their next interaction with an NPXPC.

In Fig. 1c we show the coexistence of order and chaos in the same system and observe that the chaotic and ordered trajectories enter the region of the ordered trajectories. This is due to the explicit time dependence of the Bohmian velocity field, i.e. two trajectories may pass through the same points of the configuration space but always at different times.

The nonautonomous character of the Bohmian flow complicates the study of the scattering events between particles and NPXPCs, since everything changes with time. However, when the velocity of N is small and the Bohmian flow around it varies slowly in time, we can fix the time t and treat the flow as autonomous. Then we can introduce a new (fictitious) time s for this autonomous field and draw the invariant curves of the X-point in the frame of reference of N. By doing so we can see the geometry of the scattering events for a certain time window. This is the so called 'adiabatic approximation' [15, 16]. In Fig. 2 we show two NPXPCs at two different times. In the first case N has a small velocity while in the second case we have a fast moving N. We observe the drastical reduction of the size of the NPXPC with

[2] We need incommensurable frequencies in order to observe chaos.

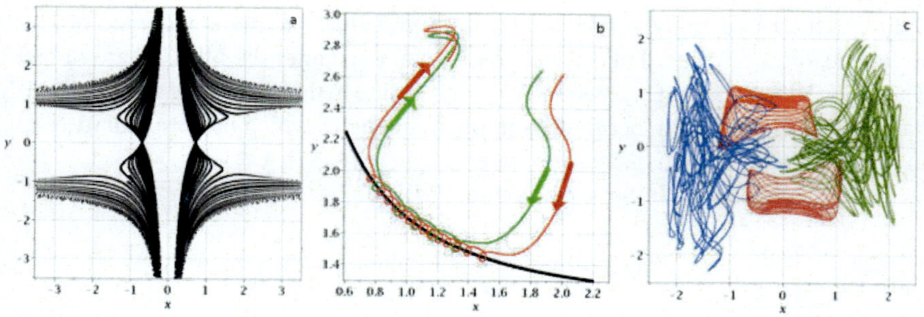

Fig. 1 **a** The trajectory of a moving nodal point: the nodal point spends some time in the broad environment of the center of the configuration space. Then accelerates and goes very fast to infinity until it enters again the configuration space from another direction an so on. **b** Two particles which get trapped around a moving nodal point and form Bohmian vortices (spiral part of the red and green trajectories). The closer the encounter with the nodal point, the denser the vortex and the longer its existence. After some time the nodal point accelerates and the particles cannot follow it any more. **c** Two chaotic (blue and green) and two ordered (red and brown) Bohmian trajectories. Ordered trajectories do not encounter the NPXPCs. The overlap between different trajectories is due to the explicit time dependence of the Bohmian field

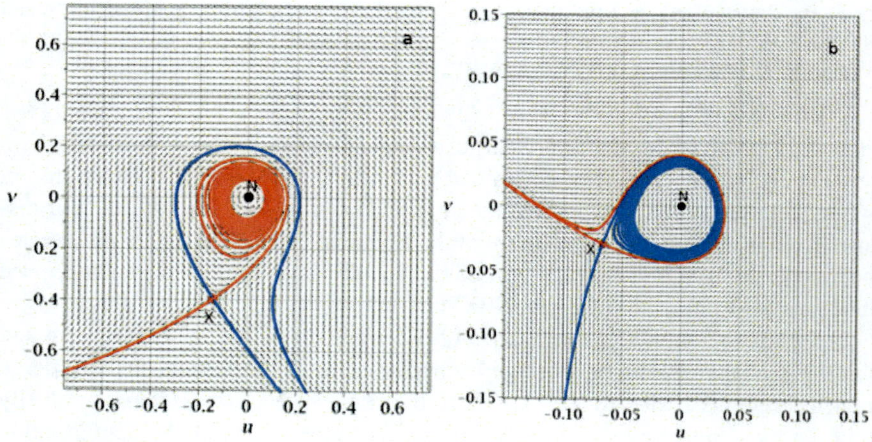

Fig. 2 Two typical NPXPCs: the moving nodal point N is centered at its own frame of reference ($u = x - x_N$, $v = y - y_N$), while the X-point lies nearby. The red and blue curves are the unstable and stable asymptotic curves of the X-point correspondingly. This structure is time dependent: in **a** we show the NPXPC of a slow moving N and in **b** of a fast moving N. The distance between N and X decreases with the velocity of N and the asymptotic curve which enters the region of N changes from unstable (red) to stable (blue)

the increase of the nodal velocity. The small arrows represent the direction of the Bohmian flow at the current time instances and show us how the Bohmian particles evolve in such a 'frozen' field (in the fictitious time s). Moreover, the red and blue curves represent the unstable and stable invariant curves emanating from X. We observe that there is a narrow channel between the two branches of the blue curve

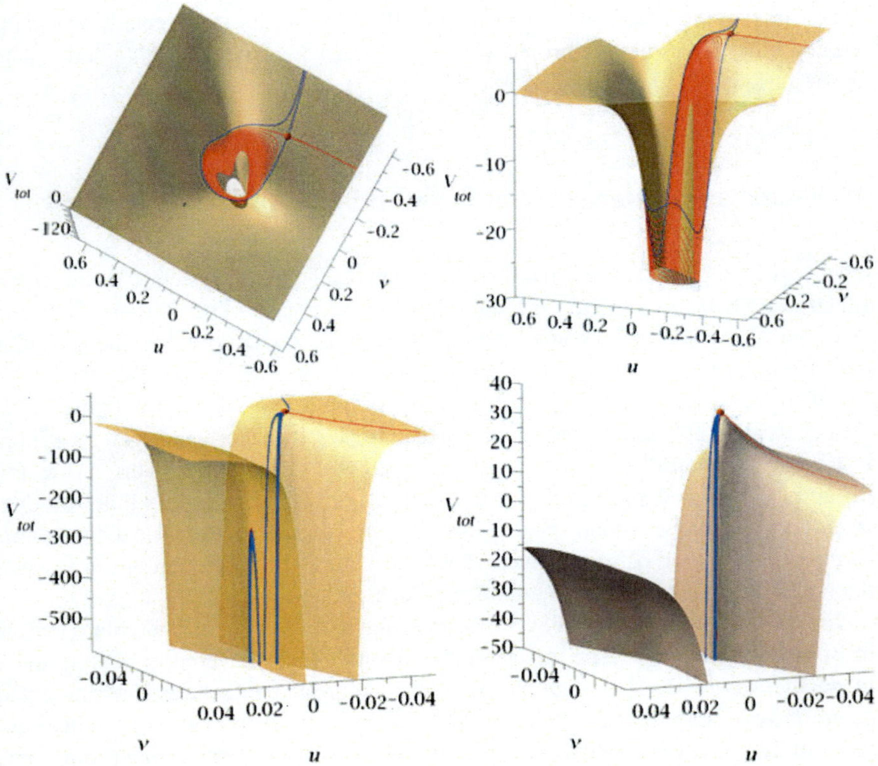

Fig. 3 Upper panel: The surface of the total potential V_{tot} from 2 different angles for a slow moving nodal point. The red dot is the X-point and is close to a local maximum of V_{tot}. The red and blue curves are the projections of the unstable and stable asymptotic curves of the X-point on the surface of V_{tot}. Lower panel: The same for a fast nodal point: in this case the X-point is on a very well distinguished maximum of V_{tot}. In both cases the nodal point is at the centre of the tube at $V_{tot} = -\infty$

which gives access to the nodal point. Particles that will enter this region will form Bohmian vortices. In any other case the incoming particles will be deflected from the X-point according to the Bohmian flow (scattering events of type I and II in [15, 16]). Therefore it is the X-point and not the nodal point which is responsible for the production of chaos.

Having in mind that N is the point of the configuration space where the quantum potential Q goes to $-\infty$ (and so does the total potential $V_{tot} = Q + V$), in our recent studies we tried to find the position of X on the surface of Q and V_{tot}. What we found is that X is always close to the local maximum of Q and V_{tot} in the close area of N [21, 22]. Moreover its distance from this local maximum is inversely proportional to the nodal velocity. This is shown in Fig. 3 where we plot V_{tot} in the cases of a slow and a fast moving nodal point correspondingly. The red dot represents X while the red and blue curves are the projections of its invariant curves (unstable and stable) on the surface of V_{tot}.

3 Chaos and Entanglement in Bohmian Qubits

Quantum entanglement is a fundamental property of quantum systems. Two systems A and B are entangled whenever their joint wavevector cannot be written in the form $|\Psi_{AB}\rangle = |\Psi_A\rangle \otimes |\Psi_B\rangle$. Entanglement lies at the heart of Quantum Mechanics and is the prerequisite of most quantum information algorithms and protocols [23].

The fundamental block of quantum information is the so called qubit (similarly to the classical bit). As a qubit we define a quantum system whose state can be written in the form $\Psi = a|0\rangle + b|1\rangle$, where $|a|^2 + |b|^2 = 1$ according to Born's rule and $|0\rangle$ and $|1\rangle$ two well separated quantum states, i.e. with a negligible inner product (overlap) $\langle 0|1\rangle$ in the Hilbert space. Thus, in sharp contrast with classical bits, qubits can be in any superposition between the basis states $|0\rangle, |1\rangle$, something that gives them unique advantanges from an informational point of view [24].

In 2019 we began our studies on the interplay between chaos and entanglement in Bohmian Mechanics. The major problem that we had to overcome was to find a model that would be simple enough and would capture all the features of an entangled quantum system. In standard Quantum Mechanics the simplest choice is to work with spin qubits. However, in BQM position representation plays a prominent role, since we always talk about particle trajectories. Thus we decided to construct qubit states in the position representation by use of coherent states of the quantum harmonic oscillator [25].

A coherent state is defined as the eigenstate of the annihilation operator of the quantum harmonic oscillator: $\hat{\alpha}|\alpha(t)\rangle = A(t)|\alpha(t)\rangle$ where $A(t) = |A(t)|\exp(i\phi(t))$. In the position representation the wavefunction of a coherent state along the axis x has the form:

$$Y(x,t) = \left(\frac{m\omega}{\pi\hbar}\right)^{\frac{1}{4}} \exp\left[-\frac{m\omega}{2\hbar}\left(x - \sqrt{\frac{2\hbar}{m\omega}}\Re[A(t)]\right)^2 + i\left(\sqrt{\frac{2m\omega}{\hbar}}\Im[A(t)]x + \xi(t)\right)\right], \tag{5}$$

with $\Re[A(t)] = a_0\cos(\sigma - \omega t)$, $\Im[A(t)] = a_0\sin(\sigma - \omega t)$, $\xi(t) = \frac{1}{2}\Big[a_0^2\sin(2(\omega t - \sigma)) - \omega t\Big]$, where $a_0 \equiv |A(0)|$ and $\sigma = \phi(0)$ is the initial phase of A. Thus we studied entangled states of two non interacting 1-d oscillators[3] described by wavefunctions of the form:

$$\Psi(x,y,t) = c_1 Y_R(x,t) Y_L(y,t) + c_2 Y_L(x,t) Y_R(y,t) \tag{6}$$

[3] Namely our Hamiltonian is $H = \frac{p_x^2}{2m_x} + \frac{p_y^2}{2m_y} + \frac{1}{2}m_x\omega_x^2x^2 + \frac{1}{2}m_y\omega_y^2y^2$.

with $|c_1|^2 + |c_2|^2 = 1$, and where Y_R and Y_L represent a one-dimensional coherent state moving to the right or to the left from the center of the oscillation along a certain axis (x or y) (Fig. 1 of [25]). We worked with common large amplitude $a_0 = 5/2$ so as the wavefunctions Y_R and Y_L have a negligible overlap in Hilbert space, i.e. they define the basis states of a qubit. Consequently the wavefunction $\Psi(x, y, t)$ refers to an entangled state of two non interacting qubits made of coherent states along the x and y directions (see Fig. 1 of [25]). The degree of the entaglement is controlled by the value of c_2, with $c_2 \in [0, \sqrt{2}/2]$, where $c_2 = 0$ corresponds to a product state while $c_2 = \sqrt{2}/2$ corresponds to a maximally entangled state (Bell state).[4]

This model was proven to be ideal for our goal since:

- It is completely analogous to a spin based two-qubit system and its entanglement can be found analytically [25–27].
- It has infinitely many nodal points (due to the infinitely many energy eigenstates contributing in the coherent state) whose position is

$$x_N = \frac{\sqrt{2}\left(k\pi \cos\left(\omega_y t\right) + \sin\left(\omega_y t\right) \ln\left(\left|\frac{c_1}{c_2}\right|\right)\right)}{4\sqrt{\omega_x} a_0 \sin\left(\omega_{xy} t\right)} \tag{7}$$

$$y_N = \frac{\sqrt{2}\left(k\pi \cos\left(\omega_x t\right) + \sin\left(\omega_x t\right) \ln\left(\left|\frac{c_1}{c_2}\right|\right)\right)}{4\sqrt{\omega_y} a_0 \sin\left(\omega_{xy} t\right)} \tag{8}$$

  with $k \in Z$, k even for $c_1 c_2 < 0$ or odd for $c_1 c_2 > 0$ and $\omega_{xy} \equiv \omega_x - \omega_y$. If we plot the (x_N, y_N) for various k's we will see that they form straight lattices (Fig. 4 of [19]) which move in the configuration space.[5] Their footprint for $t \in [0, 200]$ is given in Fig. 4a.
- The form of the probability density is very simple. It has two well defined almost gaussian blobs which rotate and oscillate in the configuration space and collide from time to time. For weak entanglement one blob is large (leading blob) and the other is small (secondary blob). With the increase of the entanglement these blobs tend to become identical (see Figs. 1 and 10 of [29]).

We found that:

1. The ordered trajectories of this model are deformed Lissajous figures. The initial conditions which produce ordered trajectories are confined on the region of the leading blob of the wavefunction (see Fig. 15 of [29]).
2. Chaos is produced due to the collisions between the blobs of $P = |\Psi|^2$ which take place at the central area of the configuration space. During the collisions the straight lattices of the NPXPCs become dense in the central area and scatter the incoming trajectories, while between the collisions the trajectories tend to follow the Lissajous-like motion of the two blobs (see Fig. 3 of [30]).

[4] In the absence of interacting terms the conservation of QE is guaranteed.

[5] In fact their position can be found analytically for any number of qubits! [28].

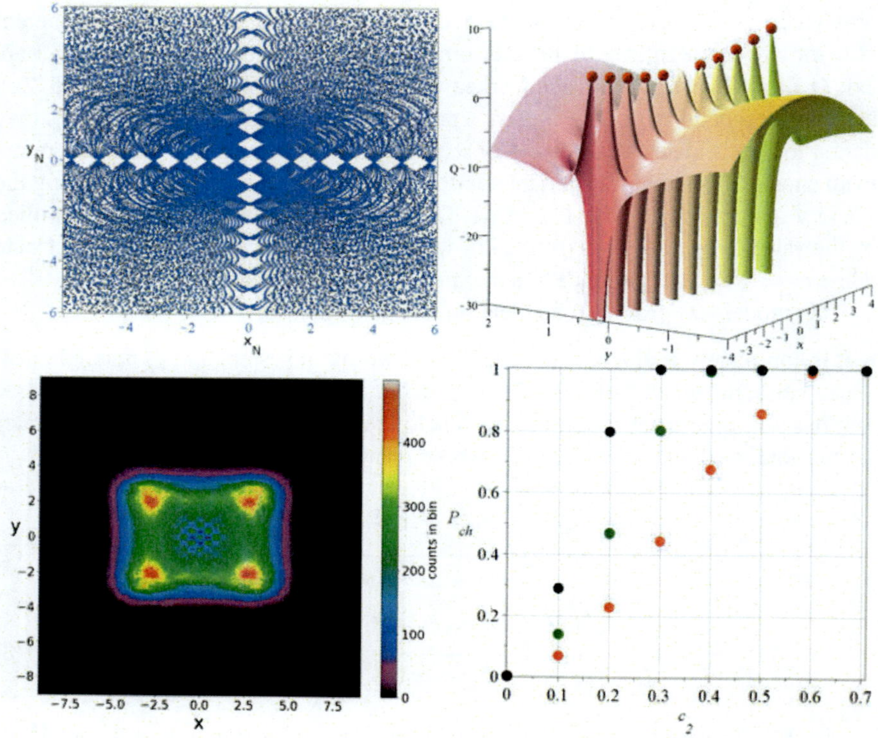

Fig. 4 Upper left: **a** The footprint of the nodal points for $t \in [0, 100]$ and for $k = -23..23$ in the maximally entangled state ($c_2 = \sqrt{2}/2$). Upper right: **b** The surface of the quantum potential Q. We observe the multiple tubes going down to $-\infty$ and the X-points (red dots) at the local maxima of surface of Q. Bottom left: **c** A typical colorplot of a chaotic trajectory in the partially entangled state with $c_2 = 0.5$. All chaotic trajectories with a given c_2 have almost the same long limit distribution of points. Thus the chaotic trajectories are practically ergodic. Bottom right: **d** The proportion of chaotic trajectories for different values of the entanglement parameter c_2 in 2, 3 and 4 entangled qubit states (red, green and black dots correspondingly). By increasing the number of qubits we find a larger number of chaotic-ergodic trajectories in the BR distribution for any given entanglement. Thus it is reasonable to make the conjecture that BR is going to be always accessible in N-qubit systems with large N

3. The higher the entanglement the larger the number of the chaotic trajectories inside the support of Ψ[6] [30, 31].
4. The X-points are on the top of the local maxima of the quantum potential as in the single nodal point case (Fig. 4b).

After studying in detail the above results with a large number of numerical simulations we tried to understand the relation between entanglement and chaos in the long time limit. Our main goal was to understand the mechanism behind the dynamical approximation of Born's rule in BQM: while in standard Quantum Mechanics

[6] The region of the configuration space where $|\Psi|^2$ is not negligible.

Born's rule $P = |\Psi|^2$ is an axiom, in BQM one can consider, in principle, initial distributions of particles with $P_0 \neq |\Psi_0|^2$. The origin of BR is a fundamental question in the Bohmian framework [32–36].

Our main result was that, in this model, all chaotic trajectories have the same long time distribution of points in the support of the wavefunction,[7] for any given nonzero value of entanglement,[8] i.e. they are ergodic. This property was found by introducing a dense grid of square cells inside the support of the wavefunction and counting the number of the passages of the trajectories in every cell of the grid (Fig. 4c). The resulting colorplots of the chaotic trajectories are almost the same.

Ergodicity gives us the opportunity to have a reliable picture of the long limit behaviour of any chaotic trajectory[9] without having to integrate an enormous number of individual initial conditions. Combined with the result 3) it implies that the establishment of BR in this model will depend only on the ratio between ordered and chaotic trajectories in the Born distribution. Namely, BR is going to be accessible by any initial distribution with $P_0 \neq |\Psi_0|^2$ but with the same ratio between ordered and chaotic trajectories with the BR distribution [19].

These results were extended in the cases of 3 and 4 qubits. The increase of the dimension of the configuration space (every qubit is defined on a different coordinate and thus contributes to the total dimension of the configuration space, i.e. N qubits→ N coordinates) was found to be crucial for the number of chaotic trajectories inside the support of the wavefunction: the larger the number of qubits the larger the number of the chaotic-ergodic trajectories for any given nonzero amount of entanglement and thus the more accessible is BR to arbitrary initial distributions of Bohmian particles. Consequently, we expect that in N-qubit systems, with N large (say more than 10), BR is going to be practically reachable by any initial distribution of Bohmian particles (Fig. 4d).

4 Summary

BQM is a trajectory based quantum theory which gives us the opportunity to study chaotic behaviour of quantum systems with all the techniques of classical dynamical systems. Order and chaos play a fundamental role in the evolution of Bohmian particles. In this paper we made a quick review of some of our most important results in Bohmian chaos in 2 dimensional systems based on the NPXPC mechanism.

Aiming at the extension of the NPXPC mechanism in 3-d Bohmian systems in [37–39] we found cases where the wavefunctions of 3-d quantum harmonic oscillators exhibit the phenomenon of partial integrability, i.e. their Bohmian trajectories evolve on 2-d integral surfaces embedded in the 3-d configuration space. After gaining a lot of information about order and chaos in 3 dimensions from these systems we

[7] I.e. the region of the configuration space where $|\Psi(t)|^2$ is not negligible.

[8] For zero entanglement the Bohmian system is decoupled and all trajectories are ordered.

[9] Which is important in the study of Born's rule.

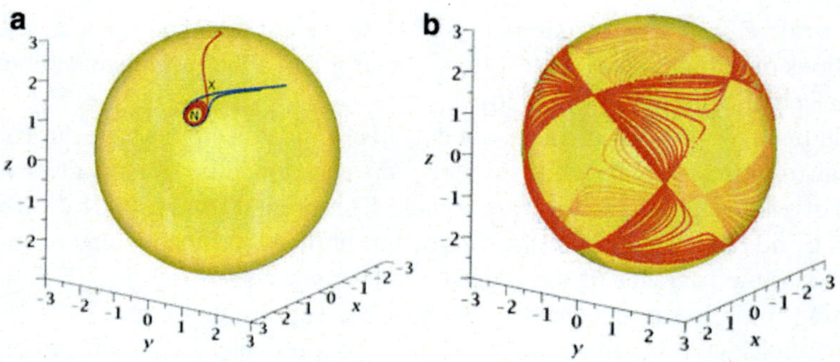

Fig. 5 **a** A NPXPC on the spherical surface of a partially integrable 3d quantum harmonic oscillator $\Psi(x, y, z, t) = \frac{1}{\sqrt{3}} \left(\Psi_{1,0,0}(x, y, z, t) + \Psi_{0,1,0}(x, y, z, t) + \Psi_{0,0,1}(x, y, z, t)\right)$. **b** the footprint of the nodal trajectory on the sphere

finally managed to extend our mechanism in arbitrary 3-d systems in [40]. The 3d NPXPC mechanism helped us to understand the generation of chaos in 3-qubit systems. Moreover, recently we studied the case of multinodal Bohmian systems whose multiple nodal points do not have a certain geometry on the configuration space (as in the case of qubits) but they are randomly scattered on the $x - y$ plane [22].

At the present moment we are trying to understand the dynamical establishing of BR in the case of a simple partially integrable system with spherical integral surface (Fig. 5) and its possible differences from the 2-d systems with planar configuration space. From this work we expect to gain new information about the relation between entanglement and chaos in multipartite Bohmian systems.

Acknowledgements This research was conducted in the framework of the program of the RCAAM of the Academy of Athens, "Study of the Dynamical Evolution of the Entanglement and Coherence in Quantum Systems." The author wants to thank Prof. George Contopoulos for his useful comments.

References

1. Bohm, D.: A suggested interpretation of the quantum theory in terms of "hidden" variables. i. Phys. Rev. **85**, 166 (1952)
2. Bohm, D.: A suggested interpretation of the quantum theory in terms of "hidden" variables. ii. Phys. Rev. **85**, 180 (1952)
3. Holland, P.R.: The Quantum Theory of Motion: An Account of the de Broglie-Bohm Causal Interpretation of Quantum Mechanics. Cambridge University Press (1995)
4. Dürr, D., Teufel, S.: Bohmian Mechanics: The Physics and Mathematics of Quantum Theory. Springer (2009)
5. Pladevall, X.O., Mompart, J.: Applied Bohmian Mechanics: From Nanoscale Systems to Cosmology. CRC Press (2012)
6. Benseny, A., Albareda, G., Sanz, Á.S., Mompart, J., Oriols, X.: Applied Bohmian mechanics. Eur. Phys. J. D **68**, 1 (2014)

7. Goldstein, S., Struyve, W.: On quantum potential dynamics. J. Phys. A **48**, 025,303 (2014)
8. Licata, I., Fiscaletti, D.: Quantum Potential: Physics, Geometry and Algebra. Springer (2014)
9. Fiscaletti, D.: Geometry Of Quantum Potential, The: Entropic Information of the Vacuum. World Scientific (2018)
10. Riggs, P.J.: Reflections on the de Broglie-Bohm quantum potential. Erkenntnis **68**, 21–39 (2008)
11. Haake, F.: Quantum Signatures of Chaos. Springer (1991)
12. Gutzwiller, M.C.: Chaos in Classical and Quantum Mechanics, vol. 1. Springer (2013)
13. Efthymiopoulos, C., Contopoulos, G., Tzemos, A.: Chaos in de Broglie - Bohm quantum mechanics and the dynamics of quantum relaxation. Ann. Fond. de Broglie **42**, 133 (2017)
14. Contopoulos, G., Tzemos, A.C.: Chaos in Bohmian quantum mechanics: a short review. Regul. Chaotic Dyn. **25**, 476–495 (2020)
15. Efthymiopoulos, C., Kalapotharakos, C., Contopoulos, G.: Nodal points and the transition from ordered to chaotic Bohmian trajectories. J. Phys. A **40**, 12,945 (2007)
16. Efthymiopoulos, C., Kalapotharakos, C., Contopoulos, G.: Origin of chaos near critical points of quantum flow. Phys. Rev. E **79**, 036,203 (2009)
17. Contopoulos, G., Efthymiopoulos, C.: Ordered and chaotic Bohmian trajectories. Celest. Mech. Dyn. Astron. **102**, 219 (2008)
18. Contopoulos, G., Delis, N., Efthymiopoulos, C.: Order in de Broglie–Bohm quantum mechanics. J. Phys. A **45**, 165,301 (2012)
19. Tzemos, A., Contopoulos, G.: The role of chaotic and ordered trajectories in establishing Born's rule. Phys. Scr. **96**, 065,209 (2021)
20. Parmenter, R.H., Valentine, R.W.: Deterministic chaos and the causal interpretation of quantum mechanics. Phys. Let. A **201**(1), 1 (1995)
21. Tzemos, A., Contopoulos, G.: Bohmian quantum potential and chaos. Chaos, Solitons Fractals **160**, 112,151 (2022)
22. Tzemos, A.C., Contopoulos, G.: Bohmian chaos in multinodal bound states. Found. Phy. **52**(4), 1–20 (2022)
23. Horodecki, R., Horodecki, P., Horodecki, M., Horodecki, K.: Quantum entanglement. Rev. Mod. Phys. **81**, 865 (2009)
24. Nielsen, M.A., Chuang, I.L.: Quantum Computation and Quantum Information. Cambridge Series on Information and the Naturciences. Cambridge University Press (2004)
25. Tzemos, A.C., Contopoulos, G., Efthymiopoulos, C.: Bohmian trajectories in an entangled two-qubit system. Phys. Scr. **94**, 105,218 (2019)
26. Ramšak, A.: Spin–spin correlations of entangled qubit pairs in the Bohm interpretation of quantum mechanics. J. Phys. A **45**(11), 115,310 (2012)
27. Zander, C., Plastino, A.: Revisiting entanglement within the Bohmian approach to quantum mechanics. Entropy **20**, 473 (2018)
28. Tzemos, A., Contopoulos, G.: Born's rule in multiqubit bohmian systems. Chaos, Solitons & Fractals **164**, 112,650 (2022)
29. Tzemos, A.C., Contopoulos, G.: Ergodicity and Born's rule in an entangled two-qubit Bohmian system. Phys. Rev. E **102**, 042,205 (2020)
30. Tzemos, A.C., Contopoulos, G.: Chaos and ergodicity in an entangled two-qubit Bohmian system. Phys. Scr. **95**, 065,225 (2020)
31. Tzemos, A.C., Contopoulos, G.: Ergodicity and Born's rule in an entangled three-qubit Bohmian system. Phys. Rev. E **104**, 054,211 (2021)
32. Valentini, A.: Signal-locality, uncertainty, and the subquantum h-theorem. i. Phys. Lett. A **156**, 5 (1991)
33. Valentini, A.: Signal-locality, uncertainty, and the subquantum h-theorem. ii. Phys. Lett. A **158**, 1 (1991)
34. Dürr, D., Goldstein, S., Zanghi, N.: Quantum equilibrium and the origin of absolute uncertainty. J. Stat. Phys. **67**, 843 (1992)
35. Valentini, A., Westman, H.: Dynamical origin of quantum probabilities. Proc. Roy. Soc. A **461**, 253 (2005)

36. Towler, M., Russell, N., Valentini, A.: Time scales for dynamical relaxation to the Born rule. Proc. Roy. Soc. A **468**, 990 (2011)
37. Tzemos, A.C., Contopoulos, G., Efthymiopoulos, C.: Origin of chaos in 3-d bohmian trajectories. Phys. Lett. A **380**(45), 3796–3802 (2016)
38. Contopoulos, G., Tzemos, A.C., Efthymiopoulos, C.: Partial integrability of 3d bohmian trajectories. J. Phys. A **50**(19), 195,101 (2017)
39. Tzemos, A.C., Contopoulos, G.: Integrals of motion in 3d bohmian trajectories. J. Phys. A **51**(7), 075,101 (2018)
40. Tzemos, A.C., Efthymiopoulos, C., Contopoulos, G.: Origin of chaos near three-dimensional quantum vortices: a general Bohmian theory. Phys. Rev. E **97**, 042,201 (2018)

A Brief Introduction to Quantum Chaos of Generic Systems

Marko Robnik

Abstract This article is an updated revised version of a recent review paper on quantum chaos in mixed-type systems, between regularity and chaos (Robnik 2020), covering the topics presented at the 28th Summer School-Conference on Dynamical Systems and Complexity, held in Chania, Crete, Greece, in July 2022, dedicated to the 70th birthday of Professor Athanassios (Thanasis) Fokas. Chaos (chaotic behaviour) can emerge in deterministic systems of classical dynamics. It is due to the sensitive dependence on initial conditions, meaning that two nearby initial states of a system develop in time such that their positions (states) separate very fast (exponentially) in time. After a finite time (Lyapunov time) the accuracy of orbit characterizing the state of the system is entirely lost, the system could be in any allowed state. The system can be also ergodic, meaning that one single orbit describing the evolution of the system visits any other neighbourhood of all other states of the system. In this sense, chaotic behaviour in time evolution does not exist in quantum mechanics. However, if we look at the structural and statistical properties of the quantum system, we do find clear analogies and relationships with the structures of the corresponding classical systems. This is manifested in the eigenstates and energy spectra of various quantum systems (mesoscopic solid state systems, molecules, atoms, nuclei, elementary particles) and other wave systems (electromagnetic, acoustic, elastic, seismic, water surface waves and gravitational waves), which are observed in nature and in the experiments. Here we review the basic aspects of quantum chaos in Hamiltonian systems. We shall focus on the most general (generic) systems, also called mixed-type systems, as their classical counterparts in the phase space exhibit regular regions coexisting with the chaotic regions for complementary initial conditions. We shall review the basic concepts of quantum chaos in the stationary picture, that is the properties of the eigenstates of the stationary Schrödinger equation, the structure of wave functions, and of the corresponding Wigner functions in the quantum phase space, and the statistical properties of the energy spectra. Before treating the general mixed-type case we shall review the two extreme cases, the universality classes,

---

M. Robnik (✉)
CAMTP - Center for Applied Mathematics and Theoretical Physics, University of Maribor, Mladinska 3, European Union, 2000 Maribor, Slovenia
e-mail: Robnik@uni-mb.si

T. Bountis et al. (eds.), *Chaos, Fractals and Complexity*, Springer Proceedings in Complexity, https://doi.org/10.1007/978-3-031-37404-3_6

namely the regular (integrable) systems, and the fully chaotic (ergodic) systems. Then the Berry-Robnik (1984) picture will be presented, and the underlying Principle of Uniform Semiclassical Condensation (PUSC) of the Wigner functions. Next, we shall consider the effects of quantum (dynamical) localization, which set in when the classical transport time (like diffusion time) is longer than the Heisenberg time scale (defined as the Planck constant divided by the mean energy level spacing). It will be shown phenomenologically that in the case of chaotic eigenstates in the quantum phase space (Wigner functions) the energy spectra display Brody level spacing distribution, where the level repulsion exponent (Brody parameter) goes from zero in the strongest localization to 1 in the fully extended states. The Berry-Robnik picture is then appropriately generalized to include the localization effects. Furthermore, the localization measures of chaotic localized eigenstates have a distribution, which in the absence of stickiness structures in the classical phase space is well described by the beta distribution. We neglect, at high energies, the tunneling effects coupling the regular and chaotic levels, since they are manifested only in low-lying levels, because the coupling decreases exponentially with increasing energy (or inverse effective Planck constant).

Keywords Nonlinear dynamics · Chaotic systems · Quantum chaos · Wave chaos · Spectral statistics · Generic Hamilton systems · Quantum localization

1 Introduction

This review is based on the recent review papers by the author [1, 2], and other recent papers with coworkers [3–14]. The first part is an introduction to quantum chaos from the stationary point of view, where we shall describe the purely regular eigenstates versus purely chaotic eigenstates. In the second part we shall address the problem of the mixed-type phase space of generic systems, where regular and chaotic eigenstates coexist. The structure of the eigenstates and their Wigner functions corresponds to the structure of the classical phase space portrait, where regular classical motion on invariant tori exists for certain initial conditions, while the motion is chaotic for the complementary initial conditions. The books by Stöckmann [15] and Haake [16] offer an excellent introduction to quantum chaos. Stöckmann's book presents also many experimental applications of quantum chaos, especially on microwave experiments he has been conducting since 1990 up to date, addressing and realizing practically all important questions of quantum chaos. Many subjects of this paper can be found in the reviews [1, 2, 17, 18].

Let us study the solutions of the Schrödinger equation of a point particle in the potential $V(\mathbf{q})$,

$$i\hbar\frac{\partial\psi}{\partial t} = \hat{H}\psi = -\frac{\hbar^2}{2m}\Delta\psi + V(\mathbf{q})\psi. \tag{1}$$

By $h = 2\pi\hbar$ we denote the Planck constant, $\psi(\mathbf{q}, t)$ is the wave function depending on the N-dimensional space coordinate vector $\mathbf{q}$ and on time t, m the mass, $V(\mathbf{q})$ the potential, and $\Delta = \partial^2/\partial\mathbf{q}^2$ is the N-dimensional Laplace operator. For instance, in the case N=2 we have

$$\Delta = \frac{\partial^2}{\partial x^2} + \frac{\partial^2}{\partial y^2}. \tag{2}$$

We shall mainly restrict ourselves to $N = 2$.

We study the solutions of the Schrödinger equation and relate them to the corresponding classical dynamics. The limiting behaviour $\hbar \to 0$ is of prime interest. The methods to find approximate solutions for small $\hbar$ are known under the name *semiclassical mechanics*, sometimes also *quasiclassical approximations*. They represent the connection between the classical and quantum mechanics. See the books [15, 16]. These approximations should be understood as short wavelength approximations, applicable to all wave systems.

Here we shall treat purely binding potential $V(\mathbf{q})$, in which the classical motion is bounded for all initial conditions. The particle cannot escape to infinity (no ionization threshold). Thus the energy spectrum of $\hat{H}$ (1) is purely discrete and infinite. An example is a classical billiard system, where a point particle is moving freely inside a potential box with hard walls, experiencing an elastic collision when hitting the boundary. If the potential outside the billiard domain is infinite, the Dirichlet boundary conditions of vanishing ψ on the boundary must be satisfied.

In a classical Hamilton system we can have either regular quasiperiodic motion on N-dimensional invariant tori, or chaotic motion. In the latter case we observe the sensitive dependence on initial conditions, which is characterized by the existence of the positive Lyapunov exponents. Two nearby orbits diverge exponentially with time $\propto \exp(\lambda t)$, and the relevant exponent λ is called the largest positive Lyapunov exponent. In the case of billiards, the type of dynamics is entirely determined by the geometrical shape of the boundary. However, in quantum mechanics an orbit (in phase space) and trajectory (in the configuration space) cannot be defined due to the Heisenberg uncertainty principle. Consequently, the divergence of nearby trajectories cannot be defined. Any attempt to define a meaningful quantum correspondent of the *asymptotic* Lyapunov exponent λ leads to the conclusion that it is always zero. Therefore in quantum systems the sensitive dependence on initial conditions does not exist. The time evolution of the wave function $\psi(t)$ (1) is stable, almost periodic, and reversible, in contradistinction to the classically chaotic systems where for times much larger than Lyapunov time $\tau = 1/\lambda$ it is fundamentally irreversible once the accuracy of integration is exhausted. For details see [16]. Therefore, as for the time evolution the correspondence between the classical chaotic and quantum systems does not exist.

However, as already mentioned, there is another aspect of classical chaos, namely the structure of the phase space, the so-called phase portrait. In integrable systems with N degrees of freedom, the quasiperiodic motion takes place on N-dimensional invariant tori, for all initial conditions. Integrable systems are very special and rare, but important, as we can entirely describe them analytically, and also understand what happens (to the phase portrait) if we slightly perturb them, by using a variety of perturbation methods. The opposite extreme are entirely chaotic, ergodic, systems where each orbit is dense and visits any other point in the phase space, with exception of measure zero (represented by the periodic orbits). The entire phase space is just one chaotic invariant component. Therefore the phase space average of functions and the time average are equal. In between there are the mixed-type systems, which possess extremely complex structure of the phase space. The regular islands of stability covered by the invariant tori coexist with chaotic sea surrounding them. The picture exhibits an infinite hierarchy of statistically selfsimilar structures. The celebrated fundamental KAM theorem explains what can happen to slightly perturbed integrable Hamiltonian systems. Most of the invariant tori still exist after the perturbation with the same N frequencies of the quasiperiodic motion, although they are typically slightly distorted. However, the rational tori are destroyed, and in place of them we get an even number of periodic orbits, half of them stable and half of them unstable, surrounded by chaotic region (Poincaré-Birkhoff theorem).

In quantum mechanics, to see the analogies with the classical phase portraits, we must look at the structure of the eigenfunctions, of their corresponding *Wigner functions in the quantum phase space* to be defined below, and at the properties of the corresponding energy spectra. In this stationary picture there is a very well defined correspondence: The quantum signatures of classical chaos, as the title of Haake's book [16] goes, are very well defined.

Let us consider the solutions of the Schrödinger equation (1) for the eigenstates with sharply defined eigenenergies E_n, $\psi(\mathbf{q}, t)$ is $\propto \psi_n(\mathbf{q}) \exp(-iE_n t/\hbar)$. The corresponding eigenfunctions ψ_n are satisfying the boundary conditions, obeying the normalizability of ψ_n, $\int |\psi_n(\mathbf{q})|^2 d^N\mathbf{q} < \infty$. In billiards we usually require $\psi = 0$ on the boundary, but other possibilities, e.g. the Neumann boundary conditions of vanishing normal derivative of ψ_n, are possible.

The eigenstates satisfy $\hat{H}\psi_n = E_n\psi_n$, so that the stationary (time-independent) Schrödinger equation is

$$\frac{\hbar^2}{2m}\Delta\psi_n + (E_n - V(\mathbf{q}))\,\psi_n = 0. \tag{3}$$

In the following sections we shall deal with the different types of solutions ψ_n and the associated energy spectra E_n.

As the final comment in introduction let us remark that the time-dependent and time-independent Schrödinger equations, (1) and (3), are some special examples of some wave equations. Other wave equations of mathematical physics exhibit similar behavior, such as the wave equations describing electromagnetic, acoustic, elastic, seismic waves, water surface waves, etc., where the same questions can be addressed, and the analogous conclusions can be reached. The books by Stöckmann [15] and Haake [16] cover these aspects. We see that the terminology *"quantum chaos"* is much too narrow, and instead, we should speak of *"wave chaos"*. Nevertheless, the quantum chaos is a well established name, but we should be aware of the great variety of wave phenomena that can occur in other wave systems. The wave chaos is closely related to the opposite effects of the spontaneous formation of ordered structure in certain wave systems such as e.g. reaction-diffusion systems. We have to understand under what conditions order or chaos can emerge, which is the subject of Haken's fundamental work on synergetics [19].

2 Quantum Mechanics of Classically Integrable Systems

We shall first deal with classically integrable systems. They are very exceptional, but important, because we can treat them analytically and also analytically and rigorously study what happens when we (slightly) perturb them. The total energy is conserved if their Hamilton function does not depend on time (autonomous system). Their phase space is entirely filled with invariant N-dimensional tori. Some examples are centrally symmetric potentials where the angular momentum is a conserved quantity (integral of motion). In billiards we have only two families of completely integrable systems, the rectangular billiards and the elliptic billiards. In the former case the absolute values of the momenta are conserved, while in the second case the product of the angular momenta with respect to the two foci is the integral of motion [20]. The circular billiard is a special case (zero eccenticity), with conserved angular momentum.

Here we address the question of what can be said to characterize the quantum mechanics of such systems. Do we observe some characteristic properties of the eigenfunctions and of the energy spectra? For the heuristic approach, let us first consider the 2-dimensional billiard systems. The Schrödinger equation, in appropriate units, reduces to the simple 2-dimensional Helmholtz equation

$$\frac{\partial^2 \psi}{\partial x^2} + \frac{\partial^2 \psi}{\partial y^2} + E\psi = 0, \tag{4}$$

where the index (quantum number(s)) n is suppressed. We assume $\psi = 0$ on the boundary. The answer to the above questions is yes. In both billiard families the eigenfunctions have an ordered structure, and in both cases the solutions can be analytically found, due to the separability of the systems. For the rectangle with horizontal width a and the vertical width b the solution is given by $\psi_{m,n}(x, y) = C \sin \frac{\pi m x}{a} \sin \frac{\pi n y}{b}$, where the constant C is fixed by the normalization. Here m and n are the two quantum numbers (positive integers). The nodal lines defined by the zeros of $\psi(\mathbf{q}) = 0$ are the horizontal straight lines $y = jb/n = const.$, where $j = 0, 1, \ldots, n$, and the vertical lines $x = ja/m = const.$, with $j = 0, 1, \ldots, m$.

In the circle billiard the nodal lines are defined by the zeros of the radial Bessel functions, which are circles, and by the zeros of the angular trigonometric function, which are polar rays, straight lines, emanating from the center of the circle. In the case of an ellipse, we can introduce the elliptical coordinates and thereby separate the solution of the Helmholtz equation.

If we perturb the shape of the integrable billiards, and solve the Helmholtz equation (4), this ordered nodal structure becomes destroyed by a *generic* perturbation, which breaks the separability and integrability of the system. Of course, such a transition is the faster the larger the energy of the eigenstate. In ergodic chaotic systems at sufficiently high energies the nodal structure is typically entirely random. Exceptions can exist, associated with so-called scars, but their measure is zero (as a relative fraction of all eigenstates).

Let us now consider the energy spectra E of the integrable billiards, obtained by solving (4)? They are certainly characterized by two quantum numbers. For the rectangle billiard with horizontal width a and the vertical width b, we find using the above eigenfunctions

$$E_{m,n} = \pi^2 \left(\frac{m^2}{a^2} + \frac{n^2}{b^2} \right), \tag{5}$$

where m, n are the two quantum numbers (positive integers). This energy spectrum is explicit and simple, and yet its statistical properties are not trivial. We shall see that the statistical properties of the energy spectra are deeply related to the type of the dynamics of the corresponding classical Hamiltonian system. If the system is integrable we find Poissonian statistics, while in classically chaotic ergodic systems the statistics is well described by the random matrix theory. We shall discuss this in the next section.

Before starting the statistical analysis we must eliminate the system's specific properties, such as the density of energy levels. For this purpose we perform the *spectral unfolding procedure*, which means transformation of the actual physical energy spectrum E_n to *the unfolded energy spectrum* e_n, upon which the mean spacing $\Delta e = \langle (e_{n+1} - e_n) \rangle$ of e_n is equal to 1 for all e. The unfolding is simply $e_n = E_n/\Delta E = E_n \rho(E_n)$, where $\rho(E)$ is the energy level density. The density of states is obatined by the Thomas-Fermi rule of filling the classical phase space inside the energy surface $E = H(\mathbf{q}, \mathbf{p}) = const.$ with the Planck cells of volume $(2\pi\hbar)^N$. Namely, the cumulative number $\mathcal{N}$ of the energy eigenvalues below the energy E is given by

$$\mathcal{N}(E) = \frac{1}{(2\pi\hbar)^N} \int_{H(\mathbf{q},\mathbf{p}) \leq E} d^N\mathbf{q}\, d^N\mathbf{p}. \tag{6}$$

Therefore,

$$\rho(E) = \frac{d\mathcal{N}}{dE} = \frac{1}{(2\pi\hbar)^N} \int \delta\left(E - H(\mathbf{q}, \mathbf{p})\right) d^N\mathbf{q}\, d^N\mathbf{p}, \tag{7}$$

where $\delta(x)$ is the Dirac delta function. For 2-dimensional billiards using the units defined by (4) we can even improve the above asymptotic estimate, valid at $E \to \infty$, namely

$$\mathcal{N}(E) = \frac{\mathcal{A}E}{4\pi} - \frac{\mathcal{L}\sqrt{E}}{4\pi} + c.c. \tag{8}$$

where $\mathcal{A}$ and $\mathcal{L}$ are the area and the circumference of the billiard, while $c.c.$ are some small constants (curvature and corner corrections) which for large E are unimportant. The first term in (8) stems from (6), while the second one is the so-called perimeter correction. Asymptotically when $E \to \infty$ the leading term is dominant, and is known also as the Weyl formula.

With the unfolded energy spectrum at hand we start the statistical analysis. Instead of defining all the correlation functions etc., we shall consider only two statistical measures. The first one is the gap probability $\mathcal{E}(S)$, which is the probability that an interval of length S (of the unfolded spectrum) is empty of levels. The second one is the level spacing distribution $P(S)$: The probability to have a level spacing within the interval $(S, S + dS)$, is equal to $P(S)dS$. They are related through $P(S) = d^2\mathcal{E}(S)/dS^2$ (see e.g. [16]).

For the irrational rectangular billiard (a^2 and b^2 are not rationally connected) we find the Poissonian statistics, namely

$$\mathcal{E}(S) = \exp(-S), \quad P(S) = \exp(-S). \tag{9}$$

We observe that there is no parameter involved in this formula, indicating that something similar should be found in other integrable systems, like the elliptic billiard. Indeed, this is the case, and we speak of *the universality class of the Poissonian spectral statistics of integrable systems*. Some subtleties around this problem (see below) were studied in [21], but the validity of the Poissonian statistics is confirmed in the asymptotic limit. Intuitively we can understand: In a quantum energy spectrum with two or more quantum numbers we have generically a superposition of infinitely many uncorrelated discrete number sequences, and such a superposition always results in a Poissonian sequence, where there are no correlations between the levels whatsoever.

A general N-dimensional classically integrable system is defined by the Hamilton function $H(\mathbf{q}, \mathbf{p})$. We can perform the construction of such N quantum numbers in terms of the classical action variables $\mathbf{I}$. In the semiclassical limit of small $\hbar$ the so-called EBK quantization (torus quantization) is based on the quantization of the classical actions $\mathbf{I}$. It is named after Einstein, Brillouin and Keller. The phase space has $2N$ dimensions, the energy surface $E = H(\mathbf{q}, \mathbf{p}) = const.$ has $2N - 1$ dimensions, and by definition we have N integrals of motion A_j, $j = 1, \ldots, N$. A_1 is by convention the Hamilton function, the energy $E = H$. The N-dimensional invariant surfaces labelled by $\mathbf{A}$ have the topology of N-dimensional tori. The actions - the generalized momenta - are defined by the N circuit integrals on a torus labelled by $\mathbf{A}$ or $\mathbf{I}$,

$$I_j = \frac{1}{2\pi} \oint_{C_j} \mathbf{p} \cdot d\mathbf{q} \tag{10}$$

For details see e.g. the references [22, 23]. The Hamilton function $H(\mathbf{q}, \mathbf{p})$ is only a function of $\mathbf{I}$, because after inverting $A_j = A_j(I_k)$, we have $A_1(\mathbf{q}, \mathbf{p}) = H(\mathbf{q}, \mathbf{p}) =$

$H(\mathbf{I})$. The conjugate angle variables θ_j are called cyclic variables, as they do not appear in the Hamilton function, and thus the actions are of course constants of motion.

For sufficiently small $\hbar$ we perform the torus quantization, or EBK quantization, by quantizing the actions of the tori,

$$\mathbf{I_m} = (\mathbf{m} + \frac{\alpha}{4})\hbar, \tag{11}$$

where $\mathbf{m}$ is a N-dimensional vector of nonnegative integers, and $= \alpha_1, \alpha_2, \ldots, \alpha_N$ are so-called Maslov indices. They count the number of caustics (singularities of the wave function in configuration space) encountered in the configuration space upon traversing round the fundamental circuit C_j. Thus, α_j depends on how the invariant torus lies in the phase space and on the structure of its projection singularities in the configuration space. The energy eigenvalues are then equal to the value of the Hamilton function at the quantized actions (11),

$$E_{\mathbf{m}} = H(\mathbf{I_m}) = H\left((\mathbf{m} + \frac{\alpha}{4})\hbar\right). \tag{12}$$

The formula (12) with (11) is basically the higher dimensional generalization of the 1-dimensional semiclassical quantization, taking into account also the Maslov corrections, which Einstein, Brillouin and Keller were not aware of. It is an approximate solution at small $\hbar$ of the Schrödinger eigenvalue problem (3). For further details see [15].

The Poissonian statistics for the quantal energy spectra of classically integrable systems can now be explained: Since we have N quantum numbers $\mathbf{m} = (m_1, m_2, \ldots, m_N)$, this means generically statistical independent superposition of infinitely many level sequences, which results in a Poissonian sequence, provided there are no special and nongeneric rational relationships or correlations between the individual level sequences. See e.g. [21], where the role of rational relationships in rectangle billiard with sides a and b is discussed. If a^2 and b^2 are rationally connected, we observe the number theoretic degeneracies, implying that the level spacing distribution is a sum of Dirac delta functions rather than a smooth distribution. It slowly converges to the Dirac delta function $P(S) = \delta(S)$ peaked at $S = 0$ asymptotically at large energies. For irrationally related a^2 and b^2 we observe Poissonian behaviour [21].

3 Quantum Chaos in Classically Ergodic Systems

Let us consider the fully chaotic, ergodic systems. We can expect again some kind of universality of the spectral fluctuations and their statistical properties. One example of a chaotic and ergodic system is the stadium of Bunimovich [24]. It is a rectangle of width ϵ with two semicircles on the opposite sides, of unit radius. The motion of a point particle is ergodic and chaotic for any positive ϵ, but the typical time t_T to fill the entire phase space depends strongly on ϵ. Namely, for small ϵ we find a diffusion process [25], which is very relevant for the quantum chaos of this system [6]. If ϵ is large, the diffusion time (or classical transport time) t_T is very short, of order unity, and becomes larger by orders of magnitude for smaller ϵ. This time scale has to be compared to another quantal time scale, the so-called *Heisenberg time*, which is important time scale in any quantum system with discrete energy spectrum, and is defined as $t_H = 2\pi\hbar/\Delta E$, where ΔE is the average energy level spacing $\Delta E = 1/\rho(E)$, determined by the Thomas-Fermi rule (7). It is observed empirically that the quantum diffusion follows the classical diffusion up to the Heisenberg time (also called break time), and is then stopped due to the destructive interference effects, resulting in a localization, which is called *dynamical localization* or *quantum localization*, discovered by Casati, Chirikov, Izrailev and Ford in 1979 [26] in the quantum kicked rotator, an example of a Floquet system. This important phenomenon of quantum localization manifests itself also in the structure of the Wigner functions of the stationary eigenstates in the quantum phase space. If the Heisenberg time is larger than the classical diffusion time (no localization effects present) we find the universal statistical behaviour of the wave functions and of the energy spectra. According to the equation (7) $\Delta E \propto (2\pi\hbar)^N$, and therefore as $\hbar \to 0$, for $N \geq 2$, t_H will be larger than classical t_T. Therefore, at some sufficiently small $\hbar$, in the semiclassical limit, the quantum localization effects disappear and the universality of the statistics of energy spectra and of the wave functions can appear.

Indeed, it has been found already by McDonald and Kaufman [27] that the nodal pattern of the stadium billiard with $\epsilon = 1$ is entirely irregular, in contradistinction to the integrable billiards (rectangle and ellipse billiard). Berry [28] has proposed that the wave functions of classically fully chaotic ergodic systems should behave as Gaussian random functions, such that the probability amplitude $\psi_n(x, y)$ has a Gaussian distribution, and this has been widely confirmed (see e.g. [29]). Therefore we have a good reason to expect universal statistics of energy spectra. Bohigas, Giannoni and Schmit have shown in 1984 [30] in the case of the stadium billiard that (after spectral unfolding) the level spacing distribution $P(S)$ is well described by the Wigner distribution (also called Wigner surmise), as an excellent approximation of the so-called GOE distribution to be discussed below, given by

$$P_W(S) = \frac{\pi S}{2} \exp\left(-\frac{\pi S^2}{4}\right), \tag{13}$$

and the corresponding gap probability is

$$\mathcal{E}_W(S) = 1 - \mathrm{erf}\left(\frac{\sqrt{\pi} S}{2}\right) = \mathrm{erfc}\left(\frac{\sqrt{\pi} S}{2}\right) \tag{14}$$

Here, the linear rise of $P(S)$ at small S, starting from zero, is important, meaning that the levels tend to avoid a degeneracy, and we speak of the linear level repulsion. This behavior is quite different from the Poissonian $P(S) = e^{-S}$, where there is no level repulsion and the degeneracies are quite likely. Bohigas, Giannoni and Schmit performed numerical exploration also of other fully chaotic ergodic billiards with short t_T, and proposed what is now well known as *Bohigas-Giannoni-Schmit (BGS) Conjecture*, namely that the statistical properties of the energy spectra of classically fully chaotic and ergodic systems are described by the Random Matrix Theory (RMT). This conjecture is at the heart of quantum chaos. Percival [31], and in particular Casati, Valz-Gris and Guarneri [32], have presented some qualitative ideas pointing in this direction.

Plentiful numerical calculations have confirmed that the conjecture is correct. The seminal paper by Berry in 1985 [33] has shown on the theoretical side, using the semiclassical methods developed by Gutzwiller in [34], that the spectral autocorrelation function and its Fourier transform, the so-called spectral form factor, agree with the RMT. In 2001 Sieber and Richter [35] extended Berry's work to the next order in power expansion for short times of the form factor. In 2006 Haake and his group [16, 36] have shown that the semiclassical form factor agrees with the RMT to all orders. Therefore, BGS Conjecture can be considered as theorem. The Gutzwiller method rests upon the semiclassical approximation of the Green function (propagator) for the Schrödinger equation, best approached in terms of the Feynnman path integral. The result is an expression of the density of the energy levels in terms of classical periodic orbits. See the book by Stöckmann [15]. This semiclassical theory predicts also limitations of the universality. For example, the delta statistics saturates at scales $L > L_{max} = \hbar/(T\Delta E)$, where ΔE is the mean energy level spacing and T is the period of the shortest periodic orbit. However, in the semiclassical limit $\hbar \to 0$ we see $L_{max} \to \infty$, because $\Delta E \propto \hbar^N$, thus we see the universality region for arbitrarily large L by taking sufficiently small effective $\hbar$ (or sufficiently large energy).

The RMT has been developed mainly by Wigner, Dyson, Mehta [16, 37] and others to describe level statistics of heavy complex nuclei. It was a major surprise that this theory applies also in few degrees of freedom systems, provided they are classically chaotic ergodic, even with just two degrees of freedom, such as billiard model systems, if the semiclassical condition $t_H > t_T$ is satisfied. The idea is that complexity of a system implies randomness of the matrix elements, where this complexity can be either due to many degrees of freedom, or due to chaotic classical dynamics in a low dimensional system, such as just $N = 2$. The main interest is in the statistical properties of the eigenvalues of appropriate ensembles of random matrices, that is matrices with random matrix elements each having a certain probability distribution.

Here we discuss only the main idea of RMT, assuming the Gaussian random distribution of the matrix elements, such that they are statistically independent of each other. Their distribution is invariant w.r.t. the transformations that preserve the symmetry of the Hamilton matrices of the ensemble. In the case of real symmetric Hamilton matrices the transformations are orthogonal transformations, and the ensemble of such random matrices is called Gaussian Orthogonal Ensemble (GOE). If H are complex Hermitian matrices, the group of symmetry preserving transformations are the unitary transformations, and the ensemble is called Gaussian Unitary Ensemble (GUE). The main question, among others, is what are the statistical properties of the eigenvalues of random matrix ensembles.

Instead of the general treatment, we consider just two-dimensional random matrix ensembles, and derive the level spacing distribution for them. For a general Hermitian matrix $\begin{pmatrix} x & y+iz \\ y-iz & -x \end{pmatrix}$, with x, y, z real, and $i^2 = -1$, the two eigenvalues are $\lambda = \pm\sqrt{x^2+y^2+z^2}$ and therefore the level spacing is $S = \lambda_1 - \lambda_2 = 2\sqrt{x^2+y^2+z^2}$. We assume that x, y, z have so far general distributions $g_x(x)$, $g_y(y)$, $g_z(z)$, correspondingly. The level spacing distribution is

$$P(S) = \int_{R^3} dx\, dy\, dz\, g_x(x) g_y(y) g_z(z) \delta(S - 2\sqrt{x^2+y^2+z^2}). \tag{15}$$

For the 2D GUE we assume $g_x(u) = g_y(u) = g_z(u) = \frac{1}{\sigma\sqrt{\pi}} \exp(-\frac{u^2}{\sigma^2})$, and impose the normalization $< S >= 1$, determining σ. Using the spherical coordinates $r = \sqrt{x^2+y^2+z^2}$ and φ, θ, we can do the integral, followed by the normalization $< S >= 1$, yielding 2D GUE formula

$$P(S) = \frac{32S^2}{\pi^2} \exp(-\frac{4S^2}{\pi}), \tag{16}$$

now exhibiting the quadratic level repulsion.

If we restrict the ensemble to the real symmetric class, taking $g_z(u) = \delta(u)$ while g_x and g_y are unchanged Gaussian, and employing the polar coordinates $r = \sqrt{x^2+y^2}$ and φ, 2D GOE formula follows $P(S) = \frac{\pi S}{2} \exp(-\frac{\pi S^2}{4})$, now exhibiting linear level repulsion, the result being identical to $P_W(S)$ from equation (13).

There is a clear cut criterion for the GOE or GUE case: If the system has an antiunitary symmetry exemplified by - but not limited to - the time reversal symmetry, then there exists a large and nontrivial family of basis in the Hilbert space where the representation of the Hamilton operator (matrix) is real symmetric, and GOE statistics applies. If there is no antiunitary symmetry, the system is a general complex Hermitian and the statistics is GUE [37–39]. (The systems with spin are entirely neglected in this review.)

In both RMT cases as well as in the Poissonian case **there is no free parameter: Universality**, the energy level statistics does not depend on any specific features of the classical dynamics, except that it must be ergodic in the first two RMT cases, and entirely integrable in the Poissonian case. Thus we have established universality classes of spectral statistics. According to the important paper by Hackenbroich and Weidenmüller [41] the result applies also to a very large class of other non-Gaussian random matrix ensembles, under the condition that the limiting distribution of the eigenvalues is smooth and confined to a finite interval. As those are mild conditions, this is the evidence that the universality classes are very robust. We have numerically verified this wider universality for a number of various non-Gaussian ensembles [42]. An elementary evidence for the robustness of the linear level repulsion is demonstrated [40, 42] by using (15). We assume $g_z(u) = \delta(u)$, and for general g_x, g_y we can integrate over the (x, y) plane by means of polar coordinates, yielding:

$$P(S) = \frac{S}{4}\int_0^{2\pi} g_x\left(\frac{S}{2}\cos\varphi\right)\, g_y\left(\frac{S}{2}\sin\varphi\right)\, d\varphi, \tag{17}$$

and for small S we obtain

$$P(S) = \frac{\pi S}{2} g_x(0) g_y(0). \tag{18}$$

Thus if g_x, g_y at $x = 0$ and $y = 0$ are finite and nonzero, the level repulsion will be always linear, independent of the details of $g_x(x), g_y(y)$. Indeed, for the Gaussian case, with the normalization $< S >= 1$, we get $\sigma = 1/\sqrt{\pi}$, and then $g_x(0) = g_y(0) = 1$ and see at once $P(S) = \pi S/2$ for small S, which agrees with (13). The result for the GUE case follows by using the general $g_z(z)$, yielding the quadratic level repulsion, which for small S in the special case of Gaussian $g_x,\ g_y,\ g_z$ agrees with (16).

4 Quantum Chaos in Generic Systems

Classically integrable and fully chaotic (ergodic) systems are exceptional and rare, they have measure zero in the space of Hamiltonians. Typical classical Hamiltonian systems are of the the mixed type, having divided phase space. Quite generally, they have extremely complex selfsimilar fractal structure, with the rare exceptions such as the mushroom billiards introduced by Bunimovich, where we have exactly one rigorously integrable (regular) component, and exactly one ergodic chaotic component.

In the general generic case the regular regions consisting of N-dimensional invariant tori coexist in the phase space with chaotic regions. Typically there is an infinite hierarchy of statistically selfsimilar structure consisting of islands of stability surrounded by the chaotic sea, which by itself might comprise several disconnected invariant chaotic components. The phase portrait has a very rich structure and is difficult to describe in detail. The quantum mechanics of such systems is also difficult. In the semiclassical limit $\hbar \to 0$ the quantum mechanics of the stationary Schrödinger equation must somehow tend to the classical mechanics. It was the idea of Percival in 1973 [31] who was the first to suggest, qualitatively, that one should distinguish between the *regular eigenstates* and the *chaotic eigenstates* (he called them irregular). However, the question is: *How*? It is rather obvious that we must introduce some kind of *the quantum phase space*, where the quantum structures can be compared with the classical ones. This can be achieved by introducing the Wigner functions of the quantum eigenstates.

4.1 The Wigner Functions

We define **the Wigner functions of eigenstates**, in terms of the orthonormal eigenfunctions $\psi_n(\mathbf{q})$ in configuration space, in the *quantum phase space* $(\mathbf{q}, \mathbf{p})$, as follows:

$$W_n(\mathbf{q}, \mathbf{p}) = \frac{1}{(2\pi\hbar)^N} \int d^N \mathbf{X} \exp\left(-\frac{i}{\hbar}\mathbf{p}.\mathbf{X}\right) \psi_n(\mathbf{q} - \frac{\mathbf{X}}{2})\psi_n^*(\mathbf{q} + \frac{\mathbf{X}}{2}). \quad (19)$$

They are real valued but **not positive definite**, and possess the following properties:

$(P1)$ $\int W_n(\mathbf{q}, \mathbf{p}) d^N\mathbf{p} = |\psi_n(\mathbf{q})|^2$ (= probability density in configuration space)
$(P2)$ $\int W_n(\mathbf{q}, \mathbf{p}) d^N\mathbf{q} = |\phi_n(\mathbf{p})|^2$ (= probability density in momentum space)
$(P3)$ $\int W_n(\mathbf{q}, \mathbf{p}) d^N\mathbf{q}\, d^N\mathbf{p} = 1$ (normalization)
$(P4)$ $(2\pi\hbar)^N \int d^N\mathbf{q}\, d^N\mathbf{p} W_n(\mathbf{q}, \mathbf{p}) W_m(\mathbf{q}, \mathbf{p}) = \delta_{nm}$ (orthogonality)
$(P5)$ $|W_n(\mathbf{q}, \mathbf{p})| \leq \frac{1}{(\pi\hbar)^N}$ (no singularities; Cauchy-Schwartz inequality)
$(P6 = P4)$ $\int W_n^2(\mathbf{q}, \mathbf{p}) d^N\mathbf{q}\, d^N\mathbf{p} = \frac{1}{(2\pi\hbar)^N}$ (divergence in the limit $\hbar \to 0$)
$(P7)$ $\hbar \to 0:$ $W_n(\mathbf{q}, \mathbf{p}) \to (2\pi\hbar)^N W_n^2(\mathbf{q}, \mathbf{p}) > 0$ (positivity in the limit $\hbar \to 0$)

From this we see that in the semiclassical limit $\hbar \to 0$ the Wigner function becomes predominantly positive definite, that it is supported in a volume cell of size $(2\pi\hbar)^N$, and thus *condenses* in such a cell. Since the Wigner functions are orthogonal, they must "live" in disjoint supports and therefore become statistically independent of each other. The question is, what is the geometry/structure of such a cell.

4.2 The Principle of Uniform Semiclassical Condensation (PUSC) of Wigner Functions

The Principle of Uniform Semiclassical Condensation (PUSC) of Wigner functions of eigenstates is based on the papers by Percival [31], Berry [28], Shnirelman [43], Voros [44], Robnik [17], and Veble, Robnik and Liu [45]. It states that the Wigner function $W_n(\mathbf{q}, \mathbf{p})$ condenses uniformly on a classical invariant component in the classical phase space, when $\hbar \to 0$ and if $t_H > t_T$. The invariant component can be an N-dimensional invariant torus, a chaotic component, or the entire energy surface in the case of classical ergodicity:

(C1) Invariant N-torus (integrable or KAM):

$$W_n(\mathbf{q}, \mathbf{p}) = \frac{1}{(2\pi)^N} \delta\left(\mathbf{I}(\mathbf{q}, \mathbf{p}) - \mathbf{I_n}\right). \quad (20)$$

(C2) Uniform on a chaotic region:

$$W_n(\mathbf{q},\mathbf{p}) = \frac{\delta(E_n - H(\mathbf{q},\mathbf{p}))\,\chi_\omega(\mathbf{q},\mathbf{p})}{\int d^N\mathbf{q}\,d^N\mathbf{p}\,\delta(E_n - H(\mathbf{q},\mathbf{p}))\,\chi_\omega(\mathbf{q},\mathbf{p})} \tag{21}$$

where $\chi_\omega(\mathbf{q},\mathbf{p})$ is the characteristic function on the chaotic component labeled by ω

(C3) Ergodicity (microcanonical Wigner function):

$$W_n(\mathbf{q},\mathbf{p}) = \frac{\delta(E_n - H(\mathbf{q},\mathbf{p}))}{\int d^N\mathbf{q}\,d^N\mathbf{p}\,\delta(E_n - H(\mathbf{q},\mathbf{p})} \tag{22}$$

Here we also define μ as the relative Liouville measure of the relevant classical invariant component indexed by ω:

$$\mu(\omega) = \frac{\int d^N\mathbf{q}\,d^N\mathbf{p}\,\delta(E_n - H(\mathbf{q},\mathbf{p}))\,\chi_\omega(\mathbf{q},\mathbf{p})}{\int d^N\mathbf{q}\,d^N\mathbf{p}\,\delta(E_n - H(\mathbf{q},\mathbf{p}))} \tag{23}$$

This principle has a great predictive power, as shown e.g. in [45]. The important condition for the uniformity of the Wigner functions of chaotic eigenstates is: The Heisenberg time t_H must be larger than classical transport time scales t_T.

4.3 Spectral Statistics in the Mixed-type Systems

PUSC predicts that in the semiclassical limit the eigenstates can be clearly classified as regular or chaotic, based on the criterion whether they overlap with an invariant N-dimensional torus or with a chaotic region. One can separate regular and chaotic energy level sequences, and perform their statistical analysis separately. As the Wigner functions are nonoverlapping the spectral sequences become statistically independent of each other. The regular sequences obey the Poissonian statistics, while the chaotic ones obey the RMT statistics (if the semiclassical condition $t_H \leq t_T$ is satisfied). The total spectrum can be theoretically represented as a statistically independent superposition of regular and chaotic level sequences. All the regular ones can be represented by a single Poissonian sequence, simply because a statistically random superposition of Poissonian level sequences is a Poisson sequence again. On the other hand, the chaotic sequences must be treated one by one, each of them associated with its relevant supporting classical chaotic region. Under these conditions the gap probability $\mathcal{E}(S)$ factorizes: The probability of having no level on interval of length S is the product of probability of having no level of the regular type, times probability of having no level of the chaotic types. In the special case of just one chaotic sequence with the approximate gap probability $\mathcal{E}_W(S) = \mathrm{erfc}\left(\frac{\sqrt{\pi}S}{2}\right)$ and one Poissonian sequence with the gap probability $\mathcal{E}_P(S) = \exp(-S)$ we obtain

$$\mathcal{E}(S) = \mathcal{E}_P(\mu_1 S)\, \mathcal{E}_W(\mu_2 S) = \exp(-\mu_1 S)\ \ \mathrm{erfc}\left(\frac{\sqrt{\pi}\mu_2 S}{2}\right). \tag{24}$$

By μ_1 we denote the relative fraction of the phase space volume of the classical regular regions in the classical phase space, while $\mu_2 = 1 - \mu_1$ is the complementary relative Liouville measure of the chaotic region. At the same time, μ_1 is the average relative density of the regular energy levels, while μ_2 is the average relative density of the chaotic levels. Since the general relationship $P(S) = d^2\mathcal{E}(S)/dS^2$ applies, we derive at once the so-called Berry-Robnik level spacing distribution [46]

$$P_{BR}(S) = e^{-\mu_1 S} e^{-\frac{\pi\mu_2^2 S^2}{4}} \left(2\mu_1\mu_2 + \frac{\pi\mu_2^3 S}{2}\right) + e^{-\mu_1 S}\mu_1^2 \mathrm{erfc}\left(\frac{\sqrt{\pi}\mu_2 S}{2}\right). \tag{25}$$

Naturally, it is normalized $< 1 >= 1$, and by definition (after unfolding!) has the normalized first moment $< S >= 1$. It has been tested in many various billiard systems, and in order to see it, it is very often necessary to reach the high-lying levels, even up to 10^6. Surely, the semiclassical condition (of the time scales) must be satisfied, and for this to be true usually very high energies are required, which is technically very demanding. The best confirmation so far has been achieved by Prosen and Robnik (1994, 1999) [47, 48], 10–15 years after the derivation of (25). The generalization to many chaotic components is quite straightforward [17, 46].

4.4 *Quantum Localization of the Chaotic Eigenstates*

Here we shall discuss the appearance and the consequences of the quantum (or dynamical) localization for the level statistics. If the condition $t_H > t_T$ is not satisfied, the Wigner functions of chaotic eigenstates are not uniformly spread on the relevant classical chaotic component, but are localized, meaning that their effective support is smaller than the classically available chaotic region. For example, this has been found in the stadium billiard if ϵ is sufficiently small. We have shown empirically [3–6, 49] that the aspects of quantum localization in time-independent eigenstates are quite analogous to the dynamical localization phenomena in time-dependent Floquet systems, exemplified by the quantum kicked rotator [50]. Below we shall show examples of localized chaotic states.

We can neglect the tunneling effects, which couple regular and chaotic levels and break the statistical independence assumption, at sufficiently high energies (small effective $\hbar$), because tunneling effects decrease exponentially with the energy or effective $1/\hbar$. On the other hand, when analyzing the quantum localization effects, we observe empirically [3–6, 47] even at high energies that the level spacing distribution of the localized chaotic eigenstates is very well described by the Brody distribution

$$P_B(S) = C_1 S^\beta \exp\left(-C_2 S^{\beta+1}\right), \tag{26}$$

where the two parameters C_1 and C_2 are determined by the two normalizations $< 1 >=< S >= 1$, namely we have $C_2 = \left[\Gamma\left(\frac{\beta+2}{\beta+1}\right)\right]^{\beta+1}$, and $C_1 = (\beta+1)C_2$. The corresponding gap probability is

$$E_B(S) = \frac{1}{\alpha(\beta+1)} Q\left(\frac{1}{\beta+1}, (\alpha S)^{\beta+1}\right) \tag{27}$$

where $\alpha = \Gamma\left(\frac{\beta+2}{\beta+1}\right)$ and $Q(a, x)$ is the incomplete Gamma function

$$Q(a, x) = \int_x^\infty t^{a-1} e^{-t} dt. \tag{28}$$

Here β is the level repulsion exponent in (26), which also measures the degree of localization of the chaotic eigenstates: if the localization is the strongest, the eigenstates practically do not overlap in the phase space (of the Wigner functions) and we find $\beta = 0$ and Poissonian distribution, while in the case of maximal uniform extendedness (no localization) we have $\beta = 1$, and the RMT statistics of levels applies. Thus, by replacing $E_W(S)$ with $E_B(S)$ we get the BRB (Berry-Robnik-Brody) distribution, which generalizes the Berry-Robnik distribution such that the localization effects are included [3].

5 The Poincaré-Husimi Functions in Billiards

We have established the formalism of the Wigner functions, thereby introducing a kind of quantum phase space. We will use it to separate the regular and chaotic eigenstates in mixed-type systems. To do this we simply look whether the given eigenstate $W_n(\mathbf{q}, \mathbf{p})$ overlaps with a classical regular or classical chaotic region. Moreover, we shall look whether the chaotic Wigner function is localized or not, and if so, to what degree. The method and approach is general, but technically difficult to implement in general. Therefore it is necessary to study some specific model systems, for which the billiard systems seem to be most suitable.

Let us consider 2D billiard, using the natural coordinates in the phase space (s, p): the arclength s round the billiard boundary, $s \in [0, \mathcal{L}]$, where $\mathcal{L}$ is the circumference, and the sine of the reflection angle, which also is the component of the unit velocity vector tangent to the boundary at the collision point, equal to p. These coordinates are canonically conjugate, known as the Poincaré-Birkhoff coordinates. The bounce map $(s_1, p_1) \to (s_2, p_2)$ is area preserving [20], and the phase portrait does not depend on the energy. For the quantum billiard we have to solve the stationary Schrödinger equation, which (using the apropriate units) reduces to the Helmholtz equation $\Delta\psi +$

$k^2\psi = 0$ with the Dirichlet boundary conditions $\psi|_{\partial\mathcal{B}} = 0$. The energy is $E = k^2$. The important quantity is the boundary function

$$u(s) = \mathbf{n} \cdot \nabla_{\mathbf{r}}\psi\,(\mathbf{r}(s))\,, \tag{29}$$

which is the normal derivative of the wavefunction ψ at the point s ($\mathbf{n}$ is the outward unit normal vector). The boundary function $u(s)$ satisfies the integral equation

$$u(s) = -2\oint dt\; u(t)\; \mathbf{n} \cdot \nabla_{\mathbf{r}} G(\mathbf{r}, \mathbf{r}(t)), \tag{30}$$

where $G(\mathbf{r}, \mathbf{r}') = -\frac{i}{4} H_0^{(1)}(k|\mathbf{r} - \mathbf{r}'|)$ is the Green function in terms of the Hankel function $H_0(x)$. The boundary function $u(s)$ contains complete information about the quantum system, because the wavefunction at any point $\mathbf{q}$ inside the billiard can be determined by the equation [51]

$$\psi_j(\mathbf{r}) = -\oint dt\; u_j(t)\; G\,(\mathbf{r}, \mathbf{r}(t))\,. \tag{31}$$

When going over to the quantum phase space we can calculate the Wigner functions based on $\psi_n(\mathbf{r})$ and perform the procedure developed in the previous section. However, in billiards it is advantageous to calculate the Poincaré-Husimi functions, which are generally just Gaussian smoothed Wigner functions. Such smoothing makes them positive definite, so that we can treat them as quasi-probability densities in the quantum phase space. Following Tualle and Voros [52] and Bäcker et al [53], we use [4] the properly $\mathcal{L}$-periodized coherent states centered at (q, p), as follows

$$c_{(q,p),k}(s) = \sum_{m\in\mathbf{Z}} \exp\{i\,k\,p\,(s - q + m\mathcal{L})\} \exp\left(-\frac{k}{2}(s - q + m\mathcal{L})^2\right). \tag{32}$$

The Poincaré-Husimi function is defined as the absolute square of the projection of the boundary function $u(s)$ onto the coherent state, namely

$$H_j(q, p) = \left|\int_{\partial\mathcal{B}} c_{(q,p),k_j}(s)\; u_j(s)\; ds\right|^2. \tag{33}$$

In Fig. 1 we show examples of a regular and of a chaotic eigenstate for the billiard introduced by Robnik in [54, 55], which is the conformal complex map $w = z + \lambda z^2$ of the unit disc $|z| = 1$. Here we take the shape parameter $\lambda = 0.15$.

Now we can classify the eigenstates as regular and chaotic according to their projection onto the classical surface of section. As we are very deep in the semiclassical regime we can expect with probability almost one that either an eigenstate is regular or chaotic, with exceptions having measure zero, ideally. To automate this technical

Fig. 1 Poincaré-Husimi functions of a chaotic (left) and regular (right) eigenstate, for the Robnik billiard at $\lambda = 0.15$. The values k (M) are: chaotic: k $(M) = 2000.0181794$ (0.981); regular: k $(M) = 2000.0777155(-0.821)$. The gray background is the classically chaotic invariant component. Only one quarter of the surface of section is shown, $(s, p) \in [0, \mathcal{L}/2] \times [0, 1]$, since due to the reflection symmetry and time-reversal symmetry the four quadrants are equivalent. Taken from [4]

task we have ascribed to each point on the grid a number $A_{i,j}$ whose value is either $+1$ if the grid point lies within the classical chaotic region or -1 if it belongs to a classical regular region. This has been done as follows. We have taken an initial condition in the chaotic region, and iterated it up to about 10^{10} collisions, enough for the convergence such that we are not missing any chaotic cells. Each visited cell (i, j) on the grid has then been assigned value $A_{i,j} = +1$, the remaining ones were assigned the value -1.

The Poincaré Husimi function $H(q, p)$ (33) (normalized) has been calculated on the grid points and the overlap index M has been calculated according to the definition $M = \sum_{i,j} H_{i,j} A_{i,j}$. In practice, M is not exactly $+1$ or -1, but can have a value in between. There are two reasons: the finite discretization of the phase space (the finite size grid), and the finite wavelength (not sufficiently small effective Planck constant, for which we can take just $1/k$). If so, the question is, where to cut the distribution of the M-values, at the threshold value M_t, such that all states with $M < M_t$ are declared regular and those with $M > M_t$ chaotic.

We have considered two natural criteria: **(I)** *The classical criterion:* the threshold value M_t is chosen such that we have exactly μ_1 fraction of regular levels and $\mu_2 = 1 - \mu_1$ of chaotic levels. **(II)** *The quantum criterion:* we choose M_t such that we get the best possible agreement of the chaotic level spacing distribution with the Brody distribution (26).

Using this method we can separate the regular and chaotic eigenstates and the corresponding eigenvalues, using the classical criterion **(I)**. The corresponding threshold value of the index M is found to be $M_t = 0.431$. The level spacing distributions are shown in Fig. 2. We see perfect agreement with Brody distribution with $\beta = 0.444$ for the chaotic levels and almost perfect Poisson distribution for the regular levels.

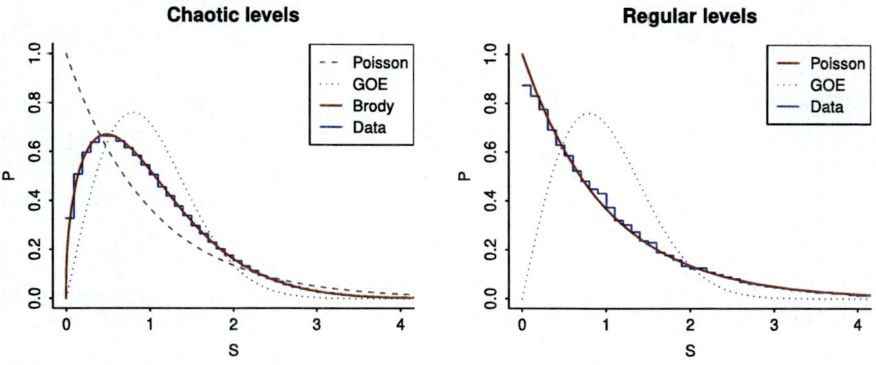

Fig. 2 Level spacing distributions for separated chaotic (left) and regular (right) levels in the Robnik billiard with $\lambda = 0.15$. We use the classical separation criterion with $M_t = 0.431$. For the chaotic subspectrum, after unfolding, there is perfect agreement with the Brody distribution with $\beta = 0.444$. For the regular part of the spectrum, after unfolding, we see excellent agreement with Poisson. Taken from [4]

6 The Phase Space Localization Measures of Chaotic Eigenstates

After the separating regular and chaotic eigenstates we want to introduce localization measures, which quantify the degree of localization of the chaotic eigenstates in the phase space [5]. We express the localization measures in terms of the discretized and normalized Husimi function. For the *entropy localization measure* denoted by A we write $A = e^{\langle I \rangle}/N_c$, where $I = -\int dq\, dp\, H(q,p) \ln\left((2\pi\hbar)^N H(q,p)\right)$ is the information entropy and N_c is a number of cells on the classical chaotic domain. The mean $\langle I \rangle$ is obtained by averaging I over a sufficiently large number of consecutive chaotic eigenstates. In the case of uniform distribution $H_{ij} = 1/N_C$ the localization measure is $A = 1$, while in the case of the strongest localization $I = 0$, and $A = 1/N_C \approx 0$.

The *correlation localization measure*, denoted by C, is defined by the overlap (correlation matrix) $C_{nm} = \frac{1}{Q_n Q_m} \sum_{ij} H_{ij}^n H_{ij}^m$, where $Q_n = \sqrt{\sum_{ij} (H_{ij}^n)^2}$ is the normalizing factor. Then $C = \langle C_{nm} \rangle$, and the averaging is over all n, m and a large number of consecutive chaotic eigenstates.

Again we use the billiard like in Sect. 5 with $\lambda = 0.15$. For a good approximation of the localization measures A and C it was sufficient to separate and extract about 1.500 consecutive chaotic eigenstates. The two localization measures are linearly equivalent as shown in Fig. 3. In order to calculate β with sufficient accuracy we need much more levels, and therefore the separation of eigenstates is then technically too demanding. We have instead calculated spectra on small intervals around $k \approx 2000$ and $k \approx 4000$, about 100.000 consecutive levels (no separation), and obtained β by fitting the $P(S)$ by the BRB distribution derived in Sect. 4.4 using the classical μ_1. The functional dependence of β on A is shown in Fig. 3. For aesthetic reasons we have

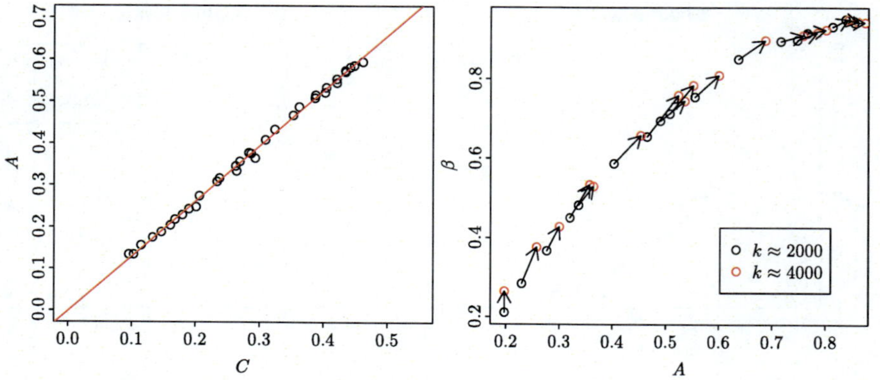

Fig. 3 Quantum localization in the Robnik billiard $\lambda = 0.15$, using about 1500 consecutive chaotic eigenstates. (Left:) Linear relation between the two entirely different localization measures, namely the entropy measure A and the correlation measure C, calculated for several different billiards at $k \approx 2000$ and $k \approx 4000$. (Right:) We show the functional relation between the Brody level repulsion parameter β and A. Arrows connect points corresponding to the same λ at two different k. Taken from [5]

Fig. 4 The Brody parameter β as a function of the localization measure A for a large number of stadium billiards of different shapes ϵ and energies $E = k^2$. Taken from [8]

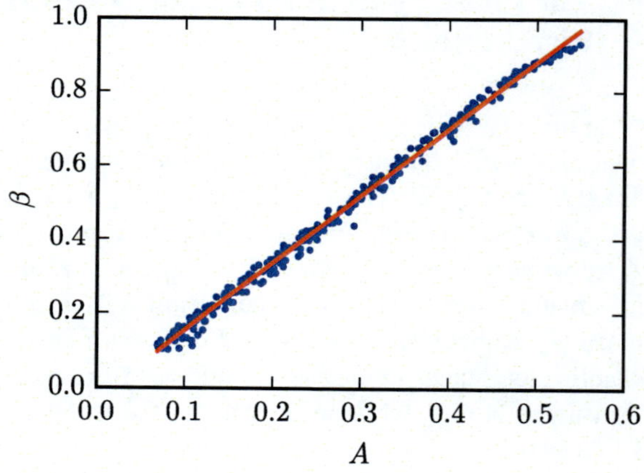

rescaled the measure $A \to A/A_{\max}$ such that it goes from 0 to 1. The maximal value of A, $A_{\max} = 0.68$, was estimated as $A_{\max} = e^{I_{\max}}/N_c$, where $I_{\max}$ is the maximum entropy of 1500 consecutive states of the almost fully chaotic $\lambda = 0.25$ billiard. Thus for fully chaotic systems the procedure always yields $A = 1$. Namely, in real chaotic eigenstates we never reach a perfectly uniform (constant) distribution $H(q, p)$, since their Poincaré -Husimi functions always have some oscillatory structure.

Let us emphasize that there is a functional relationship between A and β, as shown in Fig. 3. By increasing k from 2000 to 4000 we increase the dimensionless Heisenberg time by factor 2, therefore A must increase, but precisely in such a way, that the empirical points stay on the scaling curve, as it is observed and indicated by the arrows. We do not have yet a semiempirical functional description of the

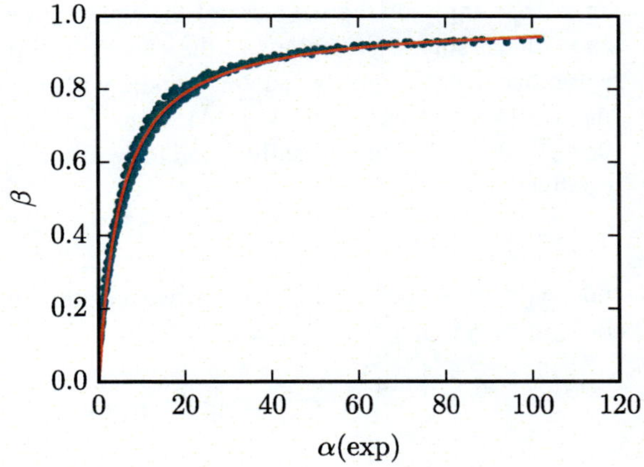

Fig. 5 The Brody parameter β is a rational function of α (34), where t_T is extracted from the exponential diffusion law, for a large number of stadium billiards of different shapes ϵ and energies $E = k^2$, as in Fig. 4. Here $\beta_\infty = 0.98$ and $s = 0.20$. Taken from [6]

relationship $\beta(A)$. In the quantum kicked rotator it is just almost linear [49, 50, 56]. Similarly it is found to be almost linear in the stadium of Bunimovich, as recently published in reference [6] and shown in Fig. 4. The level repulsion exponent β is also found to be a unique function of $\alpha = t_H/t_T$, well described empirically by the rational function

$$\beta = \beta_\infty \frac{s\alpha}{1 + s\alpha}, \tag{34}$$

as shown in Fig. 5, discussed in reference [6]. There is a great lack in theoretical understanding of the physical origin of the relationship $\beta(A)$, even in the case of (the long standing research on) the quantum kicked rotator, except for the intuitive idea, that energy spectral properties should be only a function of the degree of localization, because the localization gradually decouples the energy eigenstates and levels, switching the linear level repulsion $\beta = 1$ (extendedness) to a power law level repulsion with exponent $\beta < 1$ (localization). The full physical explanation is open for the future.

One should notice some scattering of points around the mean value in the Fig. 5, noted already by Izrailev [56] in the case of the quantum kicked rotator, which indicates that the localization measure has a certain distribution rather than being a sharp number, as has been observed recently in the kicked rotator by Manos and Robnik [57]. In the next section we address the question of the statistical properties of the localization measure A.

7 The Distribution of the Localization Measures

Recently we have shown two important results. The first one is the empirical observation [7], in the billiard family devised and studied in [54, 55], that the localization measures A of chaotic eigenstates have a distribution, which generally speaking depends strongly on the structure of the classical phase space, namely on the existence of the stickiness regions in the chaotic component. However, if the chaotic region becomes uniformly chaotic, ergodic without stickiness, we find universality: The distribution function of A, $P(A)$, is the *beta distribution* on a compact interval $[0, A_0]$, where A_0 is empirically found to be around $A_0 \approx 0.7$. The beta distribution is defined as

$$P(A) = CA^a(A_0 - A)^b, \tag{35}$$

and the two exponents a and b are positive real numbers, while C is the normalization constant such that $\int_0^{A_0} P(A)\,dA = 1$, i.e.

$$C^{-1} = A_0^{a+b+1} B(a+1, b+1), \tag{36}$$

where $B(x, y) = \int_0^1 t^{x-1}(1-t)^{y-1}dt$ is the beta function. In Fig. 6 we show the results for the distribution $P(A)$ of the chaotic eigenstates for three different mixed-type lemon billiards at three diffrent energies $E_0 = k_0^2$, and the excellent agreement with the best fitting beta distribution is obvious. For the details see [10]. In the ultimate semiclassical limit the beta distribution approaches the Dirac delta distribution $\delta(A_0 - A)$ peaked at the maximum value of $A = A_0 \approx 0.7$.

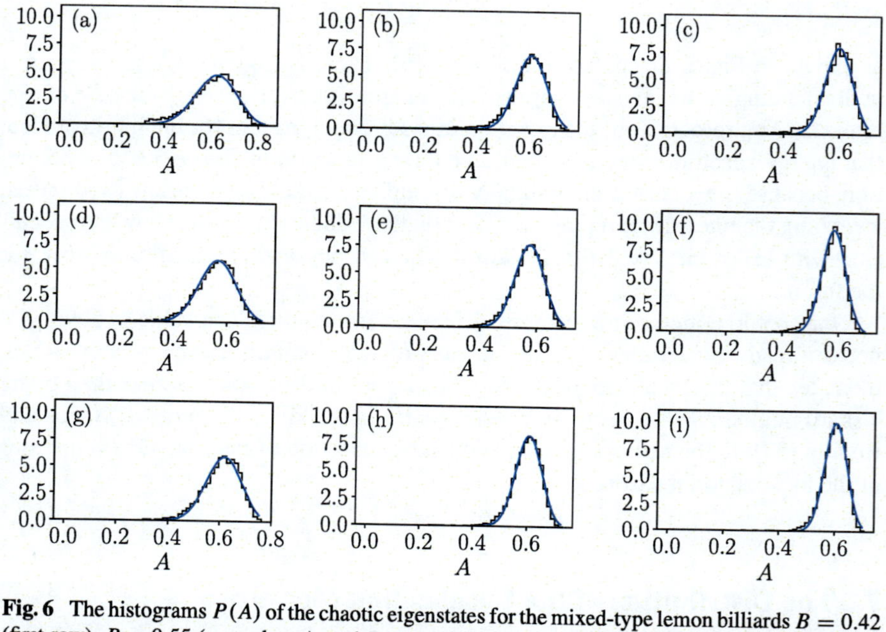

Fig. 6 The histograms $P(A)$ of the chaotic eigenstates for the mixed-type lemon billiards $B = 0.42$ (first row), $B = 0.55$ (second row), and $B = 0.6$ (third row), at various energies $E_0 = k_0^2$: (a,d,g) $k_0 = 640$, (b,e,h) $k_0 = 1760$, and (c,f,i) $k_0 = 2880$. All states are of odd-odd parity. The agreement with the best fitting beta distribution is obvious. The parameters (a, b) of the beta distribution (35) are, from (a) to (i): (20.100, 12.380), (44.209, 29.091), (59.477, 40.172), (28.982, 22.380), (52.361, 39.027), (80.315, 57.947), (31.734, 19.835), (63.872, 41.757), (90.748, 59.885). Taken from [10]

The second result is that the entire picture does not depend much, very weakly, on the definition of the localization measure A, so that e.g. the same conclusions can be met by using the normalized inverse participation ratio, rather than the (Shannon, information) entropy localization measure. It has been demonstrated [7, 8, 12] that the entropy localization measure and the normalized inverse participation ratio are almost linearly related.

These results have been reconfirmed in the stadium billiard of Bunimovich [8], and in the family of lemon billiards [10], as well as in the Dicke model [12] and in kicked top [59]. Therefore, we believe that this is a universal behaviour.

8 Semiclassical Limiting Approach to the Regime of PUSC

As discussed in the previous sections we expect that in the ultimate (sufficiently deep) semiclassical limit, according to PUSC, each quantum eigenstate in the quantum phase space (Wigner function or Husimi function) should approach either a regular state localized on an invariant torus, or should be uniformly spread on a chaotic region, such that the corresponding phase space localization measure distribution $P(A)$ is Dirac delta function $P(A) = \delta(A_0 - A)$. Therefore, in this asymptotic limit the mixed-type eigenstates should disappear, their relative fraction should tend to zero. However, before reaching this limit, we do observe mixed-type eigenstates, which has been analyzed in detail in the most recent paper [14] on the phenomenology of eigenstates in a mixed-type regime in the family of lemon billiards. There, we succeeded to demonstrate for the first time the quantitative description of this limiting monotonic disappearance of mixed-type eigenstates: It turns out that the vanishing of the relative fraction of mixed-type states is governed by a power law as a function of the unfolded energy with an exponent $\gamma \approx -0.29$. The precise mechanism for the existence of mixed-type eigenstates is still not entirely clarified. Its understanding will contribute to the theoretical explanation of the underlying power law and its exponent. Further work along these lines is in progress [60].

9 Discussion and Conclusions

Quantum chaos, or wave chaos, studies the manifestations of classical chaos in the quantum domain, or more generally, it concerns the ray dynamics in the sense of the short-wavelength approximations, by building the quantum features on this classical skeleton. It turns out, that quantum mechanics "jumps" on any structure in the classical phase space that exists, and exhibits universality if the classical dynamics has some uniformity, like in the class of integrable systems or in the class of uniformly chaotic, ergodic, systems without significant stickiness regions. While the classical chaotic dynamics with positive Lyapunov exponents and the implied sensitive dependence on initial conditions is fundamentally irreversible, the quantum evolution of bound systems with purely discrete spectrum in time is always almost periodic and stable, fundamentally reversible. Therefore quantum chaos in the time evolution does not exist. However, the eigenstates, as solutions of the stationary Schrödinger equation, of bound quantum systems exhibit features that are directly and precisely linked to the classical dynamics, to the structure of the phase portrait of the underlying and corresponding classical Hamilton system. This is revelaed in the structure of Wigner functions. If the effective Planck constant is sufficiently small, in such semiclassical limit, the Heisenberg time scale is larger than the classical transport time scale (diffusion time), the Wigner functions of chaotic eigenstates are uniformly extended over the entire available chaotic region. In this limit the regular eigenstates "live" on the invariant tori of the regular regions, while the chaotic ones cover uniformly the underlying chaotic region. The energy spectra belong then to the universality classes: the regular spectra obey the Poissonian statistics, while the chaotic ones are well described by the Gaussian random matrix theory. If the two time scales, Heisenberg time t_H and the classical transport time t_T, do **not** satisfy the semiclassical condition $t_H > t_T$, the chaotic Wigner functions (or Husimi functions) are localized due to the quantum (or dynamical) localization. The degree of localization can be quantified in various ways, but they are found to be equivalent, linearly proportional or very close to that. The Brody parameter describing the level repulsion, entering in the level spacing distribution, turns out to be a unique function of the average localization measure. By looking at the overlap of the Wigner or Husimi functions with the classically invariant components, the invariant tori and the chaotic regions, we can classify the eigenstates as regular and chaotic, respectively, and separate them. We find that the regular levels obey the Poissonian statistics, while the localized chaotic ones obey the Brody level spacing distribution. This picture goes back to the early work of Percival (1973), Berry and Robnik (1984), with the generalization capturing the localization effects by Batistić and Robnik (2010-2013). The localization measure of chaotic eigenstates has a distribution, which in the case of no stickiness in the classical phase space is the beta distribution, demonstrated to be valid for three types of billiards, in the Dicke model and in the kicked top. Moreover, the most recent result based on the phenomenology of the quantum lemon billiards shows that in the semiclassical limit the relative fraction of mixed-type eigenstates as a function of the unfolded energy decreases as a power law with the exponent $\gamma = -0.29$, thereby showing quantitatively the monotonic approach to the regime of the Principle of Uniform Semiclassical Condensation (PUSC) of the Wigner (or Husimi) functions [14].

When leaving the semiclassical limit and considering larger values of the effective Planck constant (at lower energies), we observe the tunneling effects, where the regular and chaotic eigenstates can overlap considerably, and thus no longer can be classified clearly as regular or chaotic. This line of thoughts is a subject of further research [3, 58]. However, the tunneling effects decline very fast with increasing energy, in fact exponantially, and therefore are not present at very high energies, whereas the localization effects can persist.

The fully chaotic ergodic and the regular eigenstates are meanwhile quite well understood, except in cases of very strong stickiness regions in the classical chaotic regions, where further work has to be done. Moreover, the theoretical description of the quantum localized chaotic eigenstates (their Wigner or Husimi functions), as well as of the corresponding phenomenological Brody level spacing distribution is open for the future. This includes the theoretical derivation of the apparently universal beta distribution of the localization measures. The detailed description of the mixed-type systems in general, in the strict semiclassical limit as well as at the larger values of the effective Planck constant, is topics of the current research.

Thus, several important aspects of quantum chaos are open for the future, which has meanwhile applications in many branches of theoretical physics, like e.g. solid state physics, quantum field theory, high energy physics, fluid dynamics, and of course in all wave systems.

Acknowledgements This work was supported by the Slovenian Research Agency ARRS under the grant J1-4387 "Quantum localization in few and many-body chaotic systems", and by the research program P1-0306 "Applied Mathematics, Theoretical Physics and Intelligent Systems".

References

1. Robnik, M.: Recent advances in quantum chaos of generic systems: wave chaos of mixed-type systems. In: Meyers, R.A. (ed.) Encyclopedia of Complexity and Systems Science. Springer, Berlin (2020)
2. Robnik, M.: Eur. Phys. J. Special Topics **225**, 959 (2016)
3. Batistić, B., Robnik, M.: J. Phys. A: Math. Theor. **43**, 215101 (2010)
4. Batistić, B., Robnik, M.: J. Phys. A: Math. Theor. **46**, 315102 (2013)
5. Batistić, B., Robnik, M.: Phys. Rev. E **88**, 052913 (2013)
6. Batistić, B., Lozej, Č, Robnik, M.: Nonl. Phen. Compl. Syst. (Minsk) **21**(1), 225 (2018)
7. Batistić, B., Lozej, Č, Robnik, M.: Phys. Rev. E **100**, 062208 (2019)
8. Batistić, B., Lozej, Č, Robnik, M.: Nonl. Phen. Compl. Syst. (Minsk) **23**, 17 (2020)
9. Lozej, Č, Lukman, D., Robnik, M.: Phys. Rev. E **103**, 012204 (2021)
10. Lozej, Č, Lukman, D., Robnik, M.: Nonl. Phen. Compl. Syst. (Minsk) **24**, 1 (2021)
11. Lozej, Č, Lukman, D., Robnik, M.: Physics **3**, 888 (2021)
12. Wang, Q., Robnik, M.: Phys. Rev. E **102**, 032212 (2020)
13. Wang, Q., Robnik, M.: Entropy **23**, 1347 (2021)
14. Lozej, Č, Lukman, D., Robnik, M.: Phys. Rev. E **106**, 054203 (2022)
15. Stöckmann, H.-J.: Quantum Chaos: An Introduction. Cambridge University Press, Cambridge (1999)
16. Haake, F.: Quantum Signatures of Chaos. Springer, Berlin (2010)
17. Robnik, M.: Nonl. Phen. Compl. Syst. (Minsk) **1**(1), 1 (1998)
18. Robnik, M.: In: Aizawa, Y., Miwa, Y. (eds.) Proceedigs of Waseda AICS Symposium and the 14th Slovenia-Japan Seminar, New Challenges in Complex Systems Science, 2014. Waseda University, Tokyo (2015), **B11**(3), 13–17
19. Haken, H.: Synergetics. Springer, Berlin (2004)
20. Berry, M.V.: Eur. J. Phys. **2**, 91 (1981)
21. Robnik, M., Veble, G.: J. Phys. A: Math. Gen. **49**, 4669 (1998)
22. Arnold, V.I.: Mathematical Methods of Classical Mechanics. Springer, New York (1980)
23. Ott, E.: Chaos in Dynamical Systems. Cambridge Univesity Press, Cambridge (1993)
24. Bunimovich, L.A.: Func. Anal. Appl. **8**, 254 (1974)
25. Lozej, Č, Robnik, M.: Phys. Rev. E. **97**(1), 012206–1 (2018)
26. Casati, G., Chirikov, B.V., Izrailev, F.M., Ford, J.: Lect. Notes Phys. **93**, 334 (1979)

27. McDonald, S.W., Kaufman, A.N.: Phys. Rev. Lett. **42** 1189 (1979); Phys. Rev. A **37**, 3067 (1988)
28. Berry, M.V.: J. Phys. A: Math. Gen. **10**, 2083 (1977)
29. Li, B., Robnik, M.: J. Phys. A: Math. Gen. **27**, 5509 (1994)
30. Bohigas, O., Giannoni, M.J., Schmit, C.: Phys. Rev. Lett. **52**, 1 (1984)
31. Percival, I.C.: J. Phys. B: At. Mol. Phys. **6**, L229 (1973)
32. Casati, G., Valz-Gris, F., Guarneri, I.: Lett. Nuovo Cimento **28**, 279 (1980)
33. Berry, M.V.: Proc. Roy. Soc. Lond. A **400**, 229 (1985)
34. Gutzwiller, M.C.: J. Math. Phys. **8**, 1979 (1967); **10**, 1004 (1969); **11**, 1791 (1970); **12**, 343 (1971)
35. Sieber, M., Richter, K.: Phys. Scr. **T90**, 128 (2001)
36. Müller, S., Heusler, S., Altland, A., Braun, P., Haake, F.: New J. Phys. **11**, 103025 (2009). and references therein
37. Mehta, M.L.: Random Matrices. Academic, Boston (1991)
38. Robnik, M., Berry, M.V.: J. Phys. A: Math. Gen. **19**, 669 (1986)
39. Robnik, M.: In: Seligman, T.H., Nishioka, H. (eds.) Proceedings of the International Conference Quantum Chaos and Statistical Nuclear Physics, Cuernavaca, Mexico, 1986. Lecture Notes in Physics, vol. 263, p. 120. Springer, Berlin (1986)
40. Grossmann, S., Robnik, M.: Z. Naturforsch. **62a**, 471 (2007)
41. Hackenbroich, G., Weidenmüller, H.A.: Phys. Rev. Lett. **74**, 4118 (1995)
42. Robnik, M., David, H.M., Vidmar, G., Romanovski, V.G.: Nonl. Phen. Compl. Syst. (Minsk) **13**(1), 13 (2010)
43. Shnirelman, B.: Uspekhi Matem. Nauk **29**, 181 (1974). (in Russian)
44. Voros, A.: Lect. Notes Phys. **93**, 326 (1979)
45. Veble, G., Robnik, M., Liu, J.: J. Phys. A: Math. Gen. **32**, 6423 (1999)
46. Berry, M.V., Robnik, M.: J. Phys. A: Math. Gen. **17**, 2413 (1984)
47. Prosen, T., Robnik, M.: J. Phys. A: Math. Gen. **27**, L459 (1994)
48. Prosen, T., Robnik, M.: J. Phys. A: Math. Gen. **32**, 1863 (1999)
49. Batistić, B., Manos, T., Robnik, M.: Europhys. Lett. **102**, 50008 (2013)
50. Manos, T., Robnik, M.: Phys. Rev. E **87**, 062905 (2013)
51. Berry, M.V., Wilkinson, M.: Proc. Roy. Soc. Lond. A **392**, 15 (1984)
52. Tualle, J.M., Voros, A.: Chaos Solitons Fractals **5**, 1085 (1995)
53. Bäcker, A., Fürstberger, S., Schubert, R.: Phys. Rev. E **70**, 036204 (2004)
54. Robnik, M.: J. Phys. A: Math. Gen. **16**, 971 (1983)
55. Robnik, M.: J. Phys. A: Math. Gen. **17**, 1049 (1984)
56. Izrailev, F.M.: Phys. Rep. **196**, 299 (1990)
57. Manos, T., Robnik, M.: Phys. Rev. E **91**, 042904 (2015)
58. Gehler, S., Löck, S., Shinohara, S., Bäcker, A., Ketzmerick, R., Kuhl, U., Stöckmann, H.-J.: Phys. Rev. Lett. **115**, 104101 (2015). and references therein
59. Wang, Q., Robnik, M.: Phys. Rev. E **107**, 054213 (2023)
60. Lozej, Č., Robnik, M.: Phys. Rev. E (2023), to be submitted

Dynamics and Statistics of Weak Chaos in a 4-D Symplectic Map

Tassos Bountis, Konstantinos Kaloudis, and Helen Christodoulidi

Abstract The important phenomenon of "stickiness" of chaotic orbits in low dimensional dynamical systems has been investigated for several decades, in view of its applications to various areas of physics, such as classical and statistical mechanics, celestial mechanics and accelerator dynamics. Most of the work to date has focused on two-degree of freedom Hamiltonian models often represented by two-dimensional (2D) area preserving maps. In this paper, we extend earlier results using a 4-dimensional extension of the 2D MacMillan map, and show that a symplectic model of two coupled MacMillan maps also exhibits stickiness phenomena in limited regions of phase space. To this end, we employ probability distributions in the sense of the Central Limit Theorem to demonstrate that, as in the 2D case, sticky regions near the origin are also characterized by "weak" chaos and Tsallis entropy, in sharp contrast to the "strong" chaos that extends over much wider domains and is described by Boltzmann Gibbs statistics. Remarkably, similar stickiness phenomena have been observed in higher dimensional Hamiltonian systems around unstable simple periodic orbits at various values of the total energy of the system.

Keywords Coupled MacMillan maps · Boltzmann Gibbs and Tsallis entropies · Weak and strong chaos

T. Bountis (✉)
Center for Integrable Systems, P.G. Demidov Yaroslavl State University, Yaroslavl 150003, Russia
e-mail: tassosbountis@gmail.com

K. Kaloudis
Department of Statistics and Actuarial-Financial Mathematics, University of the Aegean, Samos 83200, Greece

H. Christodoulidi
Department of Mathematics, Lincoln University, Lincoln, LN6 7TS, UK

T. Bountis et al. (eds.), *Chaos, Fractals and Complexity*, Springer Proceedings in Complexity, https://doi.org/10.1007/978-3-031-37404-3_7

1 Introduction

The behavior of nonlinear dynamical systems described by differential and difference equations has been a topic of intense interest for several decades [1–5]. As is well-known, one the most important questions in this field concerns the distinction between solutions of the equations that are called "regular", since their evolution can be predicted for long times, and those termed "chaotic", whose time evolution becomes unpredictable after relatively short times. This is typically decided by calculating the Lyapunov exponents, measuring the distance between two nearby solutions, represented by trajectories (or orbits) in the $2N-$ dimensional phase space of the system [6], with N position and N momentum variables, with time as the single independent variable. If none of the Lyapunov exponents is positive we call the orbit *regular*, while if at least one exponent is positive we call it *chaotic*.

But is this "duality" between order and chaos all there is? While there is no uncertainty about regular orbits, it has been realized that "chaos" is a lot more subtle to describe by a simple definition. One possibility is to study chaotic phase space domains from a statistical point of view, in terms of correlations and probability distributions. If these correlations decay exponentially away from a chaotic orbit, one might adopt a Boltzmann Gibbs (BG) thermodynamic description of the dynamics (as in the case of an ideal gas) and look for Gaussian probability functions (pdfs) to describe the associated statistics. What happens, however, if correlations decay by power laws and the pdfs of positions and/or momenta are no longer Gaussian? What would that imply about the corresponding chaotic behavior?

One such widely known example occurs in cases of "stickiness", where chaotic orbits of generally low-dimensional dynamical systems tend to remain confined for very long times trapped within thin chaotic layers surrounding regions of regular motion [4, 7, 9–12]. Remarkably, this phenomenon does not occur only in low dimensions. It has also been observed in multidimensional Hamiltonian lattices [8, 13–17], often in cases where chaotic regions arise around simple periodic orbits, when they have just turned unstable, as the total energy of the system is increased.

Regarding dynamical systems in discrete time, it is well–known that 2D Poincaré maps describe intersections of the orbits of a 2-degree of freedom continuous dynamical system with a 2D surface of section [4]. Thus, one may consider directly area preserving transformations of a plane onto itself to study the qualitative features of such maps [19].

One famous model in this regard is the 2D MacMillan (2DMM) area preserving, non-integrable map [18]. It may be interpreted as describing the dynamics of focusing a "flat" proton beam in a circular particle accelerator model describing the repeated passage of a "flat" beam through a periodic sequence of thin nonlinear lenses [21]:

$$
\begin{aligned}
x_{n+1} &= y_n \\
y_{n+1} &= -x_n + \frac{2Ky_n}{1+y_n^2} + \mu y_n,
\end{aligned}
\tag{1}
$$

where x_n and y_n represent a particle's position and momentum at the nth crossing of a focusing element, while μ, and K are physically important parameters. Note that the Jacobian of the transformation is unity, so that (1) is area-preserving and thus may represent the conservative (Hamiltonian) dynamics of proton beams whose radiation effects are considered negligible [21]. If $\mu = 0$ the map is integrable, as it possesses a constant of the motion given by the one parameter family of curves [20]:

$$x_n^2 + y_n^2 + x_n^2 y_n^2 - 2K x_n y_n = const.$$

In [18], (1) was studied following a nonextensive statistical mechanics approach, based on the nonadditive Tsallis entropy S_q [22]. According to this approach, the pdfs optimizing S_q, under appropriate constraints, are q–Gaussian distributions that represent quasistationary states (QSS) of the dynamics, with $1 < q < 3$ ($q = 1$ being the Gaussian). As was shown in [18], there are several cases of $K > 1$ and $\mu > 0$ parameters, where the chaotic layer around a saddle point at the origin does *not* satisfy BG statistics associated with "strong chaos", but is well described by a $q > 1$-Gaussian pdf, associated with "weak chaos".

It is, therefore, natural to ask whether similar phenomena of spatially limited, weakly chaotic dynamics occur in 4D symplectic maps, such as one encounters e.g. in 3-degree-of-freedom hamiltonian systems commonly encountered in problems of celestial mechanics, see e.g. [7, 9–11] and particle accelerator dynamics [23, 24].

In this paper, we extend for the first time the above approach to study 4D MacMillan (4DMM) maps of the form

$$\begin{aligned} x_{n+1} &= -x_{n-1} + \frac{2K_1 x_n}{1+x_n^2} + \mu x_n - \epsilon x_n y_n^2 \\ y_{n+1} &= -y_{n-1} + \frac{2K_2 y_n}{1+y_n^2} + \mu y_n - \epsilon x_n^2 y_n \end{aligned} \tag{2}$$

where x_n, y_n represent horizontal and vertical deflections of the proton beam as it passes through the nth focusing element and study the chaotic domain arising about the origin of (2), using values of K_1, K_2 and μ for which the origin is unstable. Note that (2) is symplectic, as the evolution of x_n and y_n is determined by a potential function $V(x_n, y_n)$, whose partial derivatives with x_n and y_n respectively yield the two equations of (2).

We choose suitable K_1 and/or K_2 values, for fixed $\mu > 0$, $\epsilon > 0$ small, such that the origin is (linearly) unstable and calculate the pdfs of the rescaled sums of N iterates of the map, in the sense of the Central Limit Theorem, in the large N limit for large sets of initial conditions. We then relate our results to specific properties of the phase space dynamics of the maps and distinguish cases where the pdfs represent long–lived QSS described by q-Gaussians.

We begin by describing in Sect. 2 the statistical methods used in this paper to obtain the pdfs describing our data in all cases of the 4DMM map studied here. Next, in Sect. 3, we apply this analysis to find weak chaos characterized by $q-$ Gaussian pdfs, for different parameter values connected with an unstable fixed point at the origin of our 4DMM map. We end with our conclusions in Sect. 4.

2 Statistical Analysis of Weak Chaos

Before turning to the 4DMM mapping studied here, we first carried out the same computations for the 2DMM map (1) and compared them to results depicted in Fig. 3(a) of [18]. Employing the same choices of initial conditions and the same number of iterations, we verified that we obtain practically identical results.

For the benefit of the reader, we state that the approach we follow here is to evaluate the solution x_n, y_n, $n = 0, \ldots, N$ of the 4DMM map (2) and construct probability distributions for x_n (similarly for y_n) of appropriately large rescaled sums $S_j(N)$ obtained by adding the corresponding N iterates

$$S_j(N) = \sum_{n=0}^{N} x_n^{(j)}$$

where j refers to the j-th realisation, taking values from 1 to the total number of initial conditions N_{ic}. As in [18], we generate the centered and rescaled sums

$$s_j(N) \equiv \frac{S_j(N) - \mu_j(N)}{\sigma_N} = \left(\sum_{n=0}^{N} x_n^{(j)} - \frac{1}{N_{ic}} \sum_{j=1}^{N_{ic}} \sum_{n=0}^{N} x_n^{(j)} \right) / \sigma_N \tag{3}$$

where $\mu_j(N)$ is the mean value and σ_N the standard deviation of $S_j(N)$ over N iterations

$$\sigma_N^2 = \frac{1}{N_{ic}} \sum_{j=1}^{N_{ic}} \left(S_j(N) - \mu_j(N)\right)^2 = \langle S_j^2(N) \rangle - \mu_j^2(N),$$

where $< \cdot >$ denotes averaging over N iterations. We thus find many cases, where the obtained empirical distributions are well–described by a q-Gaussian distribution of the form

$$P\left(s_j(N)\right) = \frac{\sqrt{\beta}}{C_q} \left[1 + \beta(q-1)s_j^2(N)\right]^{1/1-q} \tag{4}$$

where q is regarded as an indicator measuring the divergence from the classical Gaussian distribution, β is the 'inverse temperature' fitting parameter and C_q is a normalizing constant.

To describe the statistical properties of the above rescaled sums of the system, we employ standard parameter estimation techniques. Specifically, we are interested in identifying the q-Gaussian distribution that best describes the observed data. One of the most widely used methods for such estimations, is the Maximum Likelihood Estimator (MLE) [25]. This is a *parametric* method typically used for statistical fitting among distributions belonging to the same family, e.g. the family of Gaussians parameterized by their mean and standard deviation or the family of q-Gaussians parameterized by (q, β).

The main idea behind the MLE is that the most suitable distribution (of a given family) describing a given data set, is *the most probable* to describe the observed data. More formally, we are interested in maximizing the likelihood function, which describes "how likely" it is to observe a certain random sample, for the various values of the unknown parameters of the assumed statistical model.

To determine the likelihood function $p(\theta|\mathbf{X})$, we first calculate the joint probability function of the observed sample $\mathbf{X} = (X_1 = x_1, \ldots, X_n = x_n)$ as a function of the parameters of the problem, $\theta = (\beta, q) \in \mathbb{R}^+ \times [1, 3)$. Then, the MLE is the value of $\theta \in \Theta$ that maximizes the likelihood function, i.e. $\hat{\theta} = \arg\max_{\theta\in\Theta} p(\theta|\mathbf{X})$. For computational purposes, it is convenient to maximize the logarithmic likelihood function, which for a q-Gaussian statistical model has the form:

$$l_{\mathbf{X}}(\beta, q) = \sum_{i=1}^{n} \log \frac{\sqrt{\beta}}{C_q} \left[1 + \beta(q-1)x_i^2\right]^{1/1-q} .$$

In all simulations that follow, we perform our numerical optimization using the so-called "nlm" (nonlinear minimization) command of the *R* software for statistical computing [26].

An alternative approach to derive optimal q-Gaussian parameters is to apply nonlinear least-square fitting to binned estimates of the probability density (via histograms), using such methods as Gauss-Newton (see e.g. [27]). However, from a statistical point of view, it is more accurate to use MLE instead of curve-fitting estimates, as the MLE are theoretically guaranteed, under general (regularity) conditions, to have such desirable properties, as efficiency, consistency and asymptotic normality [28]. For an interesting discussion of the comparison between curve-based estimates and MLEs for the case of q-Exponential distributions, we refer the reader to Shalizi [29].

3 Evidence of Weak Chaos in 4DMM Maps

3.1 Weak Chaos in an Example of the 4DMM Map

We start by fixing the values of $\mu = 0.2$ and $\epsilon = 0.01$, which we will use throughout the paper, as they do not significantly affect the results. Observe now in our Fig. 1a a typical example of an optimal pdf of a q-Gaussian obtained for the choice of parameters $K_1 = 1.6$, $K_2 = 0.5$. This is a case we shall call hyperbolic–elliptic (HE), referring to the first 2D map in (2) having a hyperbolic fixed point at the origin, and

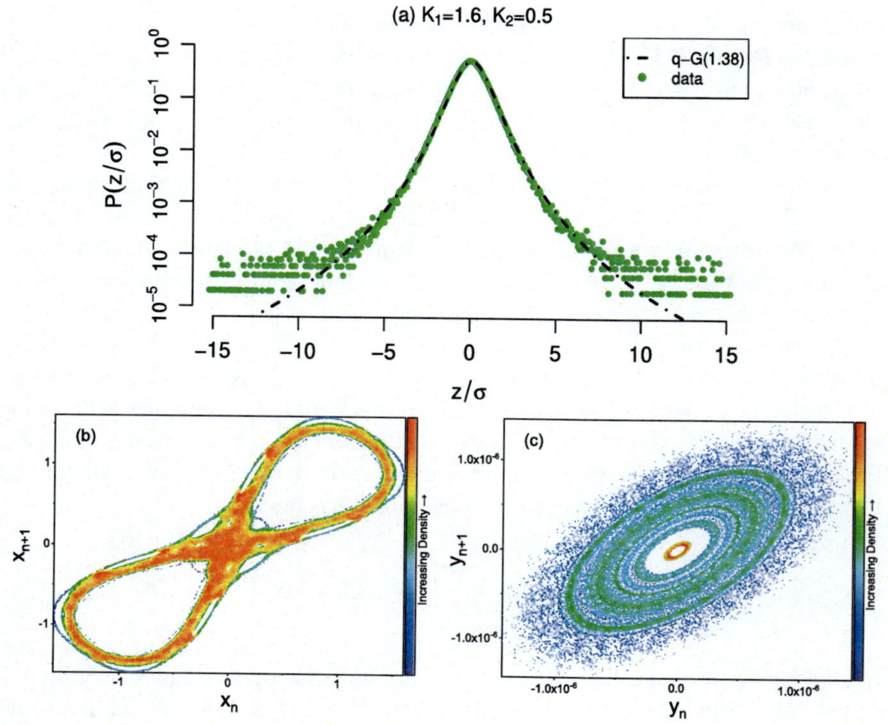

Fig. 1 **a** The computation of the pdf for the x_n variable in (2) with parameters $K_1 = 1.6$, $K_2 = 0.5$, $\mu = 0.2$, and $\epsilon = 0.01$. The dashed line represents an optimal fitting of the data by a q-Gaussian function (4) with $q = 1.38$ and $\beta = 1.19$. **b** The 2D phase space plot of the x_n, x_{n+1} projections of the 4D map (2) for the orbits and parameters used in **a**, while **c** shows the 2D phase space projection in the y_n, y_{n+1} plane variables

the second 2D map having an elliptic point. In a later subsection, we also discuss examples of the hyperbolic–hyperbolic (HH) type, where the origin is unstable in both 2D maps of (2). Note that the case EH is entirely analogous to HE due to the symmetrical form of the two 2D maps.

Throughout our study, we use 10^6 random initial conditions for each of the variables, i.e. x_0, x_1 and y_0, y_1, within the domain $(0, 10^{-6})$ close to the origin. To facilitate the visualization of stickiness phenomena, observe the phase plane picture shown in Fig. 1b. The "warm" colors represent the more dense parts of the plot, where solutions stick around for very long times, whereas "cold" colors depict orbits that scatter diffusively in phase space. We also show in Fig. 1c projections of the orbits in the y_n, y_{n+1} plane, which rotate around the origin due to our choice of $K_2 < 1$.

Each of our initial conditions is iterated $2 \cdot 10^5$ times, to achieve reliable statistics. To obtain the results shown in Fig. 1, we have employed appropriate statistical techniques (see e.g. [25, 29]) to optimize both the specific class of suitable pdfs and their parameters to obtain the best fit for such large data sets.

Clearly, a crucial role in this study is played by the fixed point at the origin and its stability properties. A simple linearization of the equations of our 4DMM map (2) about $x_n = y_n = 0$ shows that the conditions for stability of the central fixed point with respect to deviations in x_n and/or y_n are:

$$|K_i + \mu/2| < 1, \quad i = 1, 2 \tag{5}$$

Thus, we identify as EE (doubly elliptic) the case when both conditions $i = 1, 2$ in Eq. (5) hold, EH (elliptic hyperbolic) if the $i = 2$ inequality is reversed, HE (hyperbolic elliptic) if the $i = 1$ inequality in (5) is inverted, and HH (doubly hyperbolic), when both inequalities in Eq. (5) are reversed. Clearly, if the origin is doubly elliptic (EE), it will be surrounded mostly by quasiperiodic orbits and no large scale chaos will be present in its vicinity. Hence, in what follows, we will study both "partly" unstable HE and "fully" unstable cases of the HH type. We start with both K_i positive, but will also consider cases with $K_i < 0$, for $i = 1, 2$.

3.2 HE Cases of the 4DMM Map

We begin with a hyperbolic elliptic (HE) case of the 4DMM map problem (2), with the main parameters chosen so that the x-map has $K_1 > 1$, and the y-map $0 < K_2 < 1$, i.e hyperbolic in the x_n plane and elliptic in the y_n plane.

Setting $K_2 = 0.5$ and gradually increasing the value of K_1 we observe that the thin 'figure–eight' of Fig. 2a thickens around the origin as chaos slowly expands, and eventually occupies a wider "cellular" domain in phase space shown in Fig. 2d.

The pdfs for each of the panels in Fig. 2 are depicted in Fig. 3. We observe that as the trajectory winds around a thin figure–eight in Fig. 2a in a nearly organized manner, the corresponding distributions of the sums $s_N^{(j)}$ displayed in Fig. 3a follow a q–Gaussian function for two orders of magnitude, while the tails of the pdf diverge to higher values. The presence of weak chaos, however, for $K_1 = 1.5, 1.7$ in Fig. 2b, c leads to the emergence of optimal q–Gaussian distributions in Fig. 3b and c, which, for $q = 1.57, 1.67$, respectively, describe well the numerical data for five orders of magnitude!

On the other hand, for a higher $K_1 = 2$ value (see Fig. 2d) where the orbits form complex "cellular" structures, the q-Gaussian distribution that best describes the data in Fig. 3d is successful only over two orders of magnitude and corresponds to $q = 1.87$. It appears, therefore, that with increasing K_1 the value of q increases also.

3.3 HH Cases of the 4DMM Map

Let us now describe some results obtained when the origin of the map is "fully unstable", i.e. a double saddle point, which we call hyperbolic-hyperbolic (HH). To

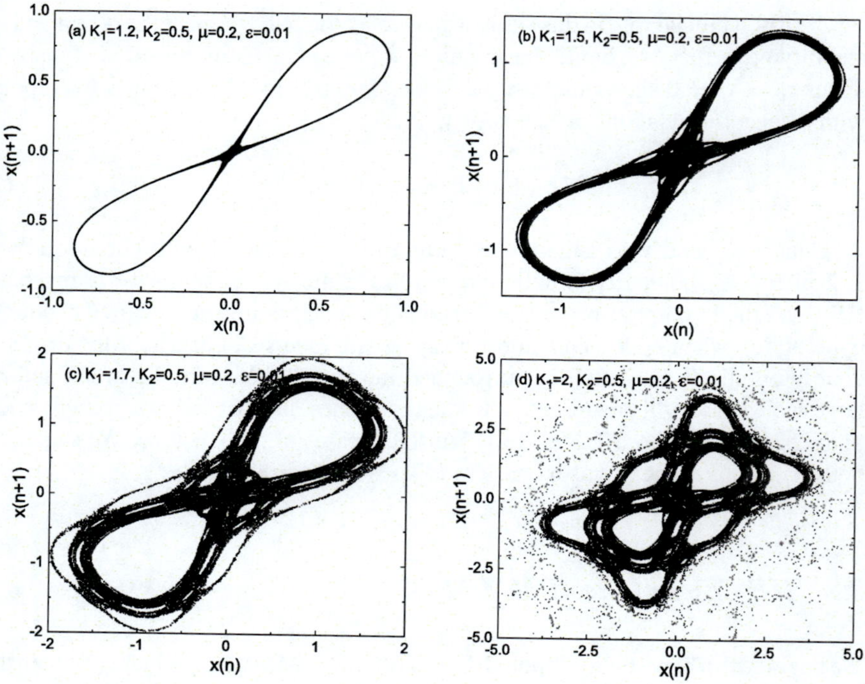

Fig. 2 2D phase space plots for the x_n plane for different K_1 values. The rest of the parameters, $K_2 = 0.5$, $\mu = 0.2$ and $\epsilon = 0.01$ remain constant for all panels. **a** $K_1 = 1.2$, **b** $K_1 = 1.5$, **c** $K_1 = 1.7$, and **d** $K_1 = 2$. The number of iterations is always 2×10^5

this end, we will take values of K_1 and K_2 that violate the condition (5) and are either positive and negative or both negative as follows:

(1) $K_1 = -1.25$, $K_2 = 1.25$: The dynamics is close to weak chaos, as the phase space plot in Fig. 4a shows, since its pdf in Fig. 4c is close to a q-Gaussian for three orders of magnitude with $q = 2.97$.

(2) With $K_1 = -1.25$, $K_2 = -1.25$: The phase space plot in Fig. 4b corresponds to what we call "strong" chaos, since its pdf, plotted in Fig. 4d is very close to a Gaussian with $q = 1.09$.

Observing Fig. 4 more closely, we suggest that the statistical results may be explained as follows: In the first column, where the orbits form a more "sparse" pattern in Fig. 4a, the associated q-Gaussian implies weak chaos, while in the second column, a more uniformly filled pattern in Fig. 4b is characterized by a true Gaussian representing strong chaos.

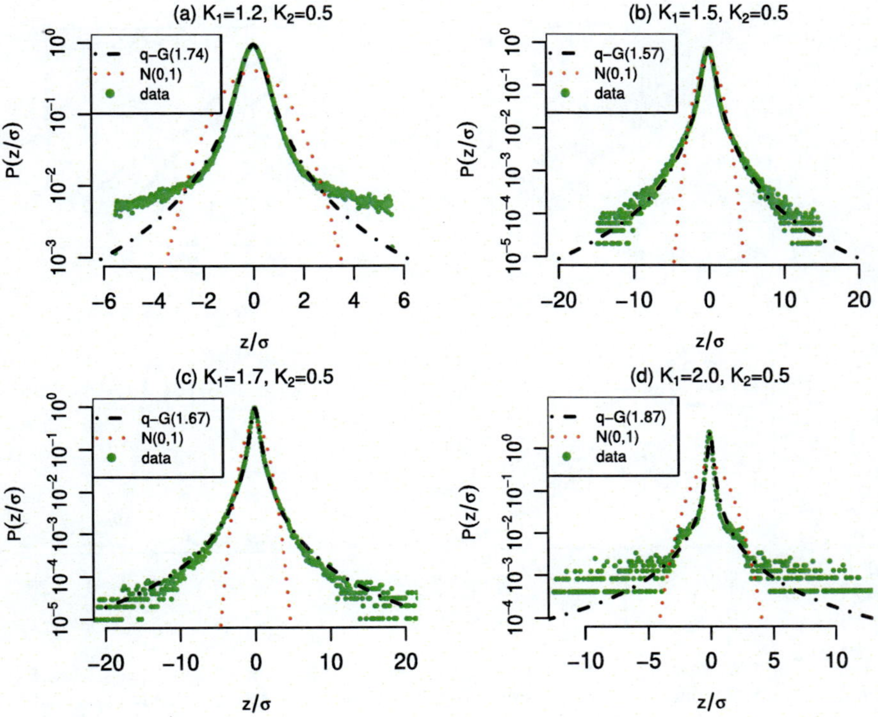

Fig. 3 The pdfs for the sums $s_N^{(j)}$ corresponding to the chaotic domains shown in Fig. 2a–d respectively. The black dashed line corresponds to the optimal fitting with the q–Gaussian distribution and the red dashed line is the normal distribution

Let us also compare, for these HH cases, the above results, with those corresponding to the y_n, y_{n+1} data as plotted in Fig. 5. Clearly, due to the $x - y$ symmetry of the map, there are strong similarities between Figs. 4 and 5, validating the conclusions of weak chaos on the left column and strong chaos on the right column of the two figures.

3.4 *Close to the Instability Transition*

We also examined a case close to the transition of instability for one of the maps. In particular, as shown in Fig. 6, we set $K_1 = 1$ and plot for $K_2 = 0.9, 1.3, 1.5$ in Fig. 6a, c, e the x_n, x_{n+1} projections of the orbits, while in Fig. 6b, d, f we present the corresponding statistical analysis. Clearly the pdfs in this case are very well described by a q-Gaussian with q increasing from 1.5 to 1.94 and 2.04, close to the value $q = 2$, which is the case of the Cauchy distribution.

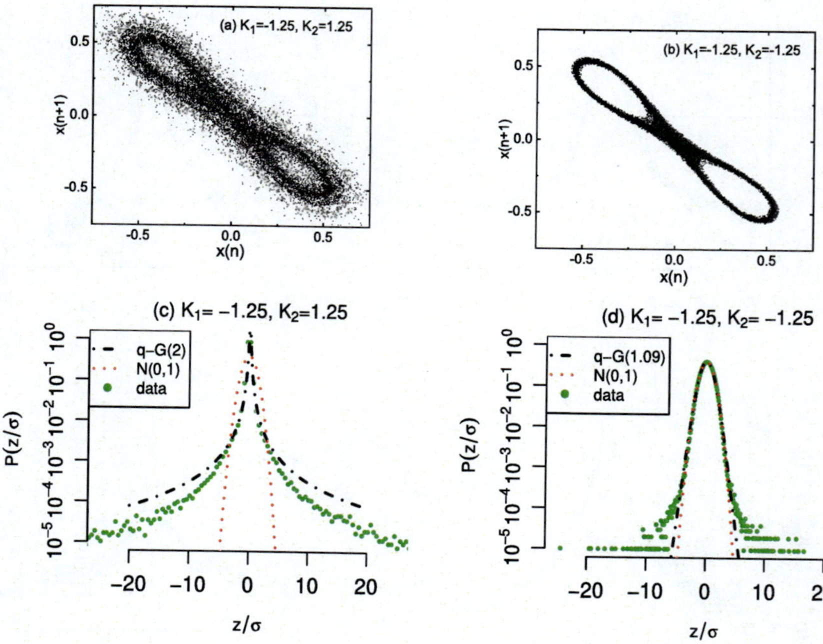

Fig. 4 Top row: Phase space plots on the x_n, x_{n+1} plane for **a** $K_1 = -1.25, K2 = 1.25$, **b** $K_1 = K2 = -1.25$. Bottom row: **c** and **d** present the pdf plots corresponding to **a** and **b** respectively

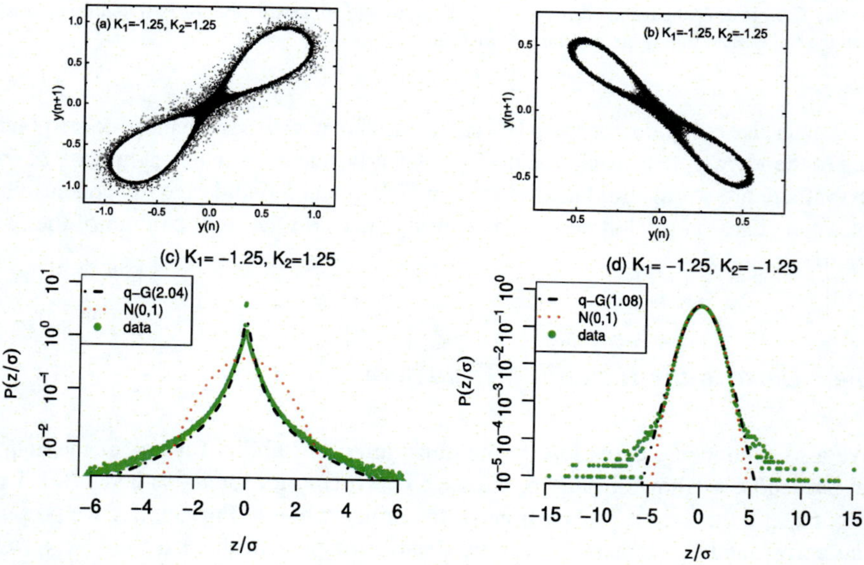

Fig. 5 **a** Top row: Phase space plots on the y_n, y_{n+1} plane for **a** $K_1 = -1.25, K2 = 1.25$ and **b** $K_1 = K2 = -1.25$. In **c** and **d** respectively we plot the pdfs corresponding to **a** and **b**. Note the similarities with Fig. 4

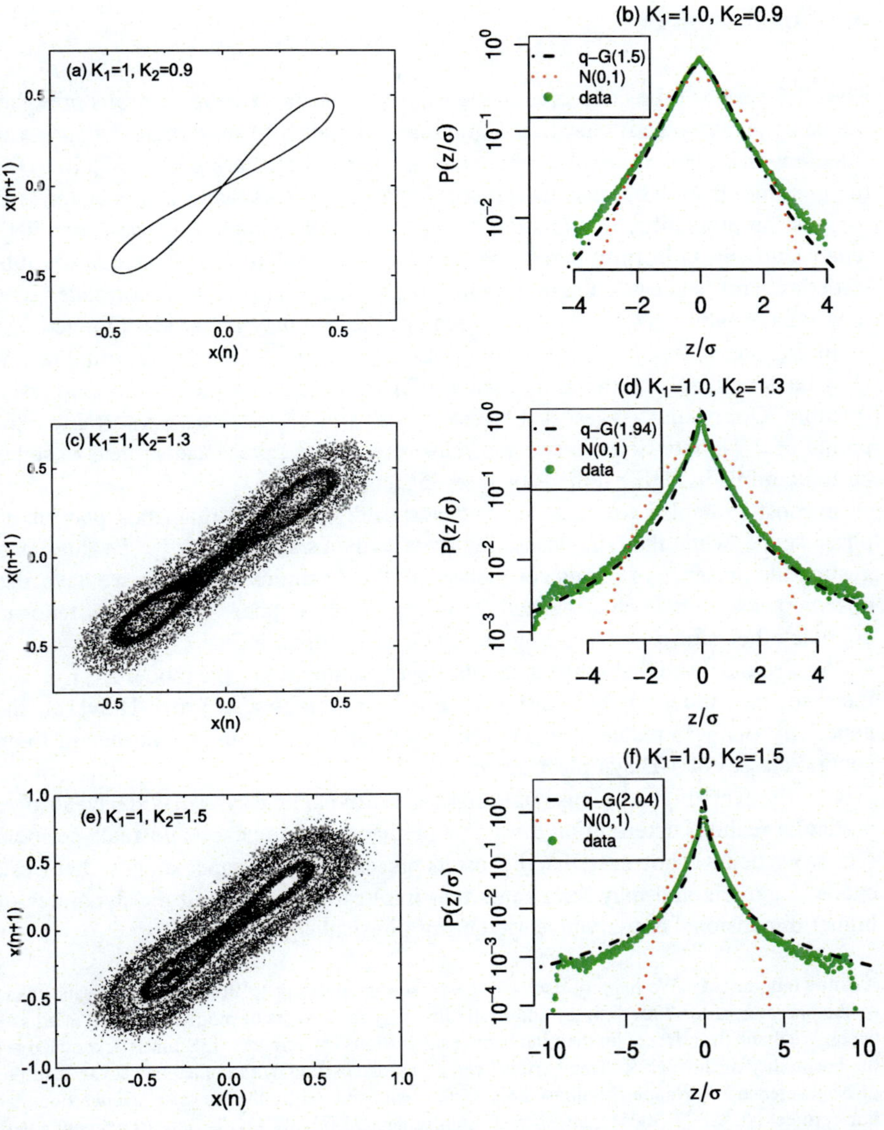

Fig. 6 Close to the instability transition: Here we set $K_1 = 1$ and present in **a-c-e** the phase space plots for x_n and in **b-d-f** the corresponding pdf plots. The first row corresponds to $K_2 = 0.9$, the second row to $K_2 = 1.3$ and the third row to $K_2 = 1.5$

4 Conclusions

The stickiness of orbits observed in the vicinity of unstable periodic orbits of higher dimensional symplectic maps, or Hamiltonian systems of more than 2 degrees of freedom, is clearly a complex phenomenon. It has been termed "weak chaos" in the literature mainly because its statistical analysis reveals that it is associated with q-Gaussian probability distributions, as opposed to the simple Gaussians one finds when studying uniformly spread stochasticity associated with Boltzmann Gibbs statistics. This is because the motion in weakly chaotic situations is correlated over long ranges, while in strongly chaotic regions the correlations are short ranged.

In this paper, we attempted to study this phenomenon, for the first time, in a 4-D symplectic map, serving as a paradigm for Hamiltonian systems of 3 degrees of freedom. Our results suggest that "weak chaos" arises typically near unstable fixed points of $2N$-dimensional maps and may very well be present also near unstable periodic orbits in higher dimensional settings.

In most examples we considered, chaos tends to form "organized" patterns in phase space, while the pdfs describing their statistics attain $1 < q < 2$ values suggesting the presence of strong correlations in the dynamics. However, we have also observed cases where chaos spreads more uniformly in phase space and q tends to approach the value $q = 1$ yielding purely Gaussian distributions.

We also observed that as the main nonlinear parameters of the model K_i, $i = 1, 2$ increase, the values of the index q of the distributions also grow. However, the genericity of these results remains open and needs to be studied further in more general classes of 4-D symplectic maps.

Clearly, every high–dimensional conservative dynamical system will have its own particular features determining the nature of chaos present near its unstable periodic orbits. We believe, however, that the results presented in this paper suggest that weak chaos is generic and may have important implications regarding the dynamics of higher dimensional conservative systems of physical significance.

Acknowledgements We happily dedicate the present manuscript to Professor Thanassis Fokas on the occasion of his 70th birthday and wish him many more years of pioneering work in all sciences. We thank the referees for their interesting suggestions and remarks. T. Bountis acknowledges the hospitality of the NCSR "Demokritos" and many discussions with colleagues at the Institute of Nanoscience and Nanotechnology. This work was supported by the Russian Science Foundation (project No. 21-71-30011), https://rscf.ru/en/project/21-71-30011/. T. Bountis acknowledges financial support for Sections 1, 3.2, 3.3 and 4 of the paper.

References

1. Guckenheimer, J., Holmes, P.: Nonlinear oscillations, dynamical systems, and bifurcations of vector fields. Appl. Math. Sci. **42** (1983)
2. Wiggins, St.: Introduction to Nonlinear Dynamical Systems and Chaos. Texts in Applied Mathematics, vol. 2. Springer (1990)

3. Ott, E.: Chaos in Dynamical Systems, 2nd edn. Cambridge University Press, Cambridge (2002)
4. Lichtenberg, A.J., Lieberman, M.A.: Regular and Chaotic Dynamics. Applied Mathematical Sciences, vol. 38, 2nd edn. Springer (2013)
5. Strogatz, S.H.: Nonlinear Dynamics and Chaos, 2nd edn. CRC Press (Taylor and Francis) (2015)
6. Skokos, C.: The Lyapunov Characteristic Exponents and Their Computation. Lecture Notes in Physics, vol. 790. Springer, Berlin, Heidelberg (2010)
7. Contopoulos, G., Harsoula, M.: Stickiness Effects in Chaos. Celestial Mechanics and Dynamical Astronomy, vol. 107, pp. 77–92. Springer (2010)
8. Bountis, T, Skokos, H.: Complex Hamiltonian Dynamics. Springer Series in Complexity. Springer (2012)
9. Katsanikas, M., Patsis, P.A., Contopoulos, G.: Instabilities and stickiness in a 3D rotating galactic potential. Int. J. Bifurc. Chaos **23**(2), 1330005 (2013)
10. Contopoulos, G., Voglis, N., Efthymiopoulos, C., et al.: Transition spectra of dynamical systems. Celest. Mech. Dyn. Astron. **67**, 293–317 (1997)
11. Kovács, T., Érdi, B.: Transient chaos in the sitnikov problem. Celest. Mech. Dyn. Astron. **105**, 289–304 (2009)
12. Bountis, T., Manos T., Antonopoulos, Ch.: Complex statistics in Hamiltonian barred galaxy models. Celest. Mech. Dyn. Astron. **113**(1), 63–80 (2012)
13. Antonopoulos, C., Basios, V., Bountis, T.: Weak chaos and the ‘melting transition’ in a confined microplasma system. PRE **81**, 016211 (2010)
14. Antonopoulos, C., Bountis, T., Basios, V.: Quasi-stationary chaotic states in multi-dimensional Hamiltonian systems. Physica A **390**, 3290–3307 (2011)
15. Christodoulidi, H., C. Tsallis, C., Bountis, T.: Fermi-pasta-ulam model with long range interactions: dynamics and thermostatistics. Eur. Phys. J. Lett. EPL **108**, 40006 (2014)
16. Christodoulidi, H., Bountis, T., Tsallis, C., Drossos, L.: Chaotic behavior of the fermi-pasta-ulam model with different ranges of particle interactions. J. Stat. Mech. **12**(12), 123206 (2016)
17. Christodoulidi, H., Bountis, T., Drossos, L.: The effect of long-range interactions on the dynamics and statistics of 1D Hamiltonian lattices with on-site potential. Eur. Phys. J. Special Topics **227**, 563–573 (2018)
18. Ruiz, G., Bountis, T., Tsallis C.: Time–evolving statistics of chaotic orbits of conservative maps in the context of the central limit theorem. Intern. J. Bifurc. Chaos **22**(9), 12502 (2012)
19. Tirnakli, U., Borges, E.P.: The standard map: from Boltzmann-Gibbs statistics to Tsallis statistics. Sci. Rep. **6**, 23644 (2016)
20. Hietarinta, J., Joshi, N., Nijhoff, F.: Discrete Systems and Integrability. Cambridge University Press (2016)
21. Turchetti, G., Scandale, W. (eds.): Nonlinear Problems in Future Particle Accelerators. World Scientific (1991)
22. Tsallis, C.: Introduction to Nonextensive Statistical Mechanics-Approaching a Complex World. Springer, New York (2010)
23. Bountis, T., Skokos, Ch.: Applications of the SALI detection method to accelerator mappings. Nucl. Instr. Meth. Phys. Res. A **561**, 173–179 (2006)
24. Bountis, T., Skokos, Ch.: Space charges can significantly affect the dynamics of accelerator maps. Phys. Lett. A **358**(2), 126–133 (2006)
25. Rossi, R.J.: Mathematical Statistics: An Introduction to Likelihood Based Inference. Wiley (2018)
26. Team, R Core: R: A language and environment for statistical computing. (2013)
27. White, D.R., Kejzar, N., Tsallis, C., Farmer, D., White, S.: Generative model for feedback networks. Phys. Rev. E **73**(1), 016119 (2006)
28. Wasserman, L.: All of Statistics: A Concise Course in Statistical Inference, vol. 26. Springer, New York (2004)
29. Shalizi, C.R.: Maximum likelihood estimation for q-exponential (Tsallis) distributions (2007). arXiv:math/0701854

Fractals

Multifractal Analysis of SEM Images of Multiphase Materials: The Case of OPC Clinker

M. Chatzigeorgiou, V. Constantoudis, M. Katsiotis, and N. Boukos

Abstract Several properties of multiphase materials are affected by the spatial distribution of their phases and the geometry of their interfaces. Given the spatial complexity of phase distributions, their quantitative characterization is a real challenge which remains largely unexplored. A widely used technique to inspect multiphase materials is the top-down Scanning Electron Microscope (SEM) images obtained in Back Scattered Electron (BSE) mode on which the different phases are depicted with different pixel intensities. On these images, the challenge is to quantify the spatial distribution of image segments characterized by different pixel intensities. Since the areas of phase segments range in multiple scales, multifractal analysis seems to be a reasonable approach to provide an insightful quantification of phase distributions. In this paper, first we demonstrate the benefits of multifractal analysis to characterize the spatial scaling behavior of different phases in a widely used multiphase material in cement industry such as clinker. Then, we focus on two issues related to the reliability of the obtained multifractal spectrum and the impact of measurement artifacts on it. A recently proposed alternative method to multifractal spectrum calculation is employed to meet the first challenge while concerning the second one we detail the effects of image artifacts on the two branches of the calculated multifractal spectrum.

Keywords Multifractality · Singularity spectrum · Back-scattered electron imaging · Multiphase material · Clinker

M. Chatzigeorgiou · V. Constantoudis (✉) · N. Boukos
Institute of Nanoscience and Nanotechnology, National Centre for Scientific Research Demokritos, Patr., Gregoriou E and 27 Neapoleos Str, 15341 Agia Paraskevi, Greece
e-mail: v.constantoudis@inn.demokritos.gr

M. Chatzigeorgiou
e-mail: e.chatzigeorgiou@inn.demokritos.gr

M. Chatzigeorgiou
School of Chemical Engineering, National Technical University of Athens, 9 Iroon Polytechniou Street, Athens Zografou 15780, Greece

M. Katsiotis
Group Innovation and Technology, TITAN Cement S.A., 22A Halkidos Street, 111 43, Athens, Greece

T. Bountis et al. (eds.), *Chaos, Fractals and Complexity*, Springer Proceedings in Complexity, https://doi.org/10.1007/978-3-031-37404-3_8

1 Introduction

Multiphase materials such as metal alloys, concrete, polymer and glass composites are used in several industry sectors (energy, construction and automobile etc.) since they exhibit a great repertoire of combinations of phase properties enhancing their performance in each specific application. The behavior of a multiphase material is controlled by the properties of its components and the way they co-exist in the bulk of material. Two crucial factors are (a) the fraction of material occupied by each component [1] and (b) the fashion that each component is spatially distributed in the whole volume of multiphase material [2]. The latter is critical since it is related to the geometrical properties of the interfaces separating the different phases and therefore the amount and type of interactions between phases. In most cases, the interfaces exhibit complicated shapes in a wide range of scales reaching the limit of micro and nanoscale. Due to this wide-scale complexity of phase distributions and interfaces, their quantitative measurement and characterization is a real challenge which needs further investigation.

The characterization of multiphase materials can be made with both imaging [3] and non-imaging techniques. Among the former, the acquisition and inspection of top-down SEM images collecting Back-Scattered Electrons (BSE-SEM images) is of prime interest. In BSE-SEM images, different phases are depicted with pixels of distinct intensities (luminosities) and hence they can be straightforwardly identified down to SEM resolution which can be on nanometer scale [4]. The quantification of the relative fraction of phases can be achieved by means of a properly designed segmentation process of BSE-SEM images which will cluster pixels according to the phase they belong to. The segmentation method can use information alone from intensity distribution of pixels or alternatively can be empowered by the spatial positioning of pixels to enhance accuracy of segmentation output [5]. In any case, after the successful application of the chosen segmentation method, the image pixels are matched to the material phases to obtain the map of phases in the analyzed SEM image. Then, we can proceed to the quantitative characterization of this map in two steps. First, we calculate the fraction of each phase in material by simply summing the pixels corresponding to the specific phase. The second step is the characterization of the fashion that phases are distributed spatially in the image. The spatial distribution of phases affects the geometry of interfaces between phases and therefore the combination of their properties in the multiphase material. Usually the phase distributions exhibit spatial complexity with boundaries covering multiple scales. To explore the scaling aspects of phase distributions, multifractal analysis seems to be a reasonable choice taking advantage of the fact that it differentiates scaling behavior versus image intensity and therefore material phase. In literature, the spatial distribution of phases in heterogeneous multiphase materials have been mainly investigated by using spatial correlation functions such the two-point correlation function, lineal-path function or the two-point cluster correlation function and Fourier of wavelet analysis [6]. During the last years, sporadic studies have emphasized the significance of multifractal analysis in quantifying specific aspects of cement-based composites such as the tendency to quantify the cluster forming tendency of C-S-H gel or the scaling statistics of pores independently of their shapes or the relationship with crack growth and anomalous behavior [7–12]. In the most recent study, Gao et al. [13] has studied the multifractal spectrum of both ordinary and blast furnace slag blended Portland cement pastes from their X-Ray CT images to account for the degree of heterogeneity of their local porosity. In this paper, we differentiate our approach in terms of both methodology and data. Concerning methodology, we elaborate an alternative method for the estimation of multifractal spectrum which has been designed to overcome the instabilities usually occurring at the negative generalized exponents which undermine the accuracy of multifractal analysis. As regards data, we collect and analyse BSE-SEM images of clinker samples depicting phase distributions on nanoscale resolution. The aim has been first to demonstrate the benefits of multifractal analysis of BSE-SEM images for the quantification of the scaling behavior of phase distribution complexity and use it for detection of image artifacts. To this end, the paper is organized as follows. In the next Sect. 2, we describe shortly the basic steps of multifractal analysis and we focus on the alternative method we propose to provide reliable calculation of the full scale multifractal spectrum. Then, the composition of the analysed multiphase material (clinker) along with the obtained BSE-SEM images are reported in Sect. 3. Section 4 delivers the results of the multifractal analysis and discusses their meaning and significance. The paper closes with a summary of the main findings within Sect. 5.

2 Multifractal Analysis

Fractal analysis has been widely used to provide a quantitative characterization of the scaling behavior of surface roughness or image texture. A limitation that occurs in this analysis is the assumption that a single scaling behavior dominates in the whole surface morphology or image texture. In many cases this assumption is not justified. A broader analysis of scaling behavior has been developed in the framework of multi-fractal analysis. The aim of this analysis is to quantify the involvement of multiple scales in a surface in order to provide a more complete description of its morphology. For example, multifractal analysis can capture the subtle differences in the scaling behavior of peaks and surface valleys or alternatively of almost flat and steep regions. One of the most commonly used techniques for multifractality measurements is the box counting method. A fundamental quantity in this approach is the measure of mass. In this work, mass corresponds to the grayscale intensities of an SEM image I. Assuming that the desirable observation scale is s, the image should be covered by non-overlapping square boxes of side s. Subsequently the normalized mass can be calculated by the following formula:

$$p(s,v) = \frac{\sum_{i}^{s^2} I_i}{\sum_{v=1}^{N_s} \sum_{i}^{s^2} I_i}, \tag{1}$$

Where the summation of grayscale intensities I_i in the numerator takes place within the v-th box of size s while the sum in the denominator is over all image pixels to normalize $p(s, v)$. The total number of squares of size s covering the structure is denoted by N_s and v is the index that enumerates these square boxes. The partition function is then utilized for the distinction between the boxes with large and small masses (intensities) by raising every term in the sum to a power q according to the following relation (2):

$$\chi(s, q) = \sum_{v=1}^{N_s} p(s, v)^q, \tag{2}$$

Where positive exponents q emphasizes the large mass segments of image containing pixels of high intensity (bright regions) while at negative q the boxes with low intensity pixels (dark regions) dominate in the sum. If the dependence of the partition function $\chi(s, q)$ on scale s follows a power law for all q, then one can conclude that the analyzed image texture exhibits a fractal behavior, i.e.

$$\chi(s, q) \sim s^{\tau(q)}, \tag{3}$$

The multifractality emerges when the generalized fractal dimension D_q defined through $\tau(q)$ with the following formula:

$$D_q = \frac{\tau(q)}{q - 1}, \tag{4}$$

varies with exponent q. The spectrum of the generalized dimensions D_q can be used to quantify the multifractal structure of image texture. Another measure that provides an insight in the image scaling properties is the singularity spectrum, which is the Legendre transform of $\tau(q)$.

$$\alpha = \frac{d\tau(q)}{dq} \text{ and } f(\alpha) = \alpha(q) \cdot q - \tau(q) \tag{5}$$

An indicative example of the singularity spectrum of an image texture is shown in Fig. 1.

The x-axis of the singularity spectrum diagram is the local dimension α while the y-axis shows the singular spectrum $f(\alpha)$. It is easier to gain an intuitive understanding of the local dimension and the singularity spectrum with Chhabra equations [14]:

$$f(\alpha) = \lim_{s \to 0} \sum_{v=1}^{N_s} \mu(q, s, v) \frac{ln\mu(q, s, v)}{lns} \text{ and } \alpha = \lim_{s \to 0} \sum_{v=1}^{N_s} \mu(q, s, v) \frac{lnp(s, v)}{lns}, \tag{6}$$

where,

$$\mu(q, s, v) = \frac{(p(s, v))^q}{\sum_{v=1}^{N_s} (p(s, v))^q} \tag{7}$$

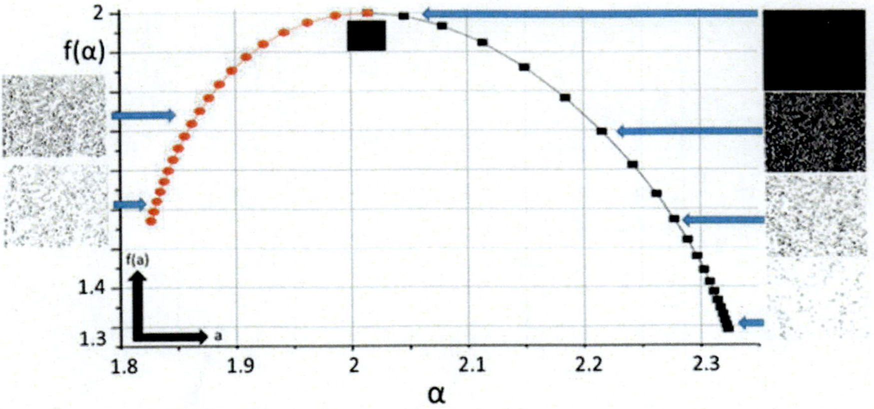

Fig. 1 A typical example of a singularity spectrum f(α) of an image texture. The right branch of spectrum (black points and line) indicates the scaling behavior of the "valleys" of image texture (dark areas) whereas the left branch (red points and line) quantifies the scaling of peaks (bright areas of image)

As Eq. 6 reveals, local dimensions α are similar to the fractal dimension with the difference that is weighted by the factor μ. At high positive q exponents and when local measures of mass (pixel intensities) are large, this factor takes a large value and hence, the local dimensions quantify the scaling behavior of the bright image areas. On the contrary, for negative q exponents the factor μ enhances in image areas dominated by small masses (dark image regions). The local dimensions which are larger than the topological dimension (i.e. $\alpha > 2$), characterize the fraction of surface that is dominated by values of small masses. On the contrary, for $\alpha < 2$ the local dimensions describe the areas of the image that are dominated by bright pixels. The singularity spectrum $f(\alpha)$ reveals how areas that are described by the same scaling law are dispersed in the surface. A large deviation from the topological dimension ($f(\alpha = 2) = 2$) indicates a sparse pattern of areas with the same local scaling behavior. A very frequent source of miscalculations in the computation of the power law exponent $\tau(q)$ is the exceedingly small masses, i.e. the image areas with very dark pixels. In the case of negative exponents q, the terms in the partition function relating with boxes within these areas often result in instabilities. These instabilities undermine our ability to extract correctly the multifractal measures i.e. D_q, α and $f(\alpha)$ by strongly biasing the power law slopes [15]. An indicative example of miscalculating multifractality is shown in Fig. 2 where the singularity spectrum of a gaussian monofractal rough texture is depicted with the black squares and line. Theoretically one expects a symmetric spectrum due to the same behavior of peaks and valleys. However, this is not the case for the spectrum calculated with the conventional box-counting method, in which the right branch of the spectrum is much more extended than the left one, despite the same scaling behavior of peaks and valleys in the Gaussian texture.

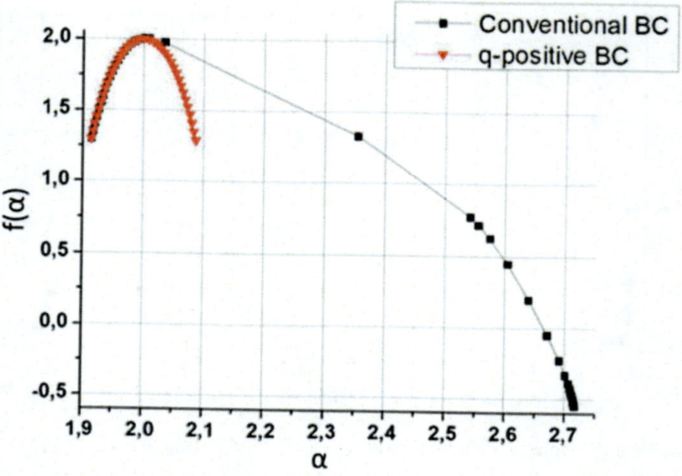

Fig. 2 The singularity spectrum $f(\alpha)$ of a rough image texture with Gaussian distribution of pixel intensities calculated with the conventional box-counting method (black squares) and the q-positive alternative method (red triangles). The spectrum of the conventional method is strongly asymmetrical with strongly elongated right branch contrary to what we expect for a texture with symmetrical bright and dark areas. On the other side, the q-positive method corrects the calculation and derives a symmetrical spectrum about (2, 2) point

This problem can be overcome by reconsidering the reason of applying negative exponents q, i.e. the scaling analysis of image dark areas. Due to the very small values that $p(s, v)$ can take on at boxes inside these areas, their contribution to the sum of the partition function at negative q values leads to infinities and instabilities. The key observation is that the calculations for positive q do not suffer from similar problems even in the boxes with very bright pixels. Therefore, one could take the complement I^- of the original image I

$$I^- = max(I) - I \tag{8}$$

and then calculate the multifractal spectrum of I^- for positive q (see Fig. 3). Since the bright areas of the complement image have the same spatial distribution at all scales with the dark areas of the original image, the above calculation can replace the q-negative spectrum of the original image eliminating the instability issue. We can name this method q-positive since it is based on the estimation of the multifractal spectrum of both the original and complement image using only positive q.

In the generalized version of the q-positive method, the normalized masses should be changed to the following equations:

$$P^+(s, v) = \frac{\sum_i^s I_i}{\sum_{v=1}^{N_s} \sum_i^s I_i} \text{ and } P^-(s, v) = \frac{\sum_i^s I_i^-}{\sum_{v=1}^{N_s} \sum_i^s I_i^-} \tag{9}$$

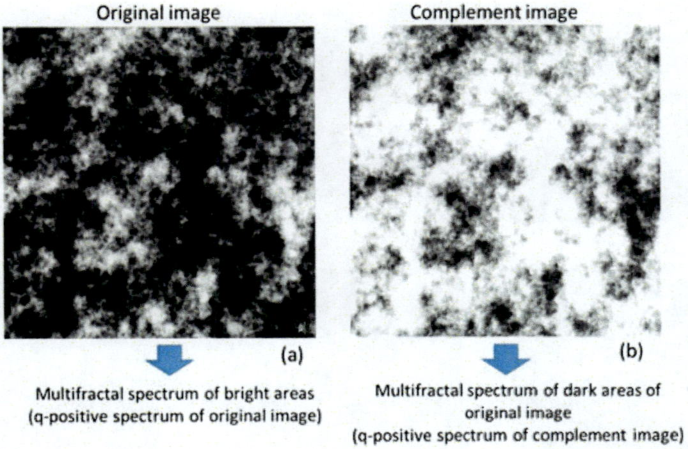

Fig. 3 An example of a grayscale image with rough texture (**a**) along with its complement in intensity (**b**). The original image is used normally for the calculation of positive q branch of multifractal spectrum while the complement one for the branch of the negative q values

and the partition function should be replaced by the following one:

$$\chi(s,q) = \frac{1+sgn(q)}{2}\sum_{v=1}^{N_s}(P^{+}(s,q))^{|q|} + \frac{1-sgn(q)}{2}\sum_{v=1}^{N_s}(P^{-}(s,v))^{|q|} \quad (10)$$

In the updated partition function, when the sign of q is positive it takes the well-known conventional formula, whereas in the case of negative q the analysis will be done in the complement image using again the positive q. The rest of multifractal analysis remains unaltered. The q-positive method that is described above is able to correctly extract the multifractality spectrum as shown in Fig. 3 where it has been applied in a Gaussian rough texture deriving the expected symmetrical spectrum.

3 Data

The images that are analyzed in this work are back-scattered electron, scanning electron microscopy images (BSE-SEM) of OPC clinker. In these images, there are 5 clinker phases visible, Periclase, Celite, Belite Alite, and Ferrite in an ascending order based on their back-scattered coefficient and hence on their grayscale intensities. The samples were prepared by impregnating clinker in a low viscosity epoxy resin in vacuum and then samples are grinded and polished with the use of Silicon Carbide papers of grid sizes 280, 400, 1200, 2500, 4000 and finally polished with diamond paste of 6 μm and 1 μm. A deodorized oil is also used to lubricate the surface after grinding and polishing process [16]. Finally, to avoid charge effects a thin carbon

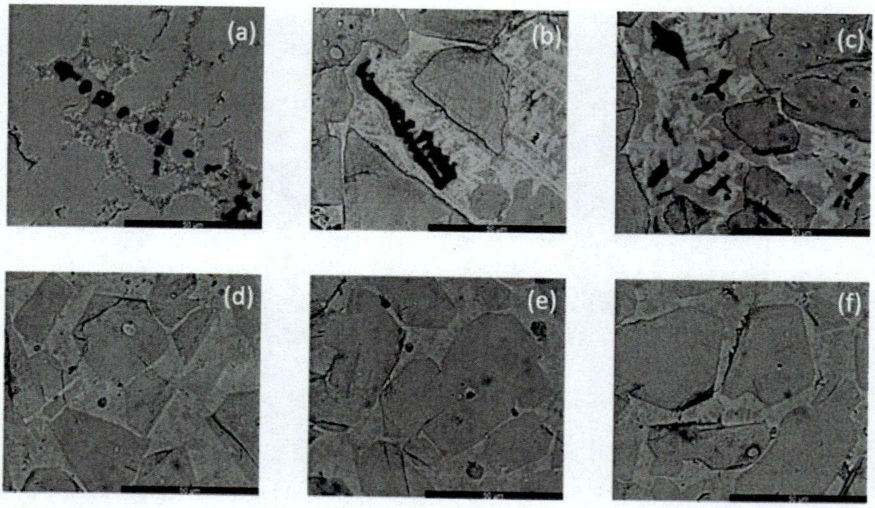

Fig. 4 The Back-Scattered Electron Scanning Electron Microscopy (BSE-SEM) images of samples of OPC clinker analyzed in this paper. Details about the preparation of samples and the acquisition of images are given in the text

layer has been deposited on the sample surface. The BSE-SEM Images have been acquired with a FEI Inspects 400 SEM under the same conditions with the same acceleration voltage (15 kV), beam current 2 nA measured with Faraday cage, the working distance was set at 10 mm whilst contrast and brightness settings are constant for all images. The resolution of images is 4096 by 3535 pixels with pixel size equal to 26 nm. In this work, six images have been acquired according to the conditions mentioned above and they are shown in Fig. 4.

4 Results

The application of multifractal analysis in the Back-scattered Electron images of clinker reveals the capability of multifractal spectrum to characterize the scaling behavior of the spatial distribution of crystallographic phases as they are depicted in a BSE-SEM image. However, before explaining this statement in the multifractal analysis of the BSE-SEM images of Fig. 4, we would like to emphasize the significance of applying the alternative implementation of multifractal analysis with the q-positive method in real BSE-SEM images. Figure 5 a illustrates the singularity spectra of the images shown in Fig. 4a–c as they have been calculated by the conventional box-counting method. One can clearly notice the strongly asymmetrical shape of the $f(\alpha)$ curves in all cases caused by the awkward enhancement of the right branches of spectra which are crossing zero to take on even negative values. As we explained in the Sect. 2, the miscalculation is due to the loss of linearity in the

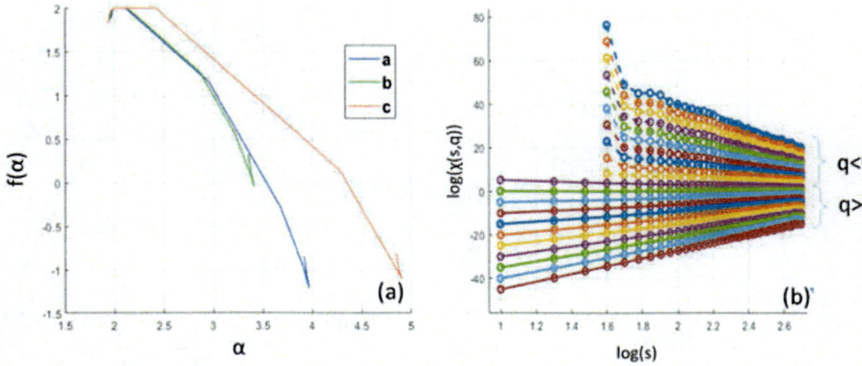

Fig. 5 **a** The singularity spectra $f(\alpha)$ of the BSE-SEM images shown in Fig. 4a–c calculated from the conventional box-counting method. One can easily notice the problematical shape of the right branches of spectra which cross zero to reach negative values of f. The reason for this is the image areas with black pixels which lead to instabilities in the calculation of $\chi(s, q)$ and deviations from power-law behavior as demonstrated by the doubly logarithmic diagrams of $\chi(s, q)$ versus s for negative q displayed in **b**

$log(\chi(s, q))$ vs log(s) plot at negative q values which correspond to the right branch of $f(\alpha)$ spectrum (see Fig. 5b). Since this part of spectrum quantifies the scaling behavior of dark image areas, the deviation from linearity is coming from the appearance of zero-intensity boxes in the calculation of $\chi(s, q)$ leading to instabilities and undermining the correctness of the total multifractal analysis. These miscalculations of partition function and therefore of the slope of log-log plots can be remedied by using the q-positive method, as described in the Sect. 2. This is demonstrated in the log-log plot shown in Fig. 5a where the linearity of $\chi(s, q)$ curves of the image of Fig. 4a is restored by the q-positive method in the whole spectrum of pixel intensities including the problematical black image areas.

After the demonstration of the need to apply the q-positive method in the analysis of real BSE-SEM images, we can proceed to the comparison of multifractal singularity spectra calculated from all BSE-SEM images shown in Fig. 4. We separate the presentation of our results in the analysis of the triplet of Fig. 4a–c first and then of the triplet of the BSE-SEM images of Fig. 4d–f. The first image (Fig. 4a) of the first triplet is dominated by Alite i.e., grayish area and periclase (dark areas). Between Alite grains and Periclase grains there are the so-called liquid phases, i.e. Ferrite phase (Bright areas) and Celite (dark gray areas). A different distribution occurs in the second image of Fig. 4b, where the same phases are present. However, in this image Periclase is more packed comparing to the previous one, and Ferrite along with Celite phases are more dominant. Finally in the third image (Fig. 4c) Periclase grains are dispersed on the image whilst Ferrite and Celite phases occupy a larger proportion of the image. The multifractal spectra of these images are shown in Fig. 4b with blue (a), green (b) and red (c) lines respectively. The left branch of the shown spectra concerns the scaling behavior of bright areas. By observing minimum

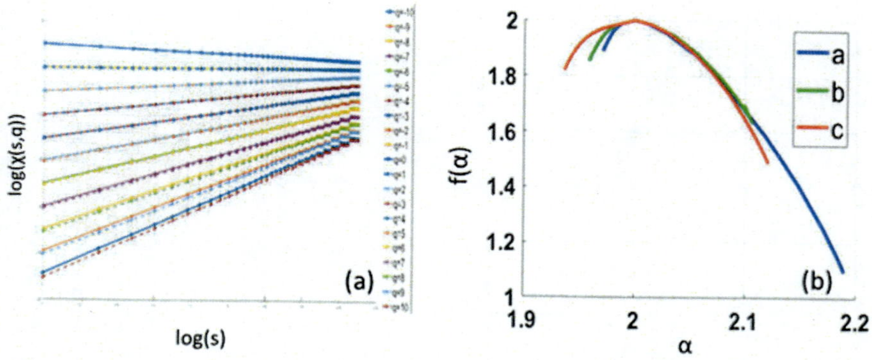

Fig. 6 **a** The $\chi(s, q)$ versus s diagrams calculated applying the q-positive method in the multifractal analysis of the image shown in Fig. 4a. One can notice the restoration of the power law behavior in all cases justified by the linearity of $\chi(s, q)$ in the log-log scale of the shown diagram, **b** The corrected singularity spectra $f(\alpha)$ of the BSE-SEM images shown in Fig. 4a, b, c. The spectra have been calculated by the q-positive method and depicted in the diagram with the blue (**a**), green (**b**) and red (**c**) line respectively

local dimensions α one can understand the size of bright aggregates. In the image a, where the bright aggregates are small, α_{min} deviates slightly from the support dimension of the image (=2) in comparison with the other images. In contrary, image c is dominated by large local bright aggregates, which leads to the larger deviation of α_{min} from the support dimension. The deviation of minimum singularity spectrum $f(\alpha_{min})$ from the support dimension, in the left branch of the diagram, highlights the spatial homogeneity of bright aggregates in an image. In the case of image a, where bright aggregates are concentrated along the diagonal of the image, the deviations of $f(\alpha_{min})$ from the support dimension are expected to be small. Indeed, this is justified by the singularity spectra of Fig. 6b since the $f(\alpha_{min})$ of the image a is closer to 2 than those of images b and c. In the dark side of the diagram maximum local dimensions reveals the size of dark aggregates in images. In image b where both the size and the number of dark aggregates is large α_{max} is considerably smaller than α_{max} of other images. In addition, the coverage of dark aggregates on the image a as f(α_{max}) uncovers is less scattered than other images. One might argue that in the image a of Fig. 4, a fraction of dark pixels does not originate from z-contrast exclusively, but includes scratches and pores of the material. This type of sample artifacts may lead the multifractal analysis to misleading results. In order to investigate more systematically the effects of these artifacts, we focus on the multifractal analysis of BSE-SEM images shown in Fig. 4d–f in which the dark pixel areas originate mainly from artifacts of the sample preparation.

By examining these images it is obvious that their dark pixels come from scratches and microporosity on sample surface. These artifacts are expected to influence the right branch of singularity spectrum which quantifies the scaling behavior of dark areas. On the other side, the results of the left branch of the singularity spectrum should be more robust to the presence of such artifacts since they are calculated

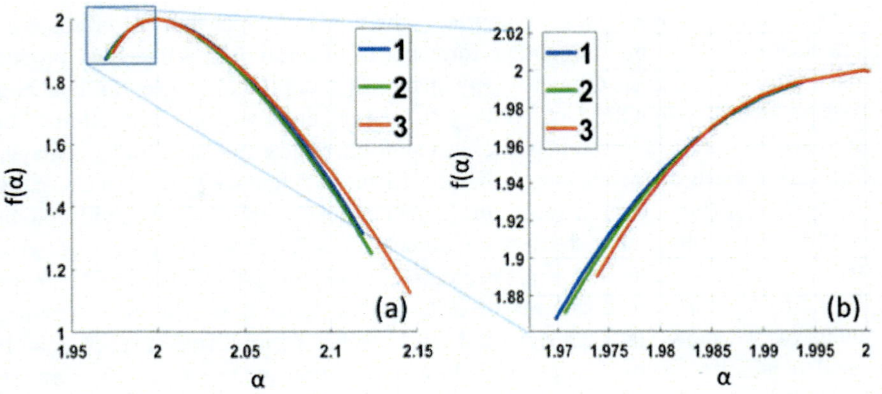

Fig. 7 **a** The corrected singularity spectra $f(\alpha)$ of the BSE-SEM images shown in Fig. 4d, e, f. The spectra have been calculated by the q-positive method and depicted in the diagram with the blue (1 for Fig. 4d image), green (2 Fig. 4e image) and red (3 Fig. 4f image) line respectively, **b** Magnification of the left portion of $f(\alpha)$ curves to highlight the differences among images concerning the multifractal analysis of their bright regions

by analyzing the scaling behavior of bright pixels depicting Ferrite phase. These observations can be shown in the singularity spectrum of these images in Fig. 4. The right branch of the singularity spectrum of the image of Fig. 4d (blue line 1 in Fig. 7a diagram) exhibits the lowest α_{max} and higher $f(\alpha_{max})$ value versus the other two images, revealing the lesser influence of scratches in the image texture. On the other hand, the image of Fig. 4f (red line 3 in Fig. 7a) is characterized by a large thick scratch, which contributes to the reduction of $f(\alpha_{max})$. The left branch of the spectra (see Fig. 7b) revealing the scaling behavior of bright pixels, indicates a very similar behavior for the images 1 (Fig. 4d) and 2 (Fig. 4e). Whilst in image 3 (Fig. 4f), α_{min} gets a value significantly closer to 2 which means that the brighter pixels in this case are less scattered over the image shaping more bulk aggregates. Nevertheless, the differences are slight revealing the overall similarity of phase distributions in these images. The above analysis reveals that the multifractal analysis is selectively sensitive to the presence of sample artifacts such as scratches. They affect only the right branch of singularity spectrum diagram which in this case does not describe the scaling behavior of low z-effective phases but the distribution of artifacts in the image.

5 Summary

Multiphase heterogeneous materials is a rapidly emerging area of research in both academia and industry since they seem to be the next step after nanomaterials era. A crucial challenge in this research field is to quantify and characterize the spatial distribution of material phases comprising the final heterogeneous material since it affects greatly the properties and performance of material. In this paper, we propose the multifractal analysis of microscopy images as a useful tool to meet this challenge and provide a systematic methodology for the quantitative characterization of the spatial distribution of different phases in a multiphase heterogeneous material paying special emphasis on their scaling behavior. To this end, first we present an alternative of the standard box-counting technique (named q-positive method) to provide reliable multifractal results even in the deep dark image areas. Then we apply this method for the estimation of the multifractal spectra in a series of top-down BSE-SEM images depicting the surfaces of samples of OPC clinker which is widely used in cement industry. We show that both right and left branches of multifractal spectra can be used to quantify the spatial aspects of phase distribution and to detect the presence of artifacts in sample preparation and/or imaging. Future work can be oriented towards two directions. First, we can apply the elaborated method in the analysis of more materials and microscope images to justify the insights that it delivers in quantitative material analysis. Secondly, we can focus more on the method itself to optimize its performance and remedy its downsides using synthesized greyscale images with predetermined multifractal characteristics.

Acknowledgements This research was supported by NCSR 'Demokritos' Industrial Research Fellowship Program [E.E.-12149], funded by the Stavros Niarchos Foundation and TITAN Cement S.A.

References

1. Paine, K.A.: Physicochemical and mechanical properties of portland cements. In: Clive, H.P., Liska, M., (eds.), Lea's Chemistry of Cement and Concrete, 5th edn. Butterworth-Heinemann (2019). https://doi.org/10.1016/B978-0-08-100773-0.00007-1
2. Erdogan, S.: Effect of clinker phase distribution within cement particles on properties of a hydrating cement paste. Construct. Building Mater. **38**, 941–949 (2013). https://doi.org/10.1016/j.conbuildmat.2012.09.051; Clive, H.P., Martin, L.: Lea's chemistry of cement and concrete (Astronomy and Astrophysics Library). Springer, Berlin (2002)
3. Nicolopoulos, S., Das, P.P., Bereciartua, P.J., Karavasili, F., Zacharias, N., Pérez, A.G., Galanis, A.S., Arenal,R., Portillo, J., Roque-Rossell, J., Kollia, M., Margiolaki, I.: Novel characterization techniques for cultural heritage using a TEM orientation imaging in combination with 3D precession diffraction tomography: a case study of green and white ancient Roman glass tesserae. Herit. Sci. **6** (2018). https://doi.org/10.1186/s40494-018-0229-7
4. Scrivener, K.L.: Backscattered electron imaging of cementitious microstructures: understanding and quantification. Cement Concr. Compos. **26**, 935–945 (2004). https://doi.org/10.1016/j.cemconcomp.2004.02.029
5. Chatzigeorgiou, M., Vrigkas, M., Beazi-Katsioti, M., Katsiotis, M., Boukos, N., Constantoudis, V.: Segmentation of SEM images of multiphase materials: when Gaussian mixture models are accurate? J. Microscopy **289**, 58–70 (2022). https://doi.org/10.1111/jmi.13150
6. Bostanabad, R., Zhang, Y., Li, X., Kearney, T., Brinson, L. C., Apley, D. W., Liu, Chen, W.: Computational microstructure characterization and reconstruction: review of the state-of-the-art techniques. Progr. Mater. Sci. **95**, 1–41 (2018)

7. Valentini, L., Artioli, G., Voltolini, M., Dalconi, M.C.: Multifractal analysis of calcium silicate hydrate (C-S-H) mapped by X-ray diffraction microtomography. J. Am. Ceram. Soc. **95**, 2647–2652 (2012). https://doi.org/10.1111/j.1551-2916.2012.05255.x
8. Gao, Y., Jiang, J., De Schutter, G., Ye, G., Sun, W.: Fractal and multifractal analysis on pore structure in cement paste. Construct. Build. Mater. **69**, 253–261 (2014). https://doi.org/10.1016/j.conbuildmat.2014.07.065
9. Carpinteri, A., Chiaia, B.: Multifractal nature of concrete fracture surfaces and size effects on nominal fracture energy. Mater. Struct. **28**, 435–443 (1995). https://doi.org/10.1007/BF02473162
10. Paggi, M., Carpinteri, A.: Fractal and multifractal approaches for the analysis of crack-size dependent scaling laws in fatigue. Chaos, Solitons Fractals **40**, 1136–1145 (2009). https://doi.org/10.1016/j.chaos.2007.08.068
11. Guo, M., Xiao, J., Zuo, S.: Multifractal analysis on pore structure of cement-based materials blended with ground limestone and its relationship with permeability. Kuei Suan Jen Hsueh Pao/J. Chinese Ceram. Soc. **47** (2019). https://doi.org/10.14062/j.issn.0454-5648.2019.05.05
12. Tian, S., Guo, Y., Dong, Z., Li, Z.: Pore microstructure and multifractal characterization of lacustrine oil-prone shale using high-resolution SEM: a case sample from natural Qungshankou Shale. Fract. Fract. **6** (2022). https://doi.org/10.3390/fractalfract6110675
13. Gao, Y., Gu, Y., Mu, S., Jiang, J., Liu, J.: The multifractal property of heterogeneous microstructure in cement paste. Fractals **29** (2021). https://doi.org/10.1142/S0218348X21400065
14. Chhabra, A.B., Meneveau, C., Jensen, R.V., Sreenivasan, K.R.: Direct determination of the $f(\alpha)$ singularity spectrum and its application to fully developed turbulance. Phys. Rev. A **40**, 5284–5294 (1989). https://doi.org/10.1103/PhysRevA.40.5284
15. Salat, H., Murcio, R., Arcaute, E.: Multifractal methodology. Phys. A **473** (2017). https://doi.org/10.1016/j.physa.2017.01.041
16. Scrivener, K., Snellings, R., Lothenbach, B.: A practical guide to microstructural analysis of cementitious materials. CRC Press (2015). ISBN: 9781498738651
17. Pirard, E., Sardini, P.: Image analysis for advanced characterization of industrial minerals and geomaterials. EMU Notes Mineral. **9**, 287–340 (2011). https://doi.org/10.1180/EMU-notes.9.3

Fractal Dimensional Analysis for Retinal Vascularization Images in Retinitis Pigmentosa: A Pilot Study

Francesca Minicucci, Fotios D. Oikonomou, and Angela A. De Sanctis

Abstract Retinal blood vessels form a complex branching pattern that has been shown to be fractal. The fractal dimension (FD) of the retinal vascular tree lies between 1 and 2. In the literature for healthy human subjects, the retinal vascularization FD was estimated at around 1.7, but it can be changed by the rarefaction or proliferation of blood vessels in the disease scenario. The aim of this paper is to investigate whether fractal analysis of retinal vascularization images can help for the early diagnosis of genetic retinal diseases as, in particular, retinitis pigmentosa (RP). This would be very useful because it represents the only defense against these illnesses. We use the results from two different imaging techniques, including Optical Coherence Tomography Angiography, to show that for retinal vascularization in patients with RP the FD is lower with respect to the corresponding healthy control group.

Keywords Retinal imaging · Fractal dimension · Retinitis pigmentosa · Wide-field swept-source optical coherence tomography angiography

F. Minicucci
Ophthalmology Clinics, Department of Medicine and Science of Ageing, University "G. d'Annunzio" of Chieti-Pescara and Department of Life, Health and Environmental Sciences, University of L'Aquila, L'Aquila, Italy

Fotios D. Oikonomou
Department of Physics, University of Patras, 26504 Rio, Greece
e-mail: pheconom@upatras.gr

Angela A. De Sanctis (✉)
Department of Management and Business Administration, University "G. d'Annunzio" of Chieti-Pescara, viale Pindaro 42, 65127 Pescara, Italy
e-mail: a.desanctis@unich.it

T. Bountis et al. (eds.), *Chaos, Fractals and Complexity*, Springer Proceedings in Complexity, https://doi.org/10.1007/978-3-031-37404-3_9

1 Introduction

Fractals are geometric objects widely studied since they were first introduced by the mathematician Benoit Mandelbrot in his book entitled "Les Objets Fractals" (1975) [1]. Fractals are shapes whose main characteristic is self-similarity, which means they repeat patterns on decreasing scales. In other words, each part of a fractal is similar to the whole shape.

Fractal analysis is a non-Euclidean geometrical framework used to assess the fractal nature of structures. The degree of complexity of a fractal is primarily described by the parameter "fractal dimension", first introduced in 1983 [2].

Fractal dimension is different from Euclidean dimension, where the dimension of a point is zero, the dimension of a line is one, the dimension of a rectangle is 2 and the dimension of a cube is 3. The fractal dimension is a real number that describes how an object's details change at different magnifications and its value is less than the dimension of the space where the shape is embedded. Thus, a fractal in two-dimensional space will have a FD between 1 and 2 and a fractal in three-dimensional space will have a FD between 2 and 3.

A simple method to measure the FD of a shape in the two-dimensional space, and the method used in this study, is to divide the shape into a grid of squares (box-counting).

Fractals are shapes often found in nature and biological systems, for example, the coastline of Britain is fractal and its fractal dimension is 1.2. In medical science, fractal structures are ramifications of the blood vessels of the human circulatory system [3, 4]. The occurrence of changes or pathologies can be signaled by variations in the fractal dimension.

The retina is of crucial importance to ophthalmologists as retinal diseases are the leading cause of blindness worldwide. The retina is a thin, light-sensitive neural layer and is supplied by a sophisticated microvascular network, that delivers nutrients and carries away waste. As part of the human circulatory system, the network's development tends to seek configurations that minimize operational energy expenditure. Often diseases will have a vascular component that can manifest as abnormalities in this network and thus the network can be studied to acquire insight into the presence (or absence) of disease.

Identification of abnormality in eye fundus can be done by examining fundus image photography using a digital fundus camera. With advancements in non-invasive ocular imaging techniques such as optical coherence tomography angiography (OCTA) permitting the segmentation of the vasculature into well-defined layers [5], the retinal vasculature has become more accessible to researchers than ever before. Hence, increasing attention has been paid to analyzing its quantitative characteristics as a potential diagnostic tool.

The retinal blood vessels form a complex branching pattern that has been shown to be fractal [6]. Therefore, a natural parameter for describing the retinal vasculature is the fractal dimension, introduced in ophthalmology in 1989 [7]. The fractal dimension of the retinal vascular tree lies between 1 and 2 [6], indicating that its branching pattern fills space more than a line, but less than a plane. Thus, the retinal fractal dimension provides a measure of the tree's global branching complexity, which can be altered by the rarefaction or proliferation of blood vessels in the disease scenario. In healthy human subjects, the retinal FD is around 1.7, which is similar to that of a 2D diffusion-limited aggregation process [6, 7]. It has been postulated that this is because the retinal vasculature grows through the diffusion of angiogenic factors in the retinal plane [8].

In 2021 [9] the authors summarize the current scientific literature on the association between FD and retinal disease. The results of the meta-analysis show decreased fractal dimension associated with the presence of glaucoma, hypertension and myopia. However, the decrease is strong with diabetic retinopathy and myopia, and weak for diabetes, glaucoma and hypertension. In particular, in 2016 [10], using the OCTA, it is proved that FD is significantly reduced in the superficial and deep capillary plexuses in eyes with diabetic retinopathy.

Due to variances in methodological setups for retinal image processing and FD calculation, it is difficult to form a consensus on this matter. Hence, before moving onto clinical applications of FD, it is necessary that a standardized protocol for image acquisition/processing be established to facilitate inter-study comparison.

The aim of this paper is to investigate how fractal analysis could help ophthalmologists for diagnosis of genetic retinal diseases, in particular, retinitis pigmentosa (RP). From a methodological point of view, this could mean that the geometry of the retina is fixed by genetics. To our knowledge, this is the first study on fractal analysis regarding a genetic retinal disease.

We will use FD of retinal images because one of the hallmarks of retinitis pigmentosa is the changes of retinal vasculature with vessel attenuation, especially in the early course of the disease [11, 12].

In the first part of the paper, we consider eye fundus images to classify RP using fractal analysis. The first phase was the image segmentation process using the green channel, and other mathematica commands aiming to extract the tree-shaped structure of blood vessels. Then the fractal dimension of the segmentation processed image was calculated using the box-counting method. Based on these results, it can be concluded that the classification of RP, using fractal analysis, can be very useful.

In the second part, we consider a recent kind of OCTA, the widefield OCTA with longer wavelengths and higher speed, which allows a better analysis of deeper tissue such as choriocapillaris and the visualization of a wider retinal field of view. In this way, it provides more details of retinal and vascular disorders not limited to the posterior pole [13]. In RP, the primary defect lies in the rod photoreceptors thus beginning in the far and mid-peripheral retina, later involving the cone photoreceptors localized more centrally [12]. In this part of the study, we consider images of retinal vascular plexuses and choriocapillaris in selected retinal areas from the foveal zone toward mid-peripheral retina in RP patients using widefield swept-source OCTA (WSS-OCTA).

2 A Brief Introduction to Fractal Dimension

In this paragraph we briefly recall the concept of fractal dimension. More details on fractals and fractal dimension can be found in [14–17]. Let us suppose that we have an "object" in d dimensions. If d = 1 this "object" could be a collection of line segments, if d = 2 a collection of parts of a plane, if d = 3 a part of the 3-dimensional space etc. It is well known that to this "object" we can assign a "measure" M. In the first case (d = 1) this "measure" is the length of the line segments, in the second case (d = 2) is the area of the plane parts, in the third case (d = 3) is the volume of the 3-dimensional part, etc.

We can cover this object with "boxes" with a small enough side l. The "boxes" in the case d = 1 are small line segments, in d = 2 are small squares and in d = 3 small cubes.

If $N(l)$ is the minimum number of boxes with side l needed to cover the object, it is obvious that,

$$M \approx N(l) l^d$$

since l^d is the length or area or volume, in general the measure, of each box [14]. If we solve the above equation relatively to d we have

$$d \approx \frac{log M}{log l} + \frac{log N(l)}{log(l^{-1})}$$

Since l is a small number and M is a constant, the term $log M / log l$ can be ignored, so,

$$d \approx \frac{log N(l)}{log(l^{-1})}$$

The number d computed from the above formula is not necessarily an integer. That is, to be precise, the number D

$$D = \lim_{l \to 0} \frac{log N(l))}{log(l^{-1})}$$

is any real number (less than the embedding dimension) and is called the "Fractal Dimension" of the above object [15–17].

We will find the Fractal Dimension (FD) of a simple image (Koch curve) as a simple implementation of the above analysis. The so-called "Koch curve" is generated from a line segment, by repeatedly removing the middle third of this (and subsequent segments) and replacing it with a triangle. So, using mathematica (Fig. 1).

Let us consider for simplicity the Koch curve at level $n = 2$ (Fig. 2). We will approximately compute the Fractal Dimension (FD) of this curve using the "box—counting" method.

(a) Level $n = 1$ (b) Level $n = 2$ (c) Level $n = 3$

Fig. 1 Koch curve

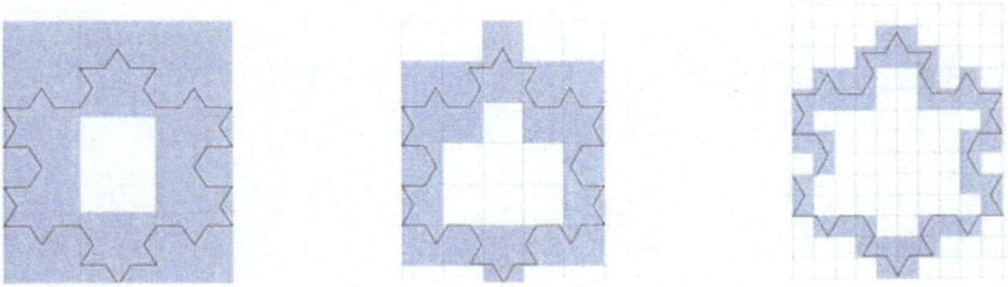

(a) $N = 8, l \approx 400$ (b) $N = 20, l \approx 240$ (c) $N = 44, l \approx 120$

Fig. 2 Box counting method

Initially we cover the whole image with boxes (squares) of decreasing side. In case (a) we use boxes of side $l \approx 400$ pixels, in case (b) of side $l \approx 240$ pixels and in (c) of side $l \approx 120$ pixels. Then, we count the minimum number N of boxes that have in their interior a part of the Koch curve, colored purple in the figure. In case (a) we have $N = 8$ boxes, in case (b) $N = 20$ boxes and in (c) $N = 44$ boxes.

Then we place the points $(logl^{-1}, logN)$ on the plane (Fig. 3—red dots) and find the line that best fits to them (least squares approximation). The equation of this line is

$$y = 1.39988x + 10.5403$$

Since we know that

$$logN \approx dlogl^{-1}$$

it is obvious that, for the Fractal Dimension d , we have

$$d \approx 1.39988 \approx 1.4$$

This result is close enough to the FD of the Koch curve as the level n goes to infinity, which is 1.26.

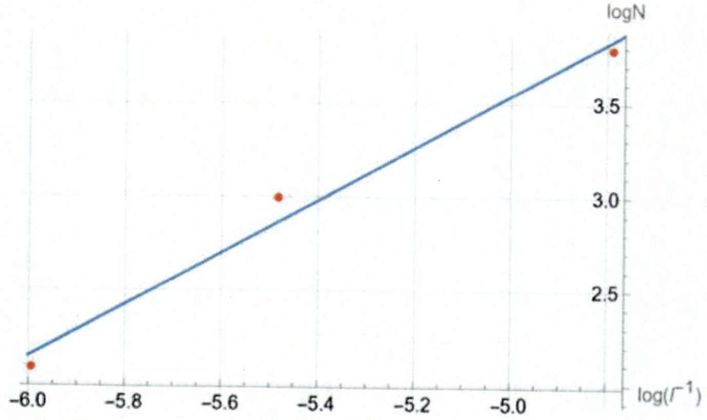

Fig. 3 Plot of the line $logN \approx dlogl^{-1}$

3 Retinitis Pigmentosa

Retinitis pigmentosa is a hereditary degenerative pathology of the chorio-retina, characterized by the presence of pigment in the nervous tunic of the eye [23]. Since it may have a wide clinical variability, it is better to consider retinitis pigmentosa as a heterogeneous group of retinal dystrophies, genetically determined and with a progressive course [26]. The common pathogenetic mechanism is represented at first by the degeneration of photoreceptors and retinal pigment epithelium cells, based on mutations of some proteins of the visual cycle. Due to the molecular dysfunctions of the rods, the cones also die secondarily [24]. It is a bilateral pathology, even if the two eyes are affected asymmetrically.

RP is a rare genetic disease. The total of clinical variants, syndromic and non-syndromic, has a variable prevalence in the different populations which were studied: in the USA, it is about 1: 3500-1: 4000, with significant variations in the various states, in Switzerland 1: 7000, in China 1: 4016, in Norway 1: 4440, in Israel 1: 4500. In Italy, it affects one out of every 3,500 inhabitants, with an expected number of about 18,000 patients. The global worldwide frequency (syndromic and non-syndromic variants) is 1 case for every 3.000-5.000 inhabitants (about 1.5 million cases in the world).

The clinical features of the disease are [25]:

- The so-called "bone spicule" pigmentation of the retina, which can be observed in ophthalmoscopy
- The dysfunction of the photoreceptors, evidenced by anomalies of the electroretinographic trace
- Night—blindness (nyctalopia), which means it is difficult or impossible to see in relatively low light
- Progressive narrowing of the peripheral visual field (tubular or telescope vision)

The most common imaging techniques used for the diagnosis of retinitis pigmentosa are [27]:

- Digital fundus photography
- Fundus fluorescein angiography (AF)
- Optical coherence tomography angiography (OCTA)
- Scanning laser ophthalmoscopy.

3.1 *First Methodology: Digital Fundus Photography*

We consider the images of Fig. 4 for a healthy eye (control group) which are found in the Messidor Database (Kindly provided by the Messidor program partners (see https://www.adcis.net/en/third-party/messidor/)) [18].

In Fig. 5 we see the images of an eye with RP,[1] in particular, Image 3 shows an initial level and the others (Image 4 and Image 5) more advanced levels of the disease.

To compute the Fractal Dimension of the above images, we have analyzed them first using mathematica. We have used the commands "ColorSeparate" choosing the "Green" channel, then "ImageAdjust" and "MorphologicalBinarize" to extract the tree-shaped structure of the blood vessels.

Then we proceed with fractal dimensional box-counting analysis which is performed using Fractalyse (ThéMA, Besançon Cedex, France). We get the results of Table 1 below, where r^2 is a measure for the quality of the linear regression analysis in our model ($r^2 = 1$ means best fitting). Fractalyse uses the p-value approach to hypothesis testing. Since p-values are very small there is strong evidence that $logN(l), log(l^{-1})$ are linearly related.

The results of table 1 are shown in the diagram of Fig. 6.

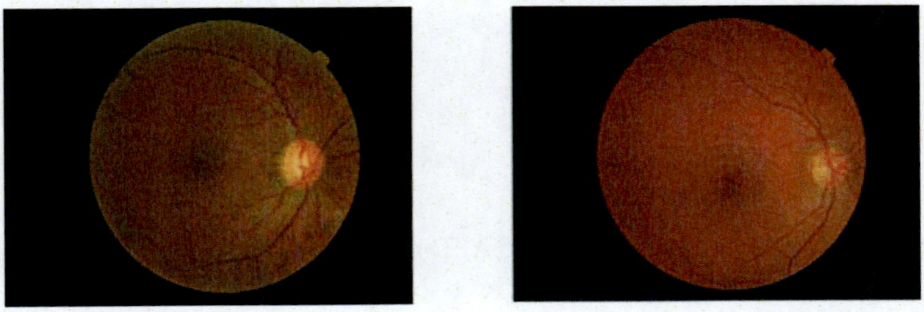

(a) Image 1 (b) Image 2

Fig. 4 Images of healthy eyes (control group)

[1] The images were kindly provided by the Ophthalmology Clinic, Department of Medicine and Science of Ageing, University "G. d'Annunzio" of Chieti-Pescara.

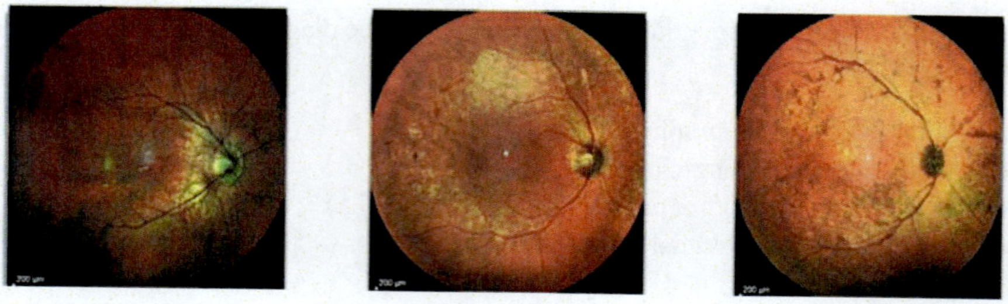

(a) Image 3 (b) Image 4 (c) Image 5

Fig. 5 Images of eyes with RP

Table 1 Fractal dimensions of images

Image	FD	r^2	Confidence (95%)	p-value
1	1.902	1.000	1.891–1.913	0.000
2	1.948	1.000	1.942–1.955	0.000
3	1.846	0.999	1.783–1.909	5.055E-10
4	1.687	1.000	1.656–1.718	1.247E-11
5	1.718	0.999	1.680–1.757	4.123E-11

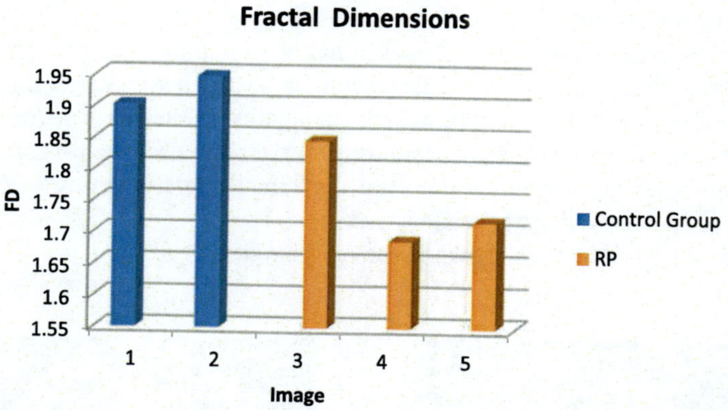

Fig. 6 Fractal dimensions of images

From this diagram, we deduce that the FD of the eye fundus is smaller for patients with RP compared to that of a healthy eye (control group). More precisely, the FD of eye fundus, in patients with RP, is smaller than that of a healthy eye from the beginning and further decreases in time, maybe due to the aging process of the eye.

3.2 *Second Methodology: Optical Coherence Tomography Angiography*

Recall that retina is the innermost of the three tunics of the eye. It has a nervous nature, due to its embryological derivation, its histological structure, and its connections with the optic nerve. The choroid, on the other hand, constitutes the most extended and posterior part of the uvea. It covers the inside of the sclera and is in turn covered by the retina. It is an electively vascular organ and has characteristics similar to corpus cavernosum, which allows for one of the highest blood flows in the human body. The choroid has the function of nourishing the outermost layers of the retina, in particular, the pigment epithelium and the outer segment of the photoreceptors [19].

The following are of particular interest for the study of chorioretinal vessels:

- Superficial capillary plexus (SCP): Layer of ganglion cells and nerve fibers
- Deep capillary plexus (DCP): Inner nuclear and outer plexiform layer
- Choriocapillary plexus (CC): Between the Bruch membrane (BM) and the Sattler layer, at the choroidal level

Optical coherence tomography angiography (OCTA) is a new imaging technique that allows indirect visualization of the chorioretinal vessels through the normal movement of blood in the capillaries [20].

The instrument, starting from a sequence of OCT images, provides a three-dimensional reconstruction of the perfused vessels of both the retina and the choroid. The OCTA allows viewing of the neuro-retinal vascular texture by layers and with a resolution of micrometers. The advantage with respect to fluorescein angiography (AF) is that it does not require the injection of a dye. The OCTA allows obtaining of separate images of the retinal and choroidal plexi in vivo. For OCTA we can use two different amplitudes of field so we distinguish:

OCTA Small Field

- Limited to the posterior pole (macula and optic disc)
- Acquisition areas of 3x3mm or 6x6mm
- Not very useful for investigating the pathologies affecting the vascularization of the peripheral retina

OCTA Wide Field

- It allows analyzing a larger retinal area
- 12x12 mm acquisition areas
- It collects more details on the retinal circulation in the middle periphery
- Useful in pathologies such as PR, which mainly affects the periphery

Let us now consider the images from OCTA Wide field of superficial capillary plexus, deep capillary plexus and choriocapillaris plexus of a healthy eye (a) and retinitis pigmentosa (RP) patient (b), which are taken from [21, Fig. 1]. The images are referred to an experiment of 12 patients with previous diagnosis of either mid-

or late-stage RP and a control group of 20 healthy age-matched subjects, at the University "G. d'Annunzio" of Chieti-Pescara, Italy.

The fractal dimensional box-counting analysis is performed again using Fractalyse. The input are these images, processed now by using ImageJ (National Institutes of Health [NIH], Bethesda, Maryland, USA) [22]. We transform the type of each image to 8 bit and then we run the macro

```
setAutoThreshold("Default dark");
setThreshold(0, a(i));
setOption("BlackBackground", false);
run("Convert to Mask")
```

for a(i) = 25i and i = 0, 1, 2, 3, 4, 5, 6, 7, 8, 9, 10 i.e. $0 \leq a(i) \leq 250$, where a(i) is a cutoff value such that every pixel with "luminosity" less than that value is considered "background", while every pixel greater than that value is considered "foreground". This procedure is necessary because the initial picture contains shades of gray. We want a binary image consisting solely of pure black pixels we have called "background" and pure white ones we have called "foreground". If an imaginary box contains white pixel(s) the box is counted during box-counting analysis. On the contrary, if contains no white pixel is not. Thus, the "setThreshold(0,a(i))" command helps us divide the image into two classes of pixels (black and white) and then perform box-counting analysis.

The above procedure is carried out for each image and we get the results shown in Table 2.

We remark again that due to the kind of images under study, the tree of blood vessels is not clearly defined (there are different levels of grey), so we have a variety

Table 2 Fractal dimensions of images

i for parameter a(i)	Healthy eye superficial plexus	RP eye superficial plexus	Healthy eye deep plexus	RP eye deep plexus	Healthy eye choriocapillaris plexus	RP eye choriocapillaris plexus
0	1.978	1.980	1.975	1.976	1.980	1.980
1	1.975	1.948	1.970	1.942	1.980	1.974
2	1.960	1.880	1.946	1.878	1.980	1.964
3	1.919	1.765	1.881	1.772	1.979	1.942
4	1.856	1.628	1.778	1.653	1.977	1.893
5	1.775	1.528	1.648	1.543	1.968	1.765
6	1.686	1.462	1.519	1.431	1.883	1.554
7	1.594	1.493	1.440	1.343	1.541	1.562
8	1.556	1.515	1.377	1.318	1.500	1.601
9	1.594	1.543	1.333	1.367	1.608	1.603
10	1.700	1.528	1.490	1.433	1.601	1.635

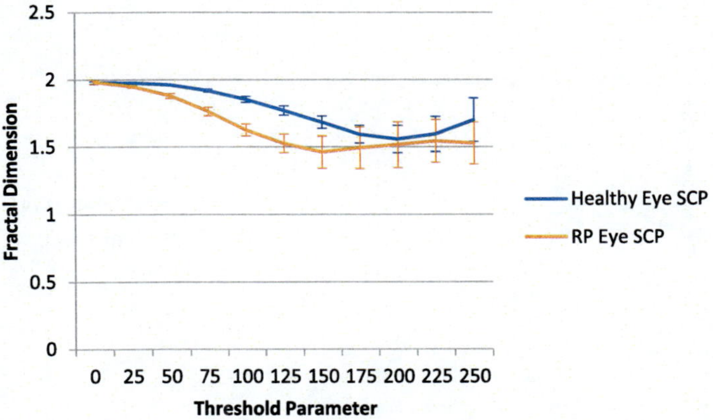

Fig. 7 Fractal dimensions of superficial plexus images, taken from [21, Fig. 1], for various values of threshold parameter

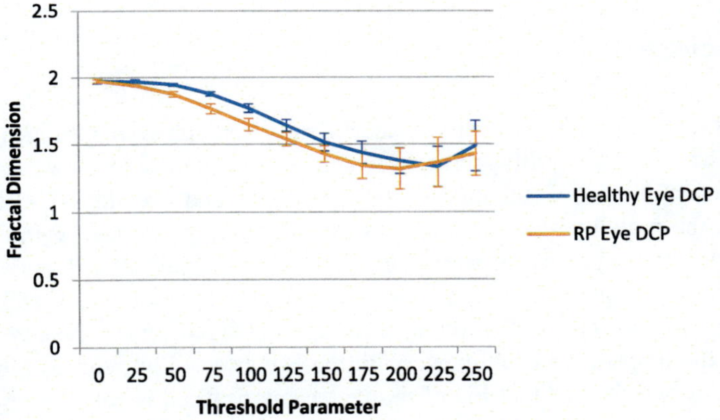

Fig. 8 Fractal dimensions of deep plexus images, taken from [21, Fig. 1], for various values of threshold parameter

of binary images, depending on the threshold parameter, with different FD. We consequently obtain curves of FD depending on the threshold parameter, which are shown in the following diagrams, for all three plexi considered here, where the blue curves refer to the healthy eye and the orange curves to the RP eye (Figs. 7, 8 and 9).

The error bars in these diagrams represent the 95% confidence intervals which provide a measure of precision for the estimated FDs.

It is encouraging to see that, for most values of the parameter a(i), the FDs of images of healthy eyes are above these of images of patients with RP, even if, possible variations indicated by the error bars are considered. This is not the case for very small or very large values of the threshold parameter, but we must expect it since in these boundaries we have a very distorted image. This numerical result is coherent with the clinical analysis observing the blood vessel attenuation [11, 12].

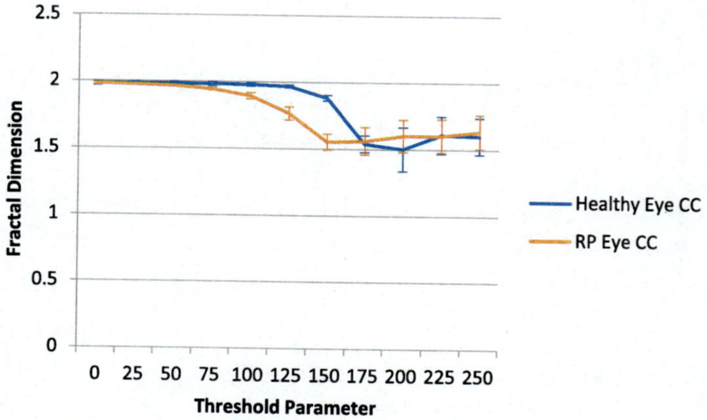

Fig. 9 Fractal dimensions of choriocapillaris plexus images, taken from [21, Fig. 1], for various values of threshold parameter

4 Discussion

Genetics determines the geometry of the retinal vascularization. Genetical pathologies, such as RP, modify this geometry, which becomes unable to allow complete and effective vision (such as tubular or telescope vision and nyctalopia). In the image analysis, the retinal vascularization curve is less dense since the beginning of the young age and the process of rarefying increases with aging.

To verify this quantitatively, we used fractal analysis of retinal vessels images, obtained with two types of different techniques (Digital fundus photography and OCTA), to compare the healthy eye with the eye affected by retinitis pigmentosa. In all cases, we showed that FD of retinal vascularization, in retinitis pigmentosa, is smaller than that of the healthy eye from the beginning and further decreases in time. For our aim, OCTA Wide Field seems to be the better imaging technique in pathologies such as RP, which mainly affects the periphery of the eyes.

For the three plexi of blood vessels in the retina, observed by OCTA Wide Field, we have constructed curves of FD depending on the threshold parameter. Such curves in RP patients are below those of the control group in most of the values, compatible with vessel attenuation clinically observed.

It would be important to compare FD in RP with FD in other genetic retinal pathologies. In general, we can suppose from our research that FD of retinal vascularization imaging in RP, in mature age, is lower as compared with other non-genetical retinal pathologies such as diabetic retinopathy or glaucoma.

Anyway, before moving onto clinical applications using FD, it is necessary that a standardized protocol for image acquisition/processing be established first to facilitate inter-study comparison. Besides, as a future work, this research aims to perform the analysis in a sample of patients with enough data to allow a quantitative estimation of FD based on statistical evidence. Therefore, this study can be considered as a pilot study.

We think that FD of retinal vascularization imaging could be very useful in the early diagnosis of RP, which represents the only defense against this illness. Indeed, an efficient and lasting cure does not exist even nowadays, guaranteeing neither complete healing nor sight recovery. The only possibility for a cure is to slow down the illness progression through a daily consumption of vitamin A, omega-3 and lutein, as well as, more recently, to apply gene therapy, by using stem cells.

Funding: This research received no external funding.

Data Availability Statement: The images of Fig. 4 for a healthy eye (control group) were provided by the Messidor Database (Kindly provided by the Messidor program partners (see https://www.adcis.net/en/third-party/messidor/)) [18].

The images of Fig. 5 were kindly provided by the Ophthalmology Clinic, Department of Medicine and Science of Ageing, University "G. d'Annunzio" of Chieti-Pescara.

Conflicts of Interest: The authors declare that there is no conflict of interest regarding the publication of this article. The funders had no role in the design of the study; in the collection, analyses, or interpretation of data; in the writing of the manuscript, or in the decision to publish the results.

References

1. Mandelbrot, B.B.: Les Objets Fractals. Flammarion, Paris (1999)
2. Mandelbrot, B.B., Wheeler, J.A.: The fractal geometry of nature. Am. J. Phys. **51**, 286–287 (1983). https://doi.org/10.1119/1.13295
3. Jayalalitha, G., Shanthoshini, D.V., Uthayakumar, R.: Fractal model for blood flow in cardiovascular system. Comput. Biol. Med. **38**, 684–693 (2008)
4. Gabryś, E., Rybaczuk, M., Kedzia, A.: Fractal models of circulatory system. Symmetrical and asymmetrical approach comparison. Chaos, Solitons Fractals **24**(3), 707–715 (2005). https://doi.org/10.1016/j.chaos.2004.09.087
5. Campbell, J.P., Zhang, M., Hwang, T.S., Bailey, S.T., Wilson, D.J., Jia, Y., Huang, D.: Detailed vascular anatomy of the human retina by projection-resolved optical coherence tomography angiography. Sci. Rep. **7**, 42201 (2017). https://doi.org/10.1038/srep42201
6. Mainster, M.A.: The fractal properties of retinal vessels: embryological and clinical implications. Eye Lond. Engl. **4**(Pt 1), 235–241 (1990). https://doi.org/10.1038/eye.1990.33
7. Family, F., Masters, B.R., Platt, D.E.: Fractal pattern formation in human retinal vessels. Phys. Nonlinear Phenom. **38**, 98–103 (1989). https://doi.org/10.1016/0167-89(89)90178-4
8. Lakshminarayanan, V., Raghuram, A., Myerson, J., Varadharajan, S.: The fractal dimension in retinal pathology. J. Mod. Opt. - J MOD Opt. **50**, 1701–1703 (2003). https://doi.org/10.1080/09500340310000069442
9. Yu, S., Lakshminarayanan, V.: Fractal dimension and retinal pathology: a meta analysis. Appl. Sci. **11**, 2376 (2021). https://doi.org/10.3390/app11052376 (Accessed 9 Sept 2021)
10. Zahid, S., Dolz-Marco, R., Freund, K.B., Balaratnasingam, C., Dansingani, K., Gilani, F., Mehta, N., Young, E., Klifto, M.R., Chae, B., Yannuzzi, L.A., Joshua, A., Young, J.A.: Fractal dimensional analysis of optical coherence tomography angiography in eyes with diabetic retinopathy. Invest Ophthalmol. Vis. Sci. **57**, 4940–4947 (2016). https://doi.org/10.1167/iovs.16-19656

11. Ma, Y., Kawasaki, R., Dobson, L.P., Ruddle, J.B., Kearns, L.S., Wong, T.Y., Mackey, D.A.: Quantitative analysis of retinal vessel attenuation in eyes with retinitis pigmentosa. Investig. Ophthalmol. Vis. Sci. **53**, 306–314 (2012)
12. Jauregui, R., Park, K.S., Duong, J.K., Mahajan, V.B., Tsang, S.H.: Quantitative progression of retinitis pigmentosa by optical coherence tomography angiography. Sci. Rep. **8**, 13130 (2018)
13. Liu, G., Yang, J., Wang, J., Li, Y., Zang, P., Jia, Y., Huang, D.: Extended axial imaging range. Widefield swept-source optical coherence tomography angiography. J. Biophoton. **10**, 1464–1472 (2017)
14. Bountis, T., Fokas, A.S., Psarakis, E.Z.: Fractal analysis of tree paintings by Piet Mondrian (1872–1944). Int. J. Arts Technol. **10**(1), 27–42 (2017)
15. Barnsley, M.F.: Fractals Everywhere, 3d edn. Academic, San Diego (1993)
16. Peitgen, H.-O., Jürgens, H., Saupe, D.: Chaos and Fractals. Springer, New York (2004)
17. Falconer, K.: Fractal Geometry Mathematical Foundations and Applications. Wiley, England (1990)
18. Decencière, F.N.: Feedback on a publicly distributed database: the Messidor database. Image Anal. Stereol. **33**(3), 231–234 (2014). ISSN 1854-5165
19. Toto, L., Borrelli, E., Mastropasqua, R., Senatore, A., Di Antonio, L., Di Nicola, M., Carpineto, P., Mastropasqua, L.: Macular features in retinitis pigmentosa: correlations among ganglion cell complex thickness, capillary density, and macular function. Investigat. Ophthalmol. Vis. Sci. (2016). https://doi.org/10.1167/iovs.16-20544
20. Mastropasqua, R., Borrelli, E., Agnifili, L., Toto, L., Di Antonio, L., Senatore, A., Palmieri, M., D'Uffizi, A., Carpineto, P.: Radial Peripapillary capillary network in Patients with retinitis Pigmentosa: an optical coherence tomography angiography study. Front. Neurol. **8**(572), 1 (2017). https://doi.org/10.3389/fneur.2017.00572
21. Mastropasqua, R., D'Aloisio, R., De Nicola, C., Ferro G., Senatore, A., Libertini, D., Di Marzio, G., Di Nicola, M., Di Martino, G., Di Antonio, L., Toto, L.: Widefield swept source OCTA in retinitis pigmentosa. Diagnostics **10**(50) (2020). https://doi.org/10.3390/diagnostics10010050
22. Rasband, W.S.: ImageJ, U. S. National Institutes of Health, Bethesda, Maryland (1997–2015). http://imagej.nih.gov/ij/. (Accessed 10 May 2021)
23. Verbakel, S.K., Van Huet, R., Boon, C., den Hollander, A.I., Collin, R., Klaver, C., et al.: Nonsyndromic retinitis pigmentosa. Prog. Retin. Eye Res. **66**, 157–86 (2018)
24. Marigo, V.: Programmed cell death in retinal degeneration: targeting apoptosis in photoreceptors as potential therapy for retinal degeneration. Cell Cycle **6**, 652–5 (2007)
25. Langham, M.E., Kramer, T.: Decreased choroidal blood flow associated with retinitis pigmentosa. Eye **4**, 374–81 (1990)
26. Ferrari, S., Di Iorio, E., Barbaro, V., Ponzin, D., Sorrentino, F.S., Parmeggiani, F.: Retinitis pigmentosa: genes and disease mechanisms. Curr. Genom. **12**, 238–49 (2011)
27. Mitamura, Y., Mitamura-Aizawa, S., Nagasawa, T., Katome, T., Eguchi, H., Naito, T.: Diagnostic imaging in patients with retinitis pigmentosa. J. Med. Invest. **59**, 1–11 (2012)

Extending the Bayesian Framework from Information to Action

Vasileios Basios, Yukio-Pegio Gunji, and Pier-Francesco Moretti

Abstract In this review, we examine an extended Bayesian inference method and its relation to biological information processing. We discuss the idea of combining two modes of Bayesian inference. The first is the standard Bayesian inference which contracts probability space. The second is its inverse, which extends and enriches the probability space of latent and observable variables. Their combination has been observed that, greatly, facilitates discovery. Moreover, this dual search during the updating process elucidates a crucial difference between biological and artificial information processing. The latter is restricted due to nonlinearities, while the former utilizes it. This duality is ubiquitous in biological information process dynamics ('flee-or-fight', 'explore-or-exploit' etc.) as is the role of fractality and chaos in its underlying nonequilibrium, nonlinear dynamics. We also propose a new experimental set up that stems from testing these ideas.

Keywords Bayesian inference · Free energy principle · Fractals · Chaos · Inverse problem

---

V. Basios (✉)
Service de Physique des Systèmes Complexes et Mécanique Statistique and Interdisciplinary Center for Nonlinear Phenomena and Complex Systems C.P.231 CeNoLi-ULB, Université Libre de Bruxelles (ULB), Brussels, Belgium
e-mail: vasileios.basios@ulb.be

Y.-P. Gunji
Department of Intermedia Arts and Science, School of Fundamental Science and Technology, Waseda University, Tokyo, Japan
e-mail: yukio@uwaseda.jp

P.-F. Moretti
CNR, National Research Council, P.le A. Moro 7, Rome, Italy
e-mail: pierfrancesco.moretti@cnr.it

T. Bountis et al. (eds.), *Chaos, Fractals and Complexity*, Springer Proceedings in Complexity, https://doi.org/10.1007/978-3-031-37404-3_10

1 Introduction

Explore different areas. The statement that one cannot be both deep and broad is a myth. Actually, the importance of being a polymath is that it allows one to make remote associations, and thus to understand the deeper essence of things. Understanding is nothing more than elucidating associations. – A. Fokas [1].

In its wide range and seminal work of Professor Athanasios Fokas one finds important contributions in an interdisciplinary research fashion on the interface between applied and pure mathematics. Among other important contributions, Professor Fokas worked and taught a lot about the challenge that inverse problems pose: how to hypothesise and determine the most plausible set of causal interconnections and identify the physical and/or statistical laws that govern the acquired data.

We too, find inspiration and try our best to follow his tall order of 'exploring different areas'. And in that spirit we approach this interdisciplinary area of biological information processing. In particular, we trace the development of the idea that chaos, affording indeterminacy and criticality, is the 'conditio sine qua non' for biological information processing and we highlight the importance of fractal basin boundaries for such an interpretation.

Moreover, we shall see how this entails a nonlinear closed feedback loop with the forward and inverse inference problems interlaced. The 'forward problem', in this case is that of categorization, i.e to propose models that calculate the results from, and reactions to, their causes. In our case it is the framework of the classical Bayesian theorem. The 'inverse problem' here deals data acquisition, i.e to calculate causes from results. And in our case connects with the converse of the Bayesian theorem, a fertile but less travelled road, which expands its original, classic, framework [2–7].

The paper is organized as follows: in Sect. 2 we review aspects that differentiate biological from artificial information processing, and dynamics, focusing on the role of chaos and fractals. In Sect. 3 we discuss the connection of Bayesian inference to the free energy principle and present an outline of our proposed extension. With Sect. 4 we conclude with a short discussion of forthcoming research plans and their outlook.

2 Biological Versus Artificial Information Processing

Life is abundant with information processing, this is commonplace, in our era though information flow is not bound to life's biological processes. We are surrounded by artificial technological contraptions serving, processing and acquiring information. Some of them are bio-inspired some are based in Turing's and Shannon's prototypical mechanistic theories of information. Evidently natural systems differ in structure and function from human-made computers as much as natural patterns differ from human-made ones. As the father of Fractal Geometry Benoît Mandelbrot famously put it *Clouds are not spheres, mountains are not cones, coastlines are not circles, and bark*

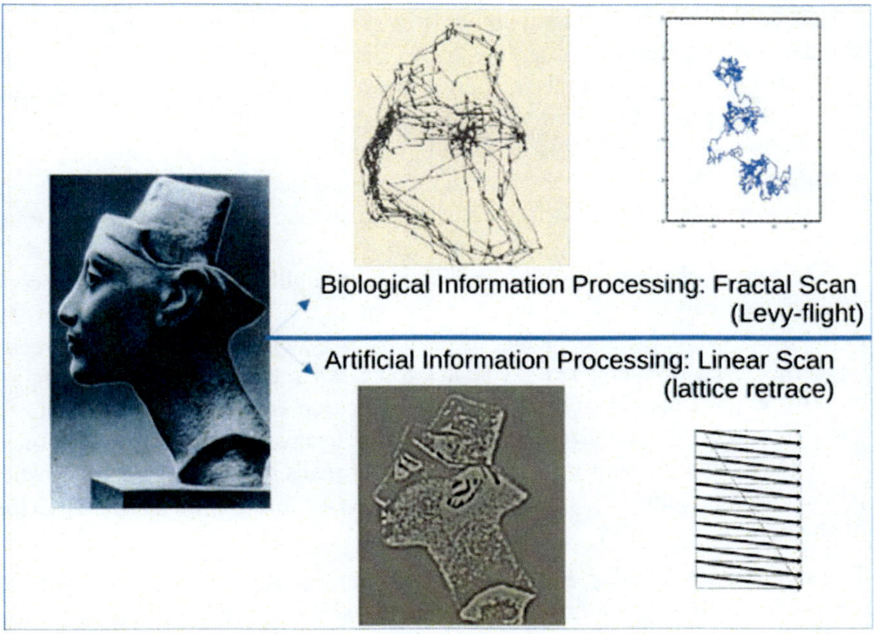

Fig. 1 Upper right: Subjects see the photograph, on the left panel, by moving their eyes in a seemingly erratic, fractal-like, Levy-flight fashion. Lower right: When machines scan the same photograph, on the left panel, they sweep the area by retracing a lattice of pixels. (The figure is our synthesis from the original Fig. 116 of [8] and its revisits in Figs. 20.1 and 20.3 of [9, 24]). Paraphrasing Benoît Madelbbrot one can say, 'clouds are not spheres, mountains are not cones ... and seeing is not scanning'

is not smooth, nor does lightning travel in a straight line, and biological information processing is not machine-information processing; and fractals do appear here, too.

From the early times of information machinery it became evident that acquisition and storage of information in biological systems happens in a radically different way (see for example the first reported study of how human visual system treats an image [8] and for an updated review see [9, 24]). In humans it happens in a fractal itinerary, what it came to be known as a Levy flight [10]. In a scanner it happens via a serial-sweep on a lattice, as Fig. 1. Moreover, biological systems process information at a multitude of levels. Or, as a pioneer of the subject, John S. Nicolis, put it [11] *The smallest biological information processor is the enzyme; the biggest is the (human) brain. They are separated by nine orders of magnitude. Yet their complexity is comparable*. Hence, fractality is obviously the most compatible property of choice for such cases of distributed probabilistic computations as the ones encountered from proteins, to organs and organisms, even to groups of organisms.

The Role of Chaos, Fractals and Complexity: It has been established,since the early days of Chaos theory that a reliable information processor must allow for chaos [12]. In particular it must afford the regimes known as 'edge of chaos' or 'self organized criticality' that are ubiquitous in systems with coexistent negative and positive nonlinear feedback circuits. Chaos Theory has shed new light in phenomena associated with biological information processors, see for example [13, 14]. Such type of chaotic dynamics allows also for adaptivity, flexibility and resilience during information processing. Moreover, since biological information is contextual, meaningful, has depth of memory and historicity. Other related types of chaotic dynamics, such as stochastic resonance, intermittency and chaotic itinerancy were also found to play key roles [12, 17].

The role of chaos in biological systems is exemplified: (i) at the macroscopic level in the analysis of chaos-order transitions in the brain and other organs (ii) at the microscopic level, modeling of neurons as systems of nonlinear differential equations, even revealing new dynamics (blue sky catastrophe and spike-trains), and (iii) at the mesoscopic level in groups of neuron communities where chimera states, non-local synchronization and modular collective dynamics were identified. So, it comes with no surprise that even the new trends of bio-inspired Information processing paradigms (e.g. artificial neural networks) discover the constructive role of chaos.

The importance of being Fractal and Chaotic: The essence of biological information processing can be expressed as the emergence of nonlinear feedback loops with two branches: The branch that provides stability and facilitates data storage, i.e. data-categorization. This is modelled via attractors in the phase space with dynamics characterized by a negative Lyapunov exponents' sum, $\Lambda < 0$. While the other branch provides instability and facilitates data acquisition actions, i.e. observation. This in turn is modeled via chaotic exploration of the phase space, with dynamics characterized by a positive Lyapunov exponents' sum, $\Lambda > 0$, [3, 12].

We must take note that the physical and biological parameters, here, are not always constant in time and a more complete treatment, albeit much more complicated and demanding, would have to account for 'chaotic itinerancy'. Chaotic itinerancy [12, 17] is a quite generic mechanism for high dimensional systems with coexisting fast-slow dynamical subsystems. It captures the complexity, plasticity and flexibility of biological systems, particularly neural dynamics, and since it is contingent upon history and parameter switching allows transitions to be stochastic [13, 18].

This emergent feedback loop is reminiscent of the 'fight or flee', 'exploit or explore' phenomena in biological systems at large. Only here the negative-feedback, contracting, branch (fight/exploit) has to do with comprehension via pre-existing categories, while the positive-feedback, expanding, branch has to do with seeking the knowledge of new data. Moreover a well functioning loop has to be well tuned, poised on the border of chaos and order. The following scheme summarizes this ubiquitous closed feedback loop pair:

Categorization: exploiting of phase space.
Stability, $\Lambda < 0 \longmapsto$ Data Storage/Memory: Inhibitory modes
Observation: exploring in phase space.
Chaos, $\Lambda > 0 \longmapsto$ Data Acquisition/Input-Output: Excitatory modes

or as J.S. Nicolis and I. Tsuda put it *To observe you need a priori categories, but to form categories you need observations* [12].

Fractal Basin Boundaries: Another key aspect of chaotic dynamics is the coexistence of attractors. And it is well known that these coexisting attractors are, most often than not, separated by fractal basin boundaries [3, 18]. One of the most well known systems with this property is the celebrated 'Newtons Fractal' a Julia set associated to Newton's iteration method for finding roots of polynomials $f(z) : z_{n+1} := z_n - \frac{f(z_n)}{f'(z_n)}, z \in \mathbb{C}$. For degree 3, $f(z) = z^3 - 1$ the attracting roots of unity and their fractal basin boundaries are shown in Fig. 2a (generated with the code available in [16]). Models with fractal basin boundaries have been proven to capture well the affinity of concepts in their apprehension and categorization and the intermittent successions among them. This kind of multistability and its stochastic switching is exemplified in the study of optical illusions. Bistable transitory dynamics, such as the ones found in the perception of the Necker Cube, or Wittgenstein's favourite linguistic paradigm of the Rabbit/Duck picture, were of the first to highlight the importance of inhibitory and excitatory connections in neural correlates during perception.

Furthermore, fractal basin boundaries of coexisitng (strange or not) attractors provide also for the probabilistic, stochastic, aspect inherent in biological processors. As it is well known a chaotic system under coarse-graining cannot be distinguished from a stochastic one [18]. This is due to two facts: Firstly, the practical impossibility to fully determine initial conditions, or any point of the phase-space for that matter, with infinite accuracy. Secondly, the inherent sensitive dependence on initial conditions of chaotic systems which renders unpredictable the course of their evolution beyond their Lyapunov time ($\tau \approx 1/\Lambda$). Hence, for any point on the fractal basin boundary, which is necessarily determined with finite accuracy, we can only assign a probability, weighted by the boundary's fractal measure, of arriving at a neighbouring attractor of the coexisting ones.

It is customary to partition the phase-space of a dynamical system by the preimages, periodic points and/or other critical points, or by a simple lattice that is refined iteratively, as in the cases of fractal-dimension determination by box-counting. For example, in Fig. 2c the highlighted region contains the fractal boundary. We can then use a rough set or a coarse- graining scheme, e.g. the partition determined by the curves R^+ and R^- marked with the white and black lines in Fig. 2c that enclose the fractal boundary of the basin of attraction of the first cubic-root of unity, in this case, ($z_1 = 1$). Every point inside R^- will end up in z_1, every point outside R^+ will *not* end up in z_1 while any point in the (highlighted region) could end up either end up in z_1 with a given probability, determined by the underlying fractal measure, or end

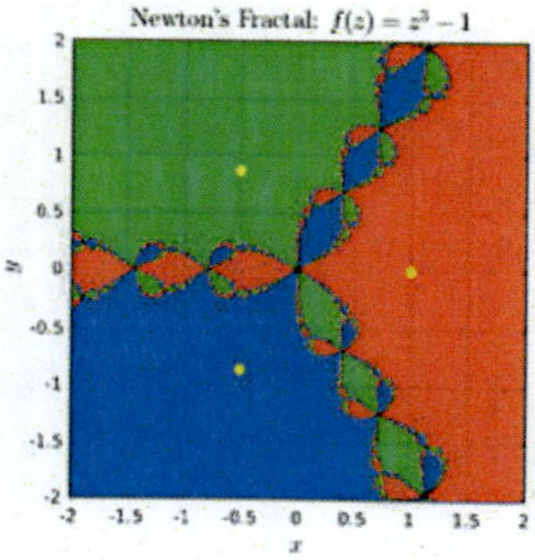

(a) An example of three coexisting attractors' basins with their fractal basin boundaries. Each cubic-root's of unity (yellow dots), basin is in different color.

(b) The 'Penrose Triangle' or the 'impossible tribar' an illusionary figure created by O. Reutersvärd, and popularized later by Sir R. Penrose, L. Penrose & M.C. Escher, (image credits [15]).

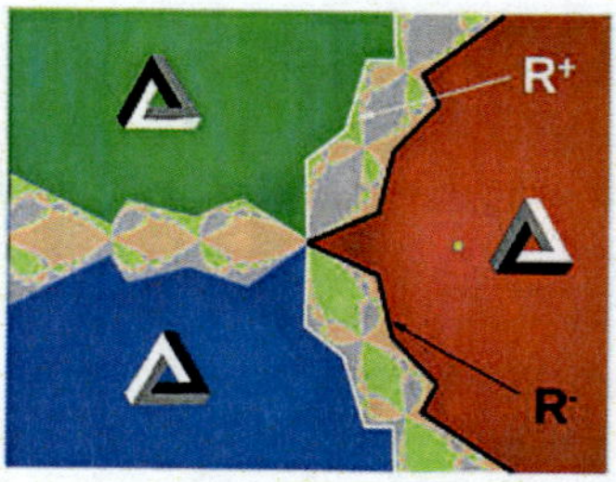

(c) The three coexisting attractors with coarse grained (rough-set) boundaries modelling the three switching perceptions of Penrose's Triangle of Fig.2b. The fractal basin boundary is within the highlighted region. The $R^{\pm}$ lines delineate roughly the basin around the root $z_1 = 1$. (yellow dot).

Fig. 2 Coexisting attractors with fractal basin boundaries in biological information processing provide models for multistable perception and semantic/categorical polyvalence

up in *either* z_2 or z_3, the other two cubic-roots of unity. Similar argument holds for the other two attractors around z_1 and z_2, with respective partitions.

Figure 2c illustrates the above mechanism that gives rise to a tri-stable visual perception of the famous Penrose Triangle shown in Fig. 2b. When the visual cortex is impinged upon with such ambiguous stimuli, it is impossible to categorize it in a single perspective. So, all three pre-existing possible and competing categories of perspective are excited. The inhibitory part of the circuit drives the system towards one of these three, but in a probabilistic fashion. Because the uncertain data on the fractal basin provides the fluctuations for stochastic transitions from one category/fixed-point to the other. Note that, such a loop puts data (stimuli) and representations (attractor-basins) on a shared basis.

These typical far-from-equilibrium processes are directly related to the ever present dissipative structures' dynamics in biological systems [18]. It results in a measurable effect with well determined transition probabilities, for a detailed model of tri-stability see [19]. Indeed ambiguous figures are found as guiding paradigmatic cases in every other textbook in cognitive sciences and neuroscience as a 'gateway to perception'. They are markedly elucidating cases of the dynamics of cognition since here the perception changes but not the signal. Which echoes another's pioneer take on the subject, Walter Freeman's, who stated that: *perception depends dominantly on expectation and marginally on sensory input*, see his contribution in [11].

Moreover, it has been established [2, 4, 20] that such transition between interrelated categories, as exemplified here, when treated within an extended Bayesian framework give rise to a logic with clear non-Boolean characteristics in accordance with the theory of Quantum Cognition [21–23].

3 Extended Bayesian Inference: Two Modes in One Loop

If biological, or even bio-inspired, information processing did not have this emergent nonlinear loop of inhibitory/excitatory dynamics as its defining characteristic, then classical Bayesian inference would suffice to describe categorization, the stable part, as it is implemented via computers' mechanical information processing. Bayesian inference is formally equivalent with a variational free-energy principle minimization problem or a 'least action variational problem' [2, 24, 25]. It is a classical optimization problem encountered in statistical mechanics and thermodynamics among other disciplines. It is often pictured as climbing up *one* mountain top (or equivalently descending in *one* valley basin depending on the choice of sign).

The Free Energy Principle in cognition and action: One way that Friston and co-workers [24, 25] express this fact is by using the following succinct formula for the free energy functional, F:

$$F(q(s), p(\mu); \eta) = Energy - Entropy = -\langle \ln p(s; \eta)\rangle_q + \langle \ln q(\mu; \eta)\rangle_q \quad (1)$$

where $p(s; \eta)$ and $q(\mu; \eta)$ are probabilistic representations (i.e. variational densities) of sensory inputs, s, from the environment, and the system's internal representations, μ, both weighted with respect to the system's external states, η and conditioned over q. Other, alternative, expressions are derived and presented in [24, 25]. One consequence of such an optimisation of the variational free-energy is that it provides a bound on surprise and enables system's resilience while driving it to an equilibrium.

Free energy principle works well when functional minimization works well. That is, when dealing with a single basin descent, as also the standard Bayesian inference procedure does. It is well known that this is true for all gradient descent methods which essentially describe approach towards equilibrium, the minimum. But, when one encounters multiple coexisting basins of attraction, stochastic terms have to enter into the picture. Then the need to extend the classic Bayesian framework arises [2]. The situation is analogous of the metastable state-transitions in statistical thermodynamics driven by stochastic fluctuations, a typical far-from-equilibrium process. Indeed, this analogy has been noted [24, 25] and, normative connections with formal analogies have been established with self-organization, autopoiesis, second order cybernetics and other theories with similar minimization challenges.

Extending the Bayesian Approach: Parallel thoughts about the Bayesian inference in the presence of missing or incomplete data led statisticians and data-scientist to consider the inverse problem of the Bayesian theorem and supply the converse

Bayesian theorem [26, 27]. In other words as the classic Bayes' theorem provides a better estimate, called 'the posterior', for an original hypothesis expressed as a probability distribution, called 'the prior', taking in consideration given data, called 'the likelihood'; so the converse Bayes theorem provides a prior distribution that is compatible with the given likelihood furnished by the data (even if there are some of them missing: hence the name 'missing data problems' [27]) for a given knowledge of the posterior.

F. T. Arrechi, another pioneer of nonlinear science, made a further breakthrough when he explained neurophysiological data from ambiguous pictures, the Necker Cube in particular, based on an argument of Quantum Cognition theory and proposed a scheme for interlacing Bayesian and Inverse Bayesian (BIB) inference [28, 29], in a closed feedback loop, see also his chapter contributed to [11]. Further studies [3, 4, 20] revealed that the underlying logic of Arrechi's experiments is a quantum-type of non-Boolean logic (an orthomodular lattice of propositions) in agreement with the tenets of the theory of Quantum Cognition [21–23].

One can start by denoting d and h the variables representing data and hypotheses, respectively. By definition, the conditional probability of an event A *given* an event B, is customary written as $P(A|B)$ while the probability of events A *and* B (i.e. $(A \cap B) = (B \cap A)$, which is the overall probability of event A and of event B occurring, but not necessarily together at once) is expressed as $P(A, B)$, obviously $P(A, B) = P(B, A)$. Thus, the conditional probabilities $P(d|h) = \frac{P(d,h)}{P(h)}$ and $P(h|d) = \frac{P(d,h)}{P(d)}$ give the celebrated Bayes Theorem:

$$P(h|d)P(d) = P(d|h)P(h)$$

Since for changing hypotheses from a set $H = \{h_k, k = 1, 2, \ldots N\}$ at each time-step t we have $P(d) = \sum_k P(d|h_k)P(h_k)$, and we obtain the iteration process, indexed with the time-step t:

$$P^t(h|d) = P^t(d|h)P^t(h)\sum_k Pt(d|h_k)P^t(h_k) \Rightarrow \; P^{t+1}(h) = P^t(h|d) \quad (2)$$

The data $D = \{d_k, k = 1, 2, \ldots N\}$ and hypotheses $H = \{h_k, k = 1, 2, \ldots N\}$, are analogous to external stimuli and their internal representations, or categories. This is the classical Bayes inference compatible with the free energy minimization principle. Expressed in an operator form as:

$$P^{t+1}(h) = \mathbf{B}P^t(h|d) \quad (3)$$

The Inverse Bayesian inference (IB) can be expressed, also formally, as (Fig. 3):

$$P^{t+1}(d|h) = \mathbf{IB}P^t(d), \text{Eq.(3) \& Bayestheorem}, \Rightarrow P^{t+1}(h) = \mathbf{B} \diamond \mathbf{IB}P^t(h) \quad (4)$$

dropping indexes for clarity and where the operator $\diamond$ denotes a special, non-deterministic, composition, respecting the tenets of the converse Bayesian theorem

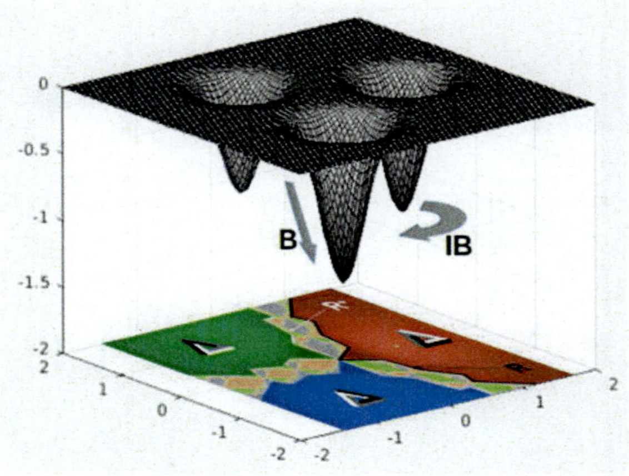

Fig. 3 Bayesian Inference (**B**, straight arrow) amounts to descending to a given basin, i.e. a category, or a hypothesis. Inverse Bayesian (**IB**, curved arrow) inference amounts to hopping and switching among different basins/hypotheses

[26, 27]. One way t do this is be letting the joint probability between a hypothesis and data to be transformed into a binary relation $R \subseteq H \times D$ such that $(h, d) \in R$ if $P(d, h) > \theta$; otherwise, $(h, d) \notin R$. Here θ is a threshold probability derived from the measure of the coarse-graining partition enclosed by the set $(R^+) - (R^-)$, e.g. as in the Figs. 2c and 3. Once a binary relation between a hypothesis and data is established, one can estimate a non-Boolean logical structure with respect to an, orthomodular, lattice. In particular, a lower approximation on hypotheses and an upper approximation on data, furnish a 'Rough Set Approximation' [20]. The iterative, non-algorithmic, part of the inverse Bayesian inference operator, **IB**, is not simply 'solving for P(h)' in Bayes' formula of Eqs. (2), (3). As with the Converse Bayes theorem, a correctly chosen functional space with proper convergence topology is crucial [26].

This picture of a landscape of data and hypotheses with multiple coexisting attractors is a challenge for any deterministic minimization process and, hence, also for free energy principle schemes. It is reminiscent of out-of-equilibrium self-organization as it brings forth the role of stochastic fluctuations in state transitions and/or other non-deterministic factors.

Nevertheless, probabilistic strategies and 'on the fly' construction of step-by-step solutions can work even if the process can be characterized as non-deterministic, or non-Turing-computable, in the normative sense of the words. Figure 4 illustrates these two Bayesian Inference processes.

Again, here, we encounter the 'exploit-explore' dual feedback loop as an instrumental and distinct feature of biological information processing. Now reflected in the probability space contraction/ expansion interplay during BIB inference:

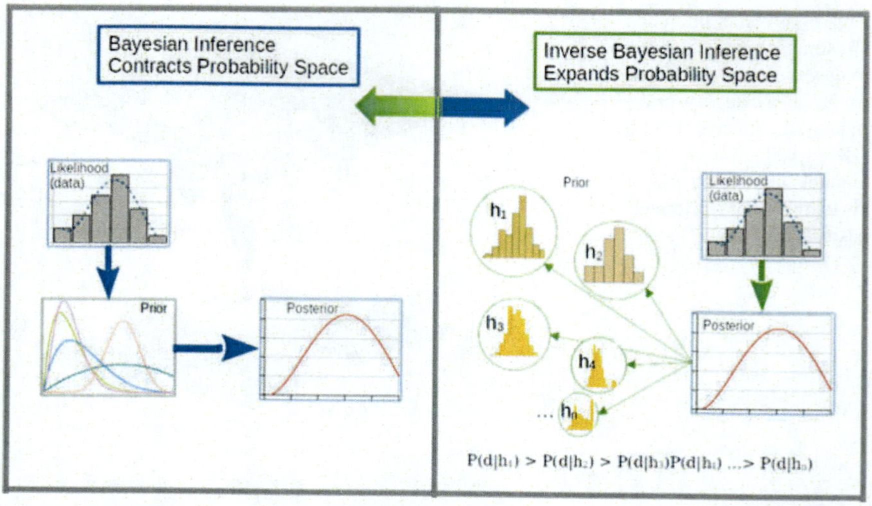

Fig. 4 The extended Bayesian (BIB) feedback loop. Left: its classical Bayesian inference branch. Right: the Inverse Bayesian inference branch with its extended probability distributions space that provide a set of competing hypotheses

Bayesian Inference: exploiting probability space
Given: Likelihood & Prior distribution $\longmapsto$ Posterior distribution
Inverse Bayesian Inference: exploring probability space
Given: Likelihood & Posterior distribution $\longmapsto$ Prior distribution

4 Outlook and Forthcoming Experiments

Apart from elucidating the logical, non-Boolean, structure of apprehension and judgement and the correspondence with the conceptual framework of Quantum Cognition, so far BIB has been successfully been implemented in explaining the appearance of fractal-type Levy flight super-diffusion and other aspects of collective behaviour of swarms [5, 6, 30, 31]. These are typical macroscopic biological processes. The extended Bayesian framework may also find application in the study of the capability of dealing with ambiguities, exploiting affordances and explore the adjacent space of possibilities, typical of natural organisms [32, 33].

Yet, as new technologies emerge, we are now able to describe the dynamics of neural morphology with spatial resolution down to the nanoscale level. Nanoelectromechanical vibrations became recently a powerful tool to investigate the role of oscillations, and noise trait, in the functioning of single neurons and the interaction with the environment [14] (e.g. effects of drugs, motor activity etc.).

In this context, experiments aiming at the analysis of such nano-vibrations of neurons have been designed and developed. We currently consider, the simplest complex system of neurons, consisting of just three neurons. This set-up allows the investigation of their collective behaviour in presence of different stimuli [34]. The ultimate goal is to infer what processes make the simplest complex alive neuronal network to act as a collective entity.

Along side and in complementing classical deterministic modeling of synchronization modes of three coupled neurons [35], BIB is expected to uncover other, not so commonly studied phenomena, in the simplest collective of neurons that nevertheless exhibit very complex behaviours, even infer new ones.

Acknowledgements GP-Y was supported by JSPS Topic-Setting Program to Advance Cutting-Edge Humanities & Social Sciences, JPJS00120351748. P-FM & VB acknowledge partial support from the Italian CNR Foresight Institute's program and from the Service de Physique des Systèmes Complexes et Mécanique Statistique of ULB. The authors would like to thank Reviewer#2 for the erudite, acute, and thorough comments that improved this manuscript.

References

1. The Seven Secrets of a Beautiful Mind, USCViterbi, 7 Sept. 2017. https://viterbischool.usc.edu/news/2017/09/seven-secrets-beautiful-mind/
2. Gunji, Y., Shinohara, S., Basios, V.: Connecting the free energy principle with quantum cognition. Front. Neurorobotics (2022). https://doi.org/10.3389/fnbot.2022.910161
3. Basios, V., Gunji, Y.: Chaotic dynamics in biological information processing: revisiting and revealing its logic. Opera Med. Phys. **3**(1), 1–13 (2017). https://doi.org/10.20388/omp2017.001.0041
4. Gunji, Y.-P., Shinohara, S., Basios, V.: Inverse bayes inference is a key of consciousness featuring macroscopic quantum logical structure. Biosystems **152**, 44–55 (2017). https://doi.org/10.1016/j.biosystems.2016.12.003
5. Gunji, Y.P., Murakami, H., Tomaru, T., Basios, V.: Inverse Bayesian inference in swarming behavior of soldier crabs. Philos. Trans. R. Soc. A. **376**, 20170370 (2018). https://doi.org/10.1098/rsta.2017.0370
6. Shinohara, S. et al.: A new method of Bayesian causal inference in non-stationary environments. PLOS, 22 May 2020 (2020). https://doi.org/10.1371/journal.pone.0233559 ($C++$ code available at: zenodo.org/record/5018080)
7. Basios, V., Gunji, Y.P.: Chaos, rhythms and processes in structure and function: extending Bayesian Inference. In: Proceedings of the Science and Technology Foresight Workshop, 'A Quest for An Interface Between Information and Action' (2021). www.foresight.cnr.it/pubblications/issn.html
8. Yarbus, A.L.: Eye Movements and Vision. Plenum Press, New York (1967). (Translated from Russian by Basil Haigh. Original Russian edition published in Moscow in 1965)
9. Purves, D., Augustine, G.J., Fitzpatrick, D., Hall, W.C., LaMantia, A-S., McNamara, J.O., White, L.E. eds.: Neuroscience, 4th ed. Sinauer Associates (1967)
10. Nicolis, J.S., Tsuda, I.: Chaotic dynamics of information processing: the 'magic number seven plus-minus two' revisited. Bull. Math. Biol. **47**(3), 343–365 (1985)
11. Nicolis, G., Basios, V.: 'Chaos Information Processing and Paradoxical Games: The legacy of J.S. Nicolis'. World Scientific (2015)
12. Nicolis, J.S., Tsuda, I.: Mathematical description of brain dynamics in perception and action. J. Conscious. Stud. **6**(11–12), 215–28 (1999)

13. Poil, S.S., et al.: Critical-state dynamics of avalanches and oscillations jointly emerge from balanced excitation/inhibition in neuronal networks. J. Neurosci.: Off. J. Soc. Neurosci. **32**(29), 9817–9823 (2012). https://doi.org/10.1523/JNEUROSCI.5990-11.2012
14. Chialvo, D.: Emergent complex neural dynamics. Nat. Phys. **6**, 744–750 (2010). https://doi.org/10.1038/nphys1803
15. Penrose triangle: In Wikipedia, 23 Jan. 2023. https://en.wikipedia.org/wiki/Penrose_triangle
16. Owais A.: Newton Fractal-Basin of Attraction (2022). (https://www.mathworks.com/matlabcentral/fileexchange/109940-newton-fractal-basin-of-attraction), MATLAB Central File Exchange. Published 12 Apr. 2022
17. Tsuda I.: Chaotic itinerancy and its roles in cognitive neurodynamics. Curr. Opin. Neurobiol. **31**, 67–71 (2015). SI: Brain rhythms and dynamic coordination
18. Nicolis, G., Nicolis, C.: Foundations of Complex Systems: Emergence, Information and Prediction. Word Scientific (2012)
19. Stewart, I., Golubitsky, M.: Symmetric networks with geometric constraints as models of visual illusions. Symmetry **11**(6), 799 (2019). https://doi.org/10.3390/sym11060799
20. Gunji, Y.P., Sonoda, K., Basios, V.: Quantum cognition based on an ambiguous representation derived from a rough set approximation. Biosystems **141**, 55–66 (2016). https://doi.org/10.1016/j.biosystems.2015.12.003
21. Khrennikov, A.: Ubiquitous Quantum Structure: From Psychology to Finances. Springer, Berlin (2010)
22. Busemeyer, J.R., Bruza, P.D.: Quantum Models of Cognition and Decision. Cambridge University Press, Cambridge (2012)
23. Aerts, D., de Bianchi, M.S.: The unreasonable success of quantum probability: part I&II'. J. Math. Psych. **67**, 51–75 and pp. 76–90 (2015). https://doi.org/10.1016/j.jmp.2015.01.003
24. Friston et al.: Perceptions as hypotheses: saccades as experiments. Front. Psychol **3**(151), 151 (2012). https://doi.org/10.3389/fpsyg.2012.00151
25. Friston, K.J.: The free-energy principle: a unified brain theory? Nat. Rev. Neurosci. **11**, 127–138 (2010). https://doi.org/10.1038/nrn2787
26. Ng, K.W.: The Converse of Bayes Theorem with Applications. Wiley (2014)
27. Tan, M.T., Tian, G.-L., Ng, K.W.: Bayesian Missing Data Problems: EM, Data Augmentation and Noniterative Computation, CRC Biostatistics Series, Chapman & Hall (2009)
28. Arecchi, F.T.: Phenomenology of consciousness from apprehension to judgment. Nonlinear Dyn. Psychol. Life Sci. **15**, 359–375 (2011)
29. Arecchi, F.T.: Chaotic neuron dynamics, synchronization and feature binding: quantum aspects. Mind Matter **1**, 15–43 (2003)
30. Gunji, Y.-P., et al.: Lévy walk in swarm models based on Bayesian and Inverse Bayesian inference. Comput. Struct. Biotechnol. J. **19**, 247–260 (2021). https://doi.org/10.1016/j.csbj.2020.11.045
31. Shinohara, S., et al.: Simulation of foraging behavior using a decision-making agent with Bayesian and inverse Bayesian inference: temporal correlations and power laws in displacement patterns. Chaos, Solitons and Fractals **157**(2022), 111976 (2022). https://doi.org/10.1016/j.chaos.2022.111976
32. Roli, A., Jaeger, J., Kauffman, S.A.: How organisms come to know the world: fundamental limits on artificial general intelligence. Front. Ecol. Evol. **9** (2022). https://doi.org/10.3389/fevo.2021.806283
33. Kauffman, S.A., Roli, A.: What is consciousness? Artificial intelligence, real intelligence, quantum mind and qualia. Biol. J. Linnean Soc. 0024–4066, blac092 (2022). https://doi.org/10.1093/biolinnean/blac092

34. Longo, G. et al.: COMA-SAN: COMplexity Analysis in the Simplest Alive Neural-network. In: Proceedings of the Science and Technology Foresight Workshop, A Quest for An Interface Between Information and Action (2021). www.foresight.cnr.it/pubblications/issn.html
35. Taylor, J.D., Chauhan, AS., Taylor, J.Y., Shilnikov, A.L., Nogaret, A.: Noise-activated barrier crossing in multiattractor dissipative neural networks. Phys. Rev. E **105**, 064203. https://doi.org/10.1103/PhysRevE.105.064203

Complexity

Fokas on Medical Imaging: Analytic Reconstructions for Emission Tomography

Nicholas E. Protonotarios, Konstantinos Kalimeris, and George A. Kastis

Abstract Mathematical problems associated with the theoretical foundations of emission tomography involve the inversion of the celebrated Radon transform of a function, defined as the set of all its line integrals, as well as the inversion of a certain generalization of the Radon transform of a function, the so-called *attenuated Radon transform*, defined as the set of all its *attenuated* line integrals. The non-attenuated and attenuated versions of the Radon transform provide the mathematical basis of emission tomography, particularly of two of the most important available medical imaging techniques, namely positron emission tomography (PET), and single-photon emission computed tomography (SPECT). Although Radon himself derived in 1917 the inversion of the transform bearing his name, seventy four years later Novikov and Fokas rederived this well-known formula by considering two classical problems in complex analysis known as the $\bar{d}$-problem and the scalar Riemann-Hilbert problem. The inversion may be obtained in a simpler manner by the use of the Fourier transform, however the derivation of Novikov and Fokas allowed Novikov to invert the attenuated Radon transform in 2002. Four years later Fokas, Iserles and Marinakis established a more straightforward derivation of this inversion. In this work, we present the seminal work of Fokas in the area of mathematical image reconstruction, based on the mathematical machinery of modern methods in complex analysis.

Keywords Radon transform · Attenuated Radon transform · Analytic image reconstruction · Medical imaging · Emission tomography

---

N. E. Protonotarios (✉)
Department of Applied Mathematics and Theoretical Physics, University of Cambridge, Cambridge CB3 0WA, UK
e-mail: np558@cam.ac.uk

N. E. Protonotarios · K. Kalimeris · G. A. Kastis
Mathematics Research Center, Academy of Athens, 11527 Athens, Greece
e-mail: kkalimeris@academyofathens.gr

G. A. Kastis
e-mail: gkastis@academyofathens.gr

G. A. Kastis
Institute of Nuclear and Radiological Science and Technology, Energy and Safety, National Center for Scientific Research "Demokritos", 15310 Agia Paraskevi, Greece

T. Bountis et al. (eds.), *Chaos, Fractals and Complexity*, Springer Proceedings in Complexity, https://doi.org/10.1007/978-3-031-37404-3_11

1 Introduction

In his seminal article for the Reports of the Saxon Academy of Sciences and Humanities of Leipzig [1, 2], entitled "Über die Bestimmung von Funktionen durch ihre Integralwerte längs gewisser Mannigfaltigkeiten," the Austrian mathematician Johann Karl August Radon (1887–1956) introduced an integral transform pair that was meant, several decades later, to give birth to the fields of computed tomography and mathematical image reconstruction. This transform would later be referred to as the *Radon transform*. Radon followed Hendrik Antoon Lorentz's (Nobel Prize in Physics 1902) unpublished work in 1905, where the initial Radon problem was introduced, and a three-dimensional function was recovered from its integrals over corresponding planes [3]. The inversion of the integral transform in three dimensions was given in 1906 by Hermann Bockwinkel, in his work on the propagation of light in biaxial crystals [4]. It is worth noting that in his work, Bockwinkel cites the inversion equation as Lorentz's [5]. The Radon transform pair provides the mathematical framework for a variety of inverse problems, not only in mathematics and medical imaging, but also in physics and certain other areas [6]. Nowadays, Radon is universally recognized as the pioneer of image reconstruction from projections. In order to honor his contributions to science, in 2003 the Austrian Academy of Sciences named the "Institute for Computational and Applied Mathematics" after Radon. Today, more than 105 years after his groundbreaking publication, Radon's work remains highly influential in the research community worldwide; see [7] for a review on recent achievements.

The Radon transform of a two-dimensional function is defined as the set of all its line integrals [8]. There exists a certain generalization of the two-dimensional Radon transform, namely the *attenuated Radon transform*, defined as the set of all line integrals of a two-dimensional function, attenuated with respect to a corresponding attenuation function. These Radon transforms provide the mathematical foundation of the most important medical imaging techniques, referred to as computed tomography (CT) [9] and as positron emission tomography (PET) [10, 11] for the non-attenuated version, and single-photon emission computed tomography (SPECT) [12, 13] for the attenuated version, respectively.

The non-attenuated Radon transform gives rise to the mathematical problem of "reconstructing" a function from its line integrals. PET consists of the numerical implementation of the inversion of the non-attenuated Radon transform. Similarly, SPECT is based in the inversion of the attenuated Radon transform, namely the reconstruction of a function from its attenuated line integrals.

In 1991, Fokas and Novikov rederived the well-known inversion of the Radon transform [14] by performing the so-called *spectral analysis* on the following eigenvalue equation:

$$\left[\frac{1}{2}\left(k+\frac{1}{k}\right)\partial_{x_1}+\frac{1}{2\mathrm{i}}\left(k-\frac{1}{k}\right)\partial_{x_2}\right]u(x_1,x_2;k)=f(x_1,x_2),\quad k\in\mathbb{C},\quad k\neq 0, \tag{1}$$

where subscripts denote partial differentiation. This analysis encompasses two certain problems in modern complex analysis known as the $\bar{d}$-problem and the scalar Riemann-Hilbert (RH) problem, respectively.

The inversion of the Radon transform can be obtained in a less complicated fashion, namely by employing the two-dimensional Fourier transform. However, the advantage of the derivation of [14] was established more than a decade later, in 2002, by Novikov [15]. In his paper, Novikov demonstrated that the inversion of the attenuated Radon transform can be obtained by applying an analysis analogous to the one performed in the eigenvalue equation (1). To this end, he performed the spectral analysis of a slight generalization of equation (1), namely of the following eigenvalue equation:

$$\left[\frac{1}{2}\left(k+\frac{1}{k}\right)\partial_{x_1}+\frac{1}{2\mathrm{i}}\left(k-\frac{1}{k}\right)\partial_{x_2}-\mu(x_1,x_2)\right]u(x_1,x_2;k)=f(x_1,x_2),$$
$$k\in\mathbb{C},\quad k\neq 0. \tag{2}$$

By employing the results of the analysis of both Eqs. (1) and (2), Fokas, Iserles and Marinakis, four years later, in 2006, derived the inverse attenuated Radon transform in a more straightforward manner [16]. The main result of the present work is the formulation of an equivalent inversion for the attenuated Radon transform, following the pioneering work of Novikov and Fokas.

2 Radon Transform and Its Attenuated Version in Two Dimensions

Via the Radon transform, a function f on $\mathbb{R}^n$ is integrated over its corresponding hyperplanes [17], it involves line integration along lines, and more precisely along families of parallel lines. The line integral of a continuous function $f : D \subset \mathbb{R}^2 \to \mathbb{R}$ along a differentiable curve $C : [a, b] \to D \subset \mathbb{R}^2$ is defined by:

$$\int_C f\mathrm{d}s=\int_a^b f(\mathbf{r}(\tau))\left\|\mathbf{r}'(\tau)\right\|_2\mathrm{d}\tau, \tag{3}$$

where $\mathbf{r} : [a, b] \to C$ is a bijective map, namely the parameterization of the curve C, and $\|\cdot\|_2$ denotes the L^2-norm in $\mathbb{R}^2$.

A line L on the x_1x_2-plane can be specified by the signed distance from the origin, ρ, $(-\infty < \rho < \infty)$, and the angle with the x_1-axis, θ $(0 \leqslant \theta < 2\pi)$, see Fig. 1. We

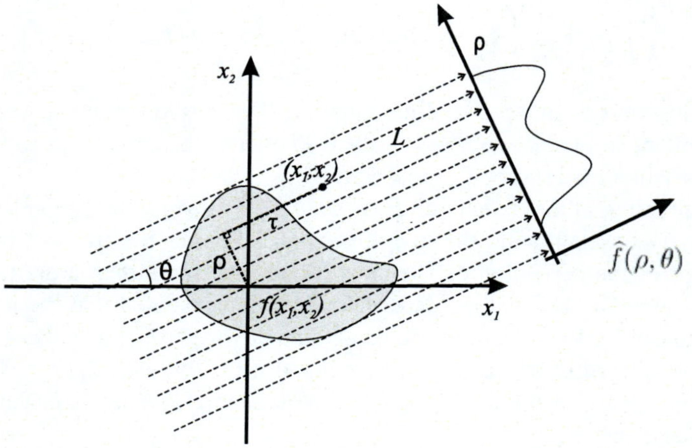

Fig. 1 Radon transform: A two-dimensional function, $f(x_1, x_2)$, and its corresponding projections, $\widehat{f}(\rho, \theta)$, in Cartesian (x_1, x_2) and local (ρ, τ) coordinates

denote the corresponding unit vectors parallel and perpendicular to L by $\mathbf{e}^{\parallel}$ and $\mathbf{e}^{\perp}$, respectively:

$$\mathbf{e}^{\parallel} = (\cos\theta, \sin\theta), \quad \text{and} \quad \mathbf{e}^{\perp} = (-\sin\theta, \cos\theta).$$

Every point $\mathbf{x} = (x_1, x_2)$ on L in Cartesian coordinates can be expressed in terms of the line coordinates (ρ, τ) via

$$\mathbf{x} = \rho\,\mathbf{e}^{\perp} + \tau\,\mathbf{e}^{\parallel},$$

where τ denotes the arc length, namely,

$$x_1 = \tau\cos\theta - \rho\sin\theta, \quad \text{and} \quad x_2 = \tau\sin\theta + \rho\cos\theta. \tag{4}$$

If we choose the following parameterization $\mathbf{r}$, resulting from Eq. (4),

$$\mathbf{r}(\tau) = \begin{bmatrix} \tau\cos\theta - \rho\sin\theta \\ \tau\sin\theta + \rho\cos\theta \end{bmatrix}, \tag{5}$$

then Eq. (4) imply that τ represents the arc length of the line L, and that Eq. (5) is a natural parameterization of the lines L. Taking into account the above, if the curve C is a line L naturally parameterized by the arc length τ, the initial line integral (3) may be rewritten as follows:

$$\int_C f\,\mathrm{d}s = \int_{-\infty}^{\infty} f\,(\tau\cos\theta - \rho\sin\theta, \tau\sin\theta + \rho\cos\theta)\,\mathrm{d}\tau. \tag{6}$$

Furthermore, Eq. (4) can be expressed in the local coordinates (ρ, τ), as follows:

$$\rho = x_2 \cos\theta - x_1 \sin\theta, \quad \text{and} \quad \tau = x_2 \sin\theta + x_1 \cos\theta. \tag{7}$$

Definition 1 The space of rapidly decreasing (Schwartz) functions on $\mathbb{R}^n$ is denoted by $\mathcal{S}(\mathbb{R}^n)$ and is defined as:

$$\mathcal{S}(\mathbb{R}^n) = \left\{ f \in C^\infty(\mathbb{R}^n) : ||f||_{\alpha,\beta} < \infty \right\} \subset C^\infty(\mathbb{R}^n), \tag{8}$$

where

$$||f||_{\alpha,\beta} = \sup_{x\in\mathbb{R}^n} \left| x^\alpha D^\beta f(x) \right|, \quad \forall \textit{ multi-index } \alpha, \beta,$$
$$\left| x^\alpha D^\beta f(x) \right| \to 0, \quad \text{as } |x| \to \infty. \tag{9}$$

Definition 2 The Radon transform, $\mathcal{R}$, is the line integral of a two-dimensional Schwartz function $f(x_1, x_2)$, $f \in \mathcal{S}(\mathbb{R}^2)$ along straight lines on the plane and is denoted by $\widehat{f}(\rho, \theta)$. It is expressed by:

$$\widehat{f}(\rho, \theta) = (\mathcal{R}f)(\rho, \theta) = \int_{-\infty}^{\infty} f(\tau\cos\theta - \rho\sin\theta, \tau\sin\theta + \rho\cos\theta) d\tau,$$
$$0 \leqslant \theta < 2\pi, \quad -\infty < \rho < \infty. \tag{10}$$

The main mathematical problem associated with the Radon transform, $\mathcal{R}$, and, by extension, with PET imaging involves the reconstruction of the the function $f(x_1, x_2)$, given the function $\widehat{f}(\rho, \theta) = (\mathcal{R}f)(\rho, \theta), 0 \leqslant \theta < 2\pi, -\infty < \rho < \infty$.

There exists a certain generalization of the Radon transform, namely the attenuated Radon transform. The attenuation notion is represented by an attenuation function $\mu(x_1, x_2)$, and is indicated in what follows by the subscript μ.

Definition 3 The attenuated Radon transform, $\mathcal{R}_\mu$, is the line integral of a two-dimensional function $f(x_1, x_2)$, attenuated with respect to the attenuation function $\mu(x_1, x_2)$. It is denoted by $\widehat{f_\mu}(\rho, \theta)$, and is expressed as follows:

$$\widehat{f_\mu}(\rho, \theta) = \left(\mathcal{R}_\mu f\right)(\rho, \theta) = \int_{-\infty}^{\infty} e^{-\int_\tau^\infty \mu(s\cos\theta - \rho\sin\theta, s\sin\theta + \rho\cos\theta)ds} \times$$
$$f(\tau\cos\theta - \rho\sin\theta, \tau\sin\theta + \rho\cos\theta)d\tau, \; 0 \leqslant \theta < 2\pi, \quad -\infty < \rho < \infty. \tag{11}$$

The mathematical problem associated with the attenuated Radon transform, $\mathcal{R}_\mu$, and, by extension, with SPECT imaging involves the reconstruction of the the function $f(x_1, x_2)$, given the functions $\widehat{f_\mu}(\rho, \theta) = (\mathcal{R}_\mu f)(\rho, \theta), 0 \leqslant \theta < 2\pi, -\infty < \rho < \infty$ and $\mu(x_1, x_2), -\infty < x_1, x_2 < \infty$, see Fig. 2.

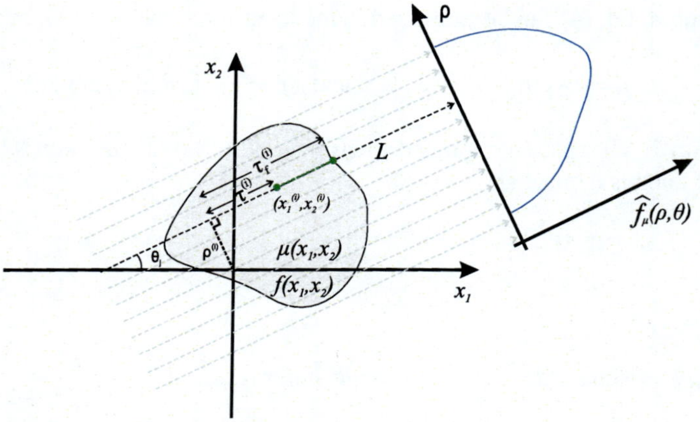

Fig. 2 Attenuated Radon transform: A two-dimensional function, $f(x_1, x_2)$, an attenuation function, $\mu(x_1, x_2)$, and its corresponding attenuated projections

3 Inversion of the Radon Transform in Two Dimensions

3.1 Fourier-Based Inversion of the Radon Transform in Two Dimensions

The most commonly attributed method for inverting the non-attenuated version of the Radon transform is the central slice theorem. This theorem provides a fundamental tool for the Fourier-based inversion of the Radon transform [9].

Theorem 1 (Central slice theorem) *The two-dimensional Fourier transform $\mathcal{F}_2$ of a function $f(x_1, x_2)$ is the one-dimensional Fourier transform $\mathcal{F}_1$ of the Radon transform $\mathcal{R}$ of the same function f, i.e.,*

$$\mathcal{F}_2\{f\} = \mathcal{F}_1\{\mathcal{R}\{f\}\}, \tag{12}$$

where $\mathcal{R}$ is defined in Eq. (10), $\mathcal{F}_2$ is defined by

$$(\mathcal{F}_2\{f\})(\xi_1, \xi_2) = \int_{-\infty}^{\infty}\int_{-\infty}^{\infty} f(x_1, x_2)e^{-2\pi i(\xi_1 x_1+\xi_2 x_2)}\mathrm{d}x_1\mathrm{d}x_2, \tag{13}$$

$\mathcal{F}_1$ is defined by

$$(\mathcal{F}_1\{g\})(r) = \int_{-\infty}^{\infty} g(\rho)e^{-2\pi i r\rho}\mathrm{d}\rho. \tag{14}$$

Proof See Section 2.2 of [9]. □

Corollary 1 *The central slice theorem implies that the Fourier-based inversion of the Radon transform may be written in the following operator form:*

$$\mathcal{R}^{-1} = \mathcal{F}_2^{-1}\mathcal{F}_1. \tag{15}$$

The inversion of the Radon transform may be accomplished without Fourier analysis, namely by employing tools arising in modern complex analysis. The computational benefit of the non-Fourier inversion are significant, as will become clear in the next Section.

3.2 *Complex Analysis Tools*

In order to highlight the seminal ideas of Fokas in the area of mathematical image reconstruction, we introduce the appropriate mathematical machinery of modern methods in complex analysis. In this direction, we will be able to solve the inverse problem defined in Eq. (10) without Fourier analysis and, ultimately, to invert the Radon transform.

Lemma 1 (Generalized Cauchy or Pompeiu's formula) *Assume that the function $f(z, \bar{z})$ is continuous and has continuous partial derivatives in a finite region D and on the simple closed boundary ∂D. Let ∂D denote the closed boundary of D (counterclockwise). Then, $f(z, \bar{z})$ can be evaluated at any interior point z via:*

$$f(z, \bar{z}) = \frac{1}{2\pi \mathrm{i}} \left(\oint_{\partial D} f(\zeta, \bar{\zeta}) \frac{\mathrm{d}\zeta}{\zeta - z} + \iint_D \frac{\partial f}{\partial \bar{\zeta}}(\zeta, \bar{\zeta}) \frac{\mathrm{d}\zeta \wedge \mathrm{d}\bar{\zeta}}{\zeta - z} \right), \tag{16}$$

where the so-called wedge product $\mathrm{d}\zeta \wedge \mathrm{d}\bar{\zeta}$, $\zeta = \xi + \mathrm{i}\eta$, $\bar{\zeta} = \xi - \mathrm{i}\eta$, *is defined as*

$$\mathrm{d}\zeta \wedge \mathrm{d}\bar{\zeta} = (\mathrm{d}\xi + \mathrm{i}\mathrm{d}\eta) \wedge (\mathrm{d}\xi - \mathrm{i}\mathrm{d}\eta) = -2\mathrm{i}\mathrm{d}\xi\mathrm{d}\eta. \tag{17}$$

Proof See Theorem 2.6.7 of [18]. □

Corollary 2 *If $f(z)$ is analytic in $\bar{D} = D \cup \partial D$, then Pompeiu's formula (16) reduces to Cauchy's integral formula, namely,*

$$f(z) = \frac{1}{2\pi \mathrm{i}} \oint_{\partial D} \frac{f(\zeta)}{\zeta - z} \mathrm{d}\zeta. \tag{18}$$

Proof If $f(z)$ is analytic in $\bar{D} = D \cup \partial D$, then $\partial f / \partial \bar{\zeta} = 0$. Hence, Pompeiu's formula (16) implies Cauchy's integral formula (18). □

Lemma 2 (Plemelj formulæ) *Let L be a smooth, simple curve, and let $\varphi(t)$ satisfy a Hölder condition on L. Then, the Cauchy-type integral*

$$\Phi(z) = \frac{1}{2\pi \mathrm{i}} \int_L \frac{\varphi(\tau)}{\tau - z} \mathrm{d}\tau, \tag{19}$$

as z approaches L from the right and the left, has the limiting values $\Phi^-(t)$ and $\Phi^+(t)$, respectively, given that t is not an endpoint of L, namely

$$\Phi^\pm(t) = \pm\frac{1}{2}\varphi(t) + \frac{1}{2\pi \mathrm{i}}\ PV \int_L \frac{\varphi(\tau)}{\tau - t} \mathrm{d}\tau, \tag{20}$$

where $PV \int$ denotes the Cauchy principal value integral defined by

$$PV \int_L g(\tau)\mathrm{d}\xi = \lim_{\epsilon \to 0} \int_{L-L_\epsilon} g(\xi)\mathrm{d}\tau, \tag{21}$$

and L_ε denotes the part of the contour L of length 2ϵ, that is centered around t.

Proof See Lemma 7.2.1 of [18]. □

Remark 1 Equation (19) yields a function which is analytic both in the interior and the exterior of the curve L, which we denote as $\Phi^\pm(z)$, respectively.

Lemma 3 *Let L be a smooth, closed simple curve, and $\Phi(z) = \Phi^\pm(z)$ analytic in the interior and exterior of L, respectively. Then, the solution of the following scalar Riemann-Hilbert problem*

$$\Phi^+(t) - \Phi^-(t) = g(t), \quad t \in L, \tag{22a}$$

$$\Phi(z) = O\left(\frac{1}{z}\right), \quad z \to \infty, \quad z \notin L, \tag{22b}$$

is given by

$$\Phi(z) = \frac{1}{2\pi \mathrm{i}} \int_L \frac{g(\tau)}{\tau - z} \mathrm{d}\tau. \tag{23}$$

Proof Equation (22a), along with Liouville's theorem [19], implies that the unique solution of the scalar Riemann-Hilbert problem represented by Eq. (22) is given by Eq. (23). □

3.3 The Inversion of the Radon Transform in Two Dimensions Without the Fourier Transform: The Fokas Approach

In 1991, Novikov and Fokas rederived the well known inversion of the Radon transform [14] by performing spectral analysis of the eigenvalue equation (1). As already noted, the Radon transform inversion can be obtained via Fourier in a more straightforward manner. However, the advantage of the derivation of [14] was manifested by Novikov in 2002 [15]. Novikov established that the inverse attenuated Radon transform can be derived by applying spectral analysis to a generalization of equation (1), namely equation (2). The spectral analysis of the eigenvalue equation (1) consists of two steps:

1. *Direct problem ($\bar{d}$-problem)*: Solve equation (1) in terms of the function f for all complex eigenvalues k. The solution must be bounded for all $k \in \mathbb{C}$.
2. *Inverse problem (Riemann-Hilbert)*: Derive an equivalent representation of u which, instead of f, depends on $\widehat{f}$, i.e., the Radon transform of f.

Definition 4 The Hilbert transform of a function $u(t)$, is defined as:

$$\mathcal{H}\{u(t)\} = \frac{1}{\pi}\left(PV\int_{-\infty}^{\infty}\frac{u(\xi)}{\xi - t}d\xi\right). \tag{24}$$

Proposition 1 *The inverse of the Radon transform $\widehat{f}(\rho, \theta)$ defined in Eq. (10), of a function $f(x_1, x_2) \in \mathcal{S}(\mathbb{R}^2)$, is given by*

$$f(x_1, x_2) = -\frac{1}{4\pi}\int_0^{2\pi}\left[\frac{\partial(\mathcal{H}\widehat{f})(\rho, \theta)}{\partial \rho}\right]_{\rho = x_2\cos\theta - x_1\sin\theta} d\theta, \tag{25}$$

with $-\infty < x_1, x_2 < \infty$, and $\mathcal{H}$ defined in Eq. (24).

Proof The spectral analysis of the eigenvalue equation (1) will unveil the inverse Radon transform, as follows.

3.3.1 $\bar{d}$-Problem (Direct)

For the direct problem, we solve equation (1) for all $k \in \mathbb{C}$. To this end, the following change of variables from (x_1, x_2) to $(z, \bar{z})$ is necessary:

$$z = \frac{1}{2i}\left(k - \frac{1}{k}\right)x_1 - \frac{1}{2}\left(k + \frac{1}{k}\right)x_2, \tag{26a}$$

$$\bar{z} = -\frac{1}{2i}\left(\bar{k} - \frac{1}{\bar{k}}\right)x_1 - \frac{1}{2}\left(\bar{k} + \frac{1}{\bar{k}}\right)x_2. \tag{26b}$$

Employing the chain rule yields

$$\partial_{x_1} = \frac{1}{2i}\left(k - \frac{1}{k}\right)\partial_z - \frac{1}{2\mathrm{i}}\left(\bar{k} - \frac{1}{\bar{k}}\right)\partial_{\bar{z}}, \tag{27a}$$

$$\partial_{x_2} = -\frac{1}{2}\left(k + \frac{1}{k}\right)\partial_z - \frac{1}{2}\left(\bar{k} + \frac{1}{\bar{k}}\right)\partial_{\bar{z}}. \tag{27b}$$

Therefore, if we introduce the function ν,

$$\nu(|k|) = \frac{1}{2\mathrm{i}}\left(\frac{1}{|k|^2} - |k|^2\right), \tag{28}$$

then, we may rewrite Eq. (1) in the following form:

$$\nu(|k|)\frac{\partial u(x_1, x_2, k)}{\partial \bar{z}} = f(x_1, x_2), \quad k \in \mathbb{C}, \quad |k| \neq 1, \tag{29}$$

Equation (29) may be simplified:

$$u_{\bar{z}} = \frac{f}{\nu}, \quad |k| \neq 1. \tag{30}$$

It is important to emphasize that if u was analytic, then $u_{\bar{z}} = 0$. Furthermore, we supplement Eq. (29) with a boundary condition at infinity, namely,

$$u = O\left(\frac{1}{z}\right), \quad z \to \infty, \quad \text{i.e. } \exists\, \alpha > 0 \text{ such that } |u| \leqslant \frac{\alpha}{|z|}. \tag{31}$$

It follows from Lemma 1 that the solution of equation (29) with the boundary condition (31) is represented by

$$u = \frac{1}{2\pi\mathrm{i}}\iint_{\mathbb{R}^2} \frac{f(x_1', x_2')}{\nu(|k|)}\frac{\mathrm{d}z' \wedge \mathrm{d}\bar{z}'}{z' - z}, \quad k \in \mathbb{C}, \quad |k| \neq 1. \tag{32}$$

Inserting

$$\mathrm{d}z' \wedge \mathrm{d}\bar{z}' = \frac{1}{2\mathrm{i}}\left|\frac{1}{|k|^2} - |k|^2\right| \mathrm{d}x_1'\mathrm{d}x_2',$$

in Eq. (32) produces another representation for u, namely,

$$u(x_1, x_2, k) = \frac{1}{2\pi\mathrm{i}}\operatorname{sgn}\left(\frac{1}{|k|^2} - |k|^2\right)\iint_{\mathbb{R}^2} f(x_1', x_2')\frac{\mathrm{d}x_1'\mathrm{d}x_2'}{z' - z}. \tag{33}$$

Equation (33) demonstrates that u depends on k only through the term

$$z - z' = \frac{1}{2\mathrm{i}}\left(k - \frac{1}{k}\right)(x_1 - x_1') - \frac{1}{2}\left(k + \frac{1}{k}\right)(x_2 - x_2'). \tag{34}$$

Therefore, $u(x_1, x_2)$ is a sectionally analytic function with a jump across the unit circle of the complex plane. In other words, Eq. (33) is perceived as the solution of the direct problem in terms of the function f, for all $k \in \mathbb{C}$.

3.3.2 Riemann-Hilbert Problem (Inverse)

For the inverse problem, we solve Eq. (1) in terms of $\widehat{f}$. Equation (33) implies

$$u = O\left(\frac{1}{k}\right), \quad k \to \infty, \tag{35}$$

hence, the solution u of equation (1) is bounded for all $k \in \mathbb{C}$. We examine the behavior of u as k approaches the unit circle, by letting

$$k^{\pm} = (1 \mp \varepsilon)e^{\mathrm{i}\theta}, \quad 0 \leqslant \theta < 2\pi, \quad \varepsilon > 0. \tag{36}$$

Therefore,

$$k^{+} \mp \frac{1}{k^{+}} = (1 - \varepsilon)e^{\mathrm{i}\theta} \mp (1 + \varepsilon)e^{-\mathrm{i}\theta} + O(\varepsilon^2), \tag{37a}$$

$$k^{-} \mp \frac{1}{k^{-}} = (1 + \varepsilon)e^{\mathrm{i}\theta} \mp (1 - \varepsilon)e^{-\mathrm{i}\theta} + O(\varepsilon^2). \tag{37b}$$

Taking into account equation (26a) yields

$$z - z' = \rho - \rho' \pm \mathrm{i}\varepsilon(\tau - \tau') + O(\varepsilon^2). \tag{38}$$

If we denote the limits of the function u as k approaches the unit circle from inside and outside, respectively, as $u^{\pm}$, then

$$u^{\pm} \equiv \lim_{\varepsilon \to 0} u(x_1, x_2, (1 \mp \varepsilon)e^{\mathrm{i}\theta}). \tag{39}$$

In Eq. (33), we replace $z - z'$ by the representation (38), i.e.,

$$u^{\pm} = \mp \frac{1}{2\pi\mathrm{i}} \lim_{\varepsilon \to 0} \iint_{\mathbb{R}^2} \frac{\varphi(\rho', \tau', \theta)\mathrm{d}\rho'\mathrm{d}\tau'}{\rho' - [\rho \pm \mathrm{i}\varepsilon(\tau' - \tau)]}, \tag{40}$$

where φ is the function f expressed in the local coordinates defined as:

$$\varphi(\rho, \tau, \theta) = f(\tau\cos\theta - \rho\sin\theta, \tau\sin\theta + \rho\cos\theta)\,. \tag{41}$$

In order to further elucidate the limit (40), we aim to control the sign of $\tau' - \tau$ by splitting the integral $\int \mathrm{d}\tau'$ in the following manner:

$$\begin{aligned} u^{\pm} = \mp\frac{1}{2\pi \mathrm{i}} \lim_{\varepsilon\to 0} \int_{-\infty}^{\infty} \Bigg\{ \int_{-\infty}^{\tau} \frac{\varphi \mathrm{d}\tau'}{\rho' - [\rho \pm \mathrm{i}\varepsilon(\tau' - \tau)]} \\ + \int_{\tau}^{\infty} \frac{\varphi \mathrm{d}\tau'}{\rho' - [\rho \pm \mathrm{i}\varepsilon(\tau' - \tau)]} \Bigg\} \mathrm{d}\rho'. \end{aligned} \tag{42}$$

In Eq. (42), $(\tau' - \tau)$ is negative in the first integral, and positive in the second integral, therefore

$$\begin{aligned} u^{\pm} = \mp\frac{1}{2\pi \mathrm{i}} \int_{-\infty}^{\tau} \left\{ \mp\pi \mathrm{i}\varphi(\rho, \tau', \theta) + (\mathcal{H}\varphi)(\rho, \tau', \theta) \right\} \mathrm{d}\tau' \\ \mp \frac{1}{2\pi \mathrm{i}} \int_{\tau}^{\infty} \left\{ \pm\pi \mathrm{i}\varphi(\rho, \tau', \theta) + (\mathcal{H}\varphi)(\rho, \tau', \theta) \right\} \mathrm{d}\tau'. \end{aligned} \tag{43}$$

In Eq. (43), we have utilized the Plemelj formulæ, see Lemma 2, and the Hilbert transform, see Definition 4. We add and subtract $\mp\frac{1}{2\pi \mathrm{i}} \int_{\tau}^{\infty} \pi \mathrm{i}\varphi(\rho, \tau', \theta)\mathrm{d}\tau'$ on the right-hand side of equation (43), to obtain

$$u^{\pm} = \mp(P^{\mp}\hat{f})(\rho, \theta) - \int_{\tau}^{\infty} \varphi(\rho, \tau', \theta)\mathrm{d}\tau', \tag{44}$$

where $P^{\pm}$ denotes the set of projectors in ρ

$$(P^{\mp}g)(\rho) = \pm\frac{g(\rho)}{2} + \frac{1}{2\pi \mathrm{i}}\, PV \int_{-\infty}^{\infty} \frac{g(r)}{r - \rho}\mathrm{d}r = \pm\frac{g(\rho)}{2} + \frac{1}{2\mathrm{i}}(\mathcal{H}g)(\rho). \tag{45}$$

Equation (44) via Eq. (45) implies

$$u^{+} - u^{-} = \mathrm{i}(\mathcal{H}\widehat{f})(\rho, \theta). \tag{46}$$

Equation (46), along with the boundary condition (35), form a scalar Riemann-Hilbert problem. Employing Lemma 3 yields the solution, namely,

$$u = \frac{1}{2\pi \mathrm{i}} \int_{|k'|=1} \frac{(u^{+} - u^{-})(\rho, \theta')\mathrm{d}k'}{k' - k}. \tag{47}$$

Equation $|k'| = 1$ may be written as $k' = e^{i\theta'}$, which implies $dk' = ie^{i\theta'}d\theta'$, i.e.,

$$u = \frac{1}{2\pi i}\int_0^{2\pi} \frac{(u^+ - u^-)(\rho, \theta')ie^{i\theta'}d\theta'}{e^{i\theta'} - k}. \tag{48}$$

In Eq. (48), we replace $u^+ - u^-$ by Eq. (46) to obtain

$$u = -\frac{1}{2\pi i}\int_0^{2\pi} \frac{e^{i\theta'}(\mathcal{H}\widehat{f})(\rho, \theta')d\theta'}{e^{i\theta'} - k}, \quad k \in \mathbb{C}, \ |k| \neq 1, \ \rho \in \mathbb{R}. \tag{49}$$

Via Eq. (49), the asymptotic analysis of the behavior of u for large k yields

$$u = \left\{\frac{1}{2\pi i}\int_0^{2\pi} e^{i\theta'}(\mathcal{H}\widehat{f})(\rho, \theta')d\theta'\right\}\frac{1}{k} + O\left(\frac{1}{k^2}\right), \quad k \to \infty. \tag{50}$$

In Eq. (1), if we substitute the above expression, we conclude that the $O(1)$ term involves f in the following manner:

$$f = \frac{1}{4\pi i}\left(\partial_{x_1} - i\partial_{x_2}\right)\int_0^{2\pi} e^{i\theta}(\mathcal{H}\widehat{f})(\rho, \theta)\Big|_{\rho = x_2\cos\theta - x_1\sin\theta} d\theta. \tag{51}$$

Since $(\partial_{x_1} - i\partial_{x_2}) = e^{-i\theta}(\partial_\tau - i\partial_\rho)$, Eq. (25) follows from Eq. (51). □

4 Inversion of the Attenuated Radon Transform in Two Dimensions

We will derive the inversion of the attenuated Radon transform via the Fokas approach [16], which is rather simpler than the one followed by Novikov in [15].

4.1 *The Inversion of the Attenuated Radon Transform in Two Dimensions Without the Fourier Transform: The Fokas Approach*

An immediate consequence of Proposition 1 is the following corollary.

Corollary 3 *Let $k^\pm$ denote the limits of k, defined in Eq.* (36)*, and let z and ν be defined in Eqs.* (26a) *and* (28)*, respectively. Then,*

$$\lim_{k \to k^{\pm}} \left(\partial_{\bar{z}}^{-1} \left\{ \frac{f(x_1, x_2)}{\nu(|k|)} \right\} \right) = \mp \left(P^{\mp} \widehat{f} \right)(\rho, \theta) - \int_{\tau}^{\infty} \varphi(\rho, s, \theta) \mathrm{d}s,$$
$$(\rho, \tau) \in \mathbb{R}^2, \quad \theta \in (0, 2\pi), \quad (52)$$

where $\widehat{f}$ denotes the Radon transform of f (defined in Eq. (10)), and $P^{\pm}$, (ρ, τ) and φ are defined in Eqs. (45), (7) and (41), respectively.

Proof It is straightforward to observe that Eq. (30) implies

$$u(x_1, x_2, k) = \partial_{\bar{z}}^{-1} \left\{ \frac{f(x_1, x_2)}{\nu(|k|)} \right\}. \quad (53)$$

By taking the limit of equation (53) as k approaches the unit circle from inside and outside, respectively, we obtain

$$u^{\pm}(x_1, x_2) = \lim_{k \to k^{\pm}} \left(\partial_{\bar{z}}^{-1} \left\{ \frac{f(x_1, x_2)}{\nu(|k|)} \right\} \right), \quad (54)$$

where $u^{\pm}$ is defined in Eq. (39). Equation (52) follows from the insertion of equation (44) in Eq. (54). □

Proposition 2 *The inverse of the attenuated Radon transform $\widehat{f}_{\mu}(\rho, \theta)$ defined in Eq. (11), of a function $f(x_1, x_2)$, attenuated with respect to the function $\mu(x_1, x_2)$ (with $f, \mu \in S(\mathbb{R}^2)$), is given by*

$$f(x_1, x_2) = \frac{1}{4\pi} (\partial_{x_1} - \mathrm{i}\partial_{x_2}) \int_0^{2\pi} e^{\mathrm{i}\theta} J(x_1, x_2, \theta) \mathrm{d}\theta, \quad -\infty < x_1, x_2 < \infty, \quad (55a)$$

with J being defined by

$$J(x_1, x_2, \theta) = e^{M(\tau, \rho, \theta)} L_{\mu}(\rho, \theta) \widehat{f}_{\mu}(\rho, \theta) \Big|_{\substack{\tau = x_2 \sin\theta + x_1 \cos\theta \\ \rho = x_2 \cos\theta - x_1 \sin\theta}}, \quad (55b)$$

where M and L_{μ} are given by

$$M(\tau, \rho, \theta) = \int_{\tau}^{\infty} \mu(s\cos\theta - \rho\sin\theta, s\sin\theta + \rho\cos\theta) \, \mathrm{d}s, \quad (55c)$$

$$L_{\mu}(\rho, \theta) = e^{P^{-}\widehat{\mu}(\rho,\theta)} P^{-} e^{P^{-}\widehat{\mu}(\rho,\theta)} + e^{-P^{+}\widehat{\mu}(\rho,\theta)} P^{+} e^{P^{+}\widehat{\mu}(\rho,\theta)}, \quad (55d)$$

with $\widehat{\mu}$ denoting the Radon transform of μ, namely

$$\widehat{\mu}(\rho, \theta) = \int_{-\infty}^{\infty} \mu\,(\tau \cos\theta - \rho \sin\theta, \tau \sin\theta + \rho\cos\theta)\mathrm{d}\tau, \quad 0 \leqslant \theta < 2\pi, \; \rho \in \mathbb{R}, \tag{55e}$$

and the projection operators $P^{\pm}$ are defined in Eq. (45).

Proof We rewrite Eq. (2) as follows:

$$u_{\bar{z}} + \frac{\mu}{\nu}u = \frac{f}{\nu}. \tag{56}$$

Furthermore, we multiply both sides by $e^{\partial_{\bar{z}}^{-1}\{\frac{\mu}{\nu}\}}$ to obtain

$$u_{\bar{z}}e^{\partial_{\bar{z}}^{-1}\{\frac{\mu}{\nu}\}} + \frac{\mu}{\nu}ue^{\partial_{\bar{z}}^{-1}\{\frac{\mu}{\nu}\}} = \frac{f}{\nu}e^{\partial_{\bar{z}}^{-1}\{\frac{\mu}{\nu}\}}. \tag{57}$$

Hence,

$$\frac{\partial}{\partial\bar{z}}\left(ue^{\partial_{\bar{z}}^{-1}\{\frac{\mu}{\nu}\}}\right) = \frac{f}{\nu}e^{\partial_{\bar{z}}^{-1}\{\frac{\mu}{\nu}\}}, \tag{58}$$

and

$$e^{\partial_{\bar{z}}^{-1}\{\frac{\mu}{\nu}\}}u = \partial_{\bar{z}}^{-1}\left\{\frac{f}{\nu}e^{\partial_{\bar{z}}^{-1}\{\frac{\mu}{\nu}\}}\right\}, \quad (x_1, x_2) \in \mathbb{R}^2, \quad k \in \mathbb{C}. \tag{59}$$

Equation (59) is an expression of the solution to the direct problem, which defines a sectionally analytic function u with a jump across the unit circle. Further investigation of the jump involves the limits of $\partial_{\bar{z}}^{-1}(f/\nu)$ as $k \to k^{\pm}$, which, in turn, involves Corollary 3. As $k \to k^{\pm}$, we employ Eq. (52) into Eq. (59) to obtain

$$e^{\left(\mp P^{\mp}\widehat{\mu} - \int_{\tau}^{\infty}\Phi(\rho,s,\theta)\mathrm{d}s\right)}u^{\pm} = \lim_{k\to k^{\pm}} \partial_{\bar{z}}^{-1}\left\{\frac{f}{\nu}e^{\left(\mp P^{\mp}\widehat{\mu} - \int_{\tau}^{\infty}\Phi(\rho,s,\theta)\mathrm{d}s\right)}\right\}, \tag{60}$$

where $\widehat{\mu}$ is defined in Eq. (55e), and Φ denotes the function μ expressed in the local coordinates, i.e., $\Phi(\rho, \tau, \theta) = \mu\,(\tau\cos\theta - \rho\sin\theta, \tau\sin\theta + \rho\cos\theta)$. Employing Corollary 3 on the right-hand side of equation (60) yields

$$\mp P^{\mp}e^{\mp P^{\mp}\widehat{\mu}}\widehat{f}_{\mu} - \int_{\tau}^{\infty} \varphi(\rho, \tau', \theta)e^{\mp P^{\mp}\widehat{\mu}e^{-\int_{\tau}^{\infty}\Phi(\rho,s,\theta)\mathrm{d}s}}\mathrm{d}\tau', \tag{61}$$

where, in Eq. (52), instead of f, we used $\left(e^{\mp P^{\mp}\widehat{\mu}e^{-\int_{\tau}^{\infty}\Phi(\rho,s,\theta)\mathrm{d}s}}\right) f$.

In Eq. (61), the term $e^{\mp P^{\mp}\widehat{\mu}}$ is independent of τ', i.e., the jump is given by

$$u^{+} - u^{-} = -J, \tag{62}$$

where J is given by Eq. (55b). Hence, it follows from Eq. (48) that

$$u = -\frac{1}{2\pi}\int_0^{2\pi}\frac{J(\rho,\tau,\theta')e^{\mathrm{i}\theta'}\mathrm{d}\theta'}{e^{\mathrm{i}\theta'}-k}, \tag{63}$$

which implies

$$u = \left[\frac{1}{2\pi}\int_0^{2\pi} e^{\mathrm{i}\theta}J(\rho,\tau,\theta)\mathrm{d}\theta\right]\frac{1}{k} + O\left(\frac{1}{k^2}\right), \quad \text{for} \quad k\to\infty. \tag{64}$$

Finally, we insert Eq. (64) in Eq. (2); the $O(1)$ term of the resulting equation yields (55a). □

4.2 A Novel Method for the Inversion of the Attenuated Radon Transform in Two Dimensions

In what follows, it is convenient to define F as half the Hilbert transform of $\widehat{\mu}$

$$F(\rho,\theta) \equiv \frac{1}{2}\mathcal{H}\{\widehat{\mu}(\rho,\theta)\} = \frac{1}{2\pi}\left(PV\int_{-\infty}^{\infty}\frac{\widehat{\mu}(r,\theta)}{r-\rho}\mathrm{d}r\right). \tag{65}$$

Proposition 3 *The inversion formula for the attenuated Radon transform, defined in Eq. (11), of a function $f(x_1,x_2)$ attenuated with respect to a function $\mu(x_1,x_2)$ is equivalent to the representation*

$$f(x_1,x_2) = -\frac{1}{2\pi}\int_0^{2\pi} e^{M(\tau,\rho,\theta)}\left[M_\rho(\tau,\rho,\theta)G(\rho,\theta)+G_\rho(\rho,\theta)\right]\Bigg|_{\substack{\tau=x_2\sin\theta+x_1\cos\theta\\ \rho=x_2\cos\theta-x_1\sin\theta}}\mathrm{d}\theta, \tag{66}$$

where M is defined in Eq. (55c) and G is defined by

$$G(\rho,\theta) = e^{-\frac{1}{2}\widehat{\mu}(\rho,\theta)}\left[\cos(F(\rho,\theta))G^C(\rho,\theta)+\sin(F(\rho,\theta))G^S(\rho,\theta)\right], \tag{67}$$

with the functions G^C and G^S defined by

$$G^C(\rho,\theta) = \frac{1}{2\pi}PV\int_{-\infty}^{\infty} e^{\frac{1}{2}\widehat{\mu}(r,\theta)}\cos F(r,\theta)\frac{\widehat{f}_\mu(r,\theta)\mathrm{d}r}{r-\rho}, \tag{68a}$$

$$G^S(\rho,\theta) = \frac{1}{2\pi}PV\int_{-\infty}^{\infty} e^{\frac{1}{2}\widehat{\mu}(r,\theta)}\sin F(r,\theta)\frac{\widehat{f}_\mu(r,\theta)\mathrm{d}r}{r-\rho}. \tag{68b}$$

Proof We apply the operator L_μ on the attenuated Radon transform $\widehat{f_\mu}$, i.e.,

$$\left(L_\mu \widehat{f_\mu}\right)(\rho,\theta) = \left\{e^{P^-\widehat{\mu}(\rho,\theta)}P^- e^{P^-\widehat{\mu}(\rho,\theta)} + e^{-P^+\widehat{\mu}(\rho,\theta)}P^+ e^{P^+\widehat{\mu}(\rho,\theta)}\right\}\widehat{f_\mu}(\rho,\theta). \quad (69)$$

Equations (45) and (65) imply

$$e^{P^\pm\widehat{\mu}} = e^{\pm\frac{\widehat{\mu}}{2}-\mathrm{i}F}. \quad (70)$$

Hence,

$$e^{P^-\widehat{\mu}}P^-\left\{e^{-P^-\widehat{\mu}}\widehat{f_\mu}\right\} = e^{-\frac{\widehat{\mu}}{2}-\mathrm{i}F}\left[-\frac{1}{2}e^{\frac{\widehat{\mu}}{2}+\mathrm{i}F}\widehat{f_\mu} + \frac{1}{2\mathrm{i}}\mathcal{H}\left\{e^{\frac{\widehat{\mu}}{2}+\mathrm{i}F}\widehat{f_\mu}\right\}\right], \quad (71a)$$

$$e^{-P^+\widehat{\mu}}P^+\left\{e^{P^+\widehat{\mu}}\widehat{f_\mu}\right\} = e^{-\frac{\widehat{\mu}}{2}+\mathrm{i}F}\left[\frac{1}{2}e^{\frac{\widehat{\mu}}{2}-\mathrm{i}F}\widehat{f_\mu} + \frac{1}{2\mathrm{i}}\mathcal{H}\left\{e^{\frac{\widehat{\mu}}{2}-\mathrm{i}F}\widehat{f_\mu}\right\}\right]. \quad (71b)$$

We simplify Eq. (69), taking into account equations (65), (70), and (71), namely,

$$\left(L_\mu \widehat{f_\mu}\right)(\rho,\theta) = \frac{1}{2\mathrm{i}}e^{-\frac{\widehat{\mu}}{2}}\left[e^{-\mathrm{i}F}\mathcal{H}\left\{e^{\frac{\widehat{\mu}}{2}+\mathrm{i}F}\widehat{f_\mu}\right\} + e^{\mathrm{i}F}\mathcal{H}\left\{e^{\frac{\widehat{\mu}}{2}\mathrm{i}F}\widehat{f_\mu}\right\}\right]. \quad (72)$$

Using Euler's formula, i.e., $e^{\mathrm{i}F} = \cos F + \mathrm{i}\sin F$, and further expanding yields

$$\left(L_\mu \widehat{f_\mu}\right)(\rho,\theta) = -2\mathrm{i}G(\rho,\theta). \quad (73)$$

It is important to note that Eq. (73) implies that the function $\left(L_\mu \widehat{f_\mu}\right)(\rho,\theta)$ is purely imaginary, i.e., $\mathrm{Re}\left\{\left(L_\mu \widehat{f_\mu}\right)(\rho,\theta)\right\} = 0$. Thus,

$$J(x_1,x_2,\theta) = -2\mathrm{i}\left[e^{M(\tau,\rho,\theta)}G(\rho,\theta)\right]_{\substack{\tau=x_2\sin\theta+x_1\cos\theta \\ \rho=x_2\cos\theta-x_1\sin\theta}}. \quad (74)$$

Calculating the action of the differential operator $(\partial_{x_1} - \mathrm{i}\partial_{x_2}) = e^{-\mathrm{i}\theta}(\partial_\tau - \mathrm{i}\partial_\rho)$ on J yields

$$(\partial_{x_1} - \mathrm{i}\partial_{x_2})J = -2e^{-\mathrm{i}\theta}e^M\left[-\mathrm{i}\mu G + M_\rho G + G_\rho\right]_{\substack{\tau=x_2\sin\theta+x_1\cos\theta \\ \rho=x_2\cos\theta-x_1\sin\theta}}, \quad (75)$$

considering that $M_\tau(\tau,\rho,\theta)\Big|_{\substack{\tau=x_2\sin\theta+x_1\cos\theta \\ \rho=x_2\cos\theta-x_1\sin\theta}} = \mu(x_1,x_2)$ and $G_\tau(\rho,\theta) = 0$ [13, 20]. Inserting the above in the right-hand side of equation (55a) and combining Eqs. (74) and (75), yields

$$f(x_1,x_2) = -\frac{1}{2\pi}\int_0^{2\pi} e^M\left[-\mathrm{i}\mu G + M_\rho G + G_\rho\right]\Bigg|_{\substack{\tau=x_2\sin\theta+x_1\cos\theta \\ \rho=x_2\cos\theta-x_1\sin\theta}} \mathrm{d}\theta. \quad (76)$$

The first term of the integral on (76) vanishes; hence, Eq. (66) is obtained. Indeed, the first term of the integral on (76) can be simplified as follows:

$$-\mathrm{i}\int_0^{2\pi} \mu(x_1,x_2)\left[e^{M(\tau,\rho,\theta)}G(\rho,\theta)\right]_{\substack{\tau=x_2\sin\theta+x_1\cos\theta\\ \rho=x_2\cos\theta-x_1\sin\theta}}\mathrm{d}\theta = \frac{1}{2}\mu(x_1,x_2)\int_0^{2\pi} J(x_1,x_2,\theta)\mathrm{d}\theta. \quad (77)$$

Furthermore, Eq. (2.9) of [16] evaluated at $\lambda = 0$ yields

$$u(x_1,x_2,0) = \frac{1}{2\pi}\int_0^{2\pi} J(x_1,x_2,\theta)\mathrm{d}\theta.$$

Taking the limit $\lambda \to 0$ in Eq. (2.2) of [16] implies $u_{\bar{z}}(x_1,x_2,0) = 0$. Therefore, u is analytic everywhere, including infinity. Proposition 2.1 of [16] imposes the boundary condition $u = O\left(z^{-1}\right)$, as $z \to \infty$. From Liouville's theorem, it follows that the entire function u must be zero, hence

$$\int_0^{2\pi} J(x_1,x_2,\theta)\mathrm{d}\theta = 0. \quad (78)$$

Equation (77) implies that

$$\int_0^{2\pi} \mu(x_1,x_2)\left[e^{M(\tau,\rho,\theta)}G(\rho,\theta)\right]_{\substack{\tau=x_2\sin\theta+x_1\cos\theta\\ \rho=x_2\cos\theta-x_1\sin\theta}}\mathrm{d}\theta = 0,$$

which, in turn, implies that (76) simplifies to (66). Hence, Eq. (55a) is equivalent to Eq. (66). □

Acknowledgements This research was partially supported by the European Union and Greek national funds through the Operational Program Competitiveness, Entrepreneurship and Innovation, under the call RESEARCH-CREATE-INNOVATE (project code: TAEDK-06189). K. Kalimeris acknowledges support by the research programme 200/984 of the Research Committee of the Academy of Athens.

References

1. Radon, J.: Über die Bestimmung von Funktionen durch ihre Integralwerte längs gewisser Mannigfaltigkeiten. Akad. Wiss. **69**, 262–277 (1917)
2. Radon, J.: On the determination of functions from their integral values along certain manifolds. IEEE Trans. Med. Imag. **5**(4), 170–176 (1986). https://doi.org/10.1109/TMI.1986.4307775
3. Cormack, A.M.: Computed tomography: some history and recent developments. Proc. Sympos. Appl. Math. **27**, 35–42 (1982)

4. Bockwinkel, H.: On the propagation of light in a biaxial crystal about a midpoint of oscillation. Verh. Konink Acad. V. Wet. Wissen. Natur. **14**, 636–651 (1906)
5. Wininger, K.L.: History of mathematics special interest group of the Mathematical Association of America, HOMSIGMAA (2011)
6. Denecker, K., Van Overloop, J., Sommen, F.: The general quadratic Radon transform. Inverse Probl. **14**(3), 615 (1998)
7. Ramlau, R., Scherzer, O.: The Radon Transform: The First 100 Years and Beyond, vol. 22. Walter de Gruyter GmbH & Co KG (2019)
8. Kuchment, P.: The Radon transform and medical imaging. In: CBMS-NSF Regional Conference Series in Applied Mathematics, Society for Industrial and Applied Mathematics (2014)
9. Barrett, H.H.: The Radon transform and its applications. Prog. Opt. **21**, 217–286 (1984). https://doi.org/10.1016/S0079-6638(08)70123-9
10. Cherry, S.R., Sorenson, J.A., Phelps, M.E.: Physics in Nuclear Medicine, 4th edn. W.B. Saunders, Philadelphia (2012). https://doi.org/10.1016/B978-1-4160-5198-5.00033-2
11. Kastis, G.A., Kyriakopoulou, D., Gaitanis, A., Fernández, Y., Hutton, B.F., Fokas, A.S.: Evaluation of the spline reconstruction technique for PET. Med. Phys. **41**(4), 042501 (2014)
12. Iniewski, K.: Medical Imaging: Principles, Detectors, and Electronics. Wiley-Interscience, New York (2009)
13. Protonotarios, N.E., Fokas, A.S., Kostarelos, K., Kastis, G.A.: The attenuated spline reconstruction technique for single photon emission computed tomography. J. R. Soc. Interface **15**(148), 20180509 (2018). https://doi.org/10.1098/rsif.2018.0509
14. Fokas, A., Novikov, R.: Discrete analogues of δ-equation and of Radon transform. Comptes Rendus Acad. Sci. Série 1, Mathématique **313**(2), 75–80 (1991)
15. Novikov, R.: An inversion formula for the attenuated X-ray transformation. Ark. Mat. **40**(1), 145–167 (2002). https://doi.org/10.1007/BF02384507
16. Fokas, A., Iserles, A., Marinakis, V.: Reconstruction algorithm for single photon emission computed tomography and its numerical implementation. J. R. Soc. Interface **3**(6), 45–54 (2006). https://doi.org/10.1098/rsif.2005.0061
17. Natterer, F., Wuebbeling, F.: Mathematical Methods in Image Reconstruction. Society for Industrial and Applied Mathematics, Monographs on Mathematical Modeling and Computation (2001)
18. Ablowitz, M.J., Fokas, A.S., Fokas, A.: Complex Variables: Introduction and Applications. Cambridge University Press (2003)
19. Fokas, A.S.: A Unified Approach to Boundary Value Problems. Society for Industrial and Applied Mathematics, vol. 78 (2008). https://doi.org/10.1137/1.9780898717068
20. Protonotarios, N.E., Kastis, G.A., Fokas, A.S.: A new approach for the inversion of the attenuated radon transform. In: Mathematical Analysis and Applications, pp. 433–457. Springer (2019). https://doi.org/10.1007/978-3-030-31339-5_16

Inverse EEG Problem, Minimization and Numerical Solutions

Georgia Parakevopoulou, Athanassios S. Fokas, Antonios Charalambopoulos, and Stavros Perantonis

Abstract Mental processes are associated with brain activation, which in turn gives rise to neuronal electric currents, generating electric fields. Electroencephalography (EEG) is based on the measurements of the electric potential on the scalp and has a variety of neurophysiological and clinical applications. In recent years, there have been efforts to use EEG for determining the underlying neuronal electric current. This gives rise to a mathematical inverse problem for which one of the authors made pioneering contributions. In particular, he established that, although the EEG inverse problem suffers from lack of uniqueness, the assumption that the primary current minimizes the L^2-norm yields a unique solution. Here, we present the completion of the analytical formulation presented earlier by analyzing a specific boundary term that was neglected until now. Furthermore, a first attempt is made towards a more efficient use of a neural network approach for the numerical solution of this problem.

Keywords Inverse problems · Electroencephalography · Machine learning · Surrogate modelling · Artificial neural networks

G. Parakevopoulou (✉) · A. Charalambopoulos
School of Applied Mathematical and Physical Sciences, Department of Mathematics, National Technical University of Athens, Zografou Campus, 15780 Athens, Greece
e-mail: g_paraskevopoulou@mail.ntua.gr

A. S. Fokas
Department of Applied Mathematics and Theoretical Physics, University of Cambridge, Cambridge CB3 0WA, UK

Department of Electrical Engineering, University of Southern California, Los Angeles 90089, USA

S. Perantonis
Institute of Informatics and Telecommunications, National Center for Scientific Research - "Demokritos", Patriarchou Gregoriou E' and 27 Neapoleos Str, 15431 Agia Paraskevi, Greece

T. Bountis et al. (eds.), *Chaos, Fractals and Complexity*, Springer Proceedings in Complexity, https://doi.org/10.1007/978-3-031-37404-3_12

1 Neurophysiology of EEG

The fundamental cells of our nervous system are the neurons. Each neuron has a soma, an axon, and a large number of dendrites. Neurons communicate to each other through action potentials which are generated via the ionic currents flowing across the sodium and potassium channels of the neuronal membrane. When an action potential reaches the presynaptic terminal it triggers the release into the synaptic cleft of an appropriate neurotransmitter (the synaptic cleft is a tiny gap between the presynaptic axon terminal and the postsynaptic dendrite). After travelling across the synaptic cleft, the neurotransmitter attaches to the postsynaptic receptors. This gives rise to postsynaptic potentials which drive the postsynaptic cell away of its resting state and alters the probability that an action potential will be produced [1]. A post synaptic potential creates an extracellular voltage near the neural dendrites that is more negative than elsewhere along the neuron or vice versa, depending on the kind of neurotransmitter involved and the position of the synapse. In both cases a dipole is created [2]. Therefore, a specific mental process is associated with brain activation of a unique form which in turn expresses itself via the generation of a specific neuronal electric current.

Michael Faraday was the first to establish that there is a relation between electricity and magnetism. The precise mathematical form of this relationship is expressed by Maxwell's equations, which capture the following fact: an electric current gives rise to both a magnetic field and an electric potential. This implies that the activation of neurons generates a magnetic field and an electric potential. Electroencephalography (EEG), is based on the measurements of the electric potential on the scalp, which is generated by the primary current density distribution which arises from neuronal post-synaptic processes. Each electrode detects the sum of charges from a large number of neurons in their vicinity [2]. Since electrodes measure the sum of both the positive and negative charges beneath them, only neurons that are activated synchronously and are arranged in a parallel form produce a measurable signal [3]. There is much literature indicating that EEG measurements depend not only on the location of electrodes, but also on the specific mental function performed by the subject during the recordings [4].

2 The Importance of EEG

EEG has a variety of neurophysiological and medical applications. It is a minimally invasive technique, that provides almost real time information about mental processes. EEG's temporal resolution is better than 1 millisecond, which is several orders of magnitude better than the imaging techniques of fMRI, PET and SPECT. In most clinical applications EEG uses 19 electrodes, but for research purposes there exists a high-density EEG cap with 256 sensors.

In normal subjects there exist several characteristic rhythms, distinguished by their frequencies, depending on the state of the patient (e.g. alert wakefulness, drowsiness, sleep). It has also been shown that electroencephalograms obtained during different types of clinical conditions exhibit characteristic patterns [3]. For the above reasons, EEG is widely used for diagnosing and predicting many abnormal neurological conditions, such as epilepsy, sleep disorders and autism. In addition, neuroscientists use EEG to elucidate various mental processes [5] including the neuronal processes characterising various emotional states and the processes of decision making.

It is important to note that EEG is increasingly explored in the field of Brain-Computer Interfaces (BCI) in order to enable people with disabilities (including those who are not able to communicate with others) to control and direct mechanical and electronic devices, see for example [6]. EEG devices have also been used for educational purposes [7] or measuring the reading ability or confusion levels of students during several tasks [5]. As EEG devices are becoming inexpensive and more accessible, it is expected that the impact of EEG in research and medical fields will increase further. In addition to price, another important factor is the precision of the information that can be obtained via EEG. This work aims to enhance this important direction.

In recent years, there have been efforts to use EEG for the ambitious purpose of estimating the underlying neuronal electric current. This gives rise to the following inverse mathematical problem: assuming that the electric potential is known everywhere on the scalp, determine the electric current that gave rise to this potential. The problem of computing the electric potential for a given head model and a given configuration of dipole sources is known as the forward problem. The solution of this problem is a prerequisite for solving the inverse problem. There has been extensive research investigating the forward problem. In particular, the use of a boundary element solver (BEM), OpenMEEG [8] is well established. It solves the forward problem for an arbitrary piecewise homogeneous conductor and a set of dipole sources by analyzing the related boundary integral equations. Regarding the solution of the inverse problem one should mention the low-resolution brain electromagnetic tomography technique (sLORETA) [9], which uses a discrete formulation. Other discrete formulation approaches are presented in [10].

Fokas and collaborators have investigated extensively the forward and the inverse EEG problem [4]. In contrast to the above discrete approaches, they have modelled the neuronal electric current as a continuous vectorial function. Furthermore, they have decomposed the primary neuronal current into its irrotational and solenoidal components. This approach, apparently, provides a more accurate representation of the underlying physics. Importantly, Fokas [11] established that although the inverse problem suffers from lack of uniqueness, the assumption that the primary current minimizes the L^2-norm, yields a unique solution. Furthermore, under the assumption that the underlying primary neuronal current vanishes on the cortical surface S_c; a hybrid analytical-numerical solution for the inverse EEG problem is implemented in [12]. This assumption leads to the absence of a specific boundary term in the associative mathematical formulation.

Although the above assumption simplifies the mathematical formulation it cannot be justified for physical considerations. Thus, here we consider this model without this assumption. Furthermore, we present a modification of the neural network used for the regression of the electric potential on the scalp. We note that this step plays an essential role for the solution of the earlier model as well as for the analysis of the more complete model presented here. It is expected that the new modification will be important for the complete numerical solution which will be presented elsewhere.

3 Head Model

The first step for solving both the direct and the inverse problems is to model the head as an appropriate conductor. A standard model consists of the nested compartments that can be seen in Fig. 1. The bounded domain Ω_c represents the cerebrum. The shells Ω_f, Ω_b and Ω_s model the spaces occupied by the cerebrospinal fluid (CSF), the skull and the scalp, respectively. These compartments are distinguished by their different values of conductivity, which are denoted by σ_c, σ_f, σ_b and σ_s. The spaces Ω_c, Ω_f, Ω_b and Ω_s are bounded by the surfaces S_c, (S_c, S_f), (S_f, S_b) and (S_b, S_s), respectively. The domain exterior to the head is denoted by Ω_e and it is assumed that it is not conductive. Table 1 presents the conductivity values of the head model as documented in [12–14]. For numerical purposes, the domain Ω_f can be ignored, since the brain-CSF interface has a negligible effect in the forward model (see the detailed analysis of [15]).

Fig. 1 The different compartments of the head model

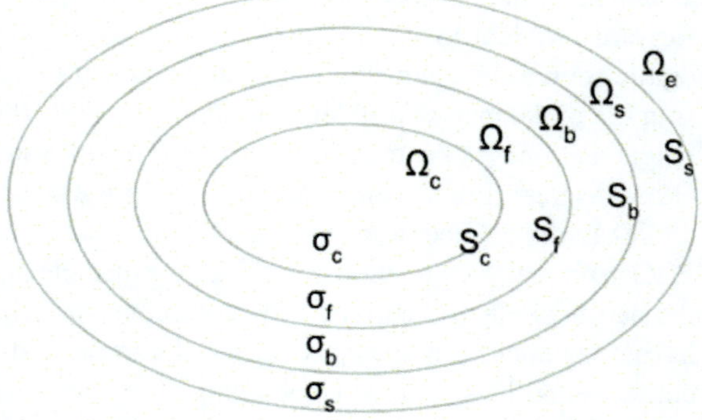

Table 1 The conductivity values for the different compartments of the head model

Domain Ω	Conductivities σ (S/m)
Cerebrum Ω_c	0.33
CSF Ω_f	1.0
Skull Ω_b	0.0042
Scalp Ω_s	0.33

4 Basic Equations for the EEG Formulation

Let $\mathbf{J}^p(\boldsymbol{\tau})$ denote the primary neuronal current, where $\boldsymbol{\tau} \in \Omega_c$. Under the assumption that J^p has sufficient smoothness, we can employ Helmholtz decomposition to express it in terms of its irrotational and solenoidal components:

$$\mathbf{J}^p(\boldsymbol{\tau}) = \nabla_{\boldsymbol{\tau}} \Psi(\boldsymbol{\tau}) + \nabla_{\boldsymbol{\tau}} \times \mathbf{A}(\boldsymbol{\tau}), \qquad \boldsymbol{\tau} \in \Omega_c, \tag{1}$$

where $\mathbf{A}(\boldsymbol{\tau})$ satisfies the constraint $\nabla_{\boldsymbol{\tau}} \cdot \mathbf{A}(\boldsymbol{\tau}) = 0$. The function $\Psi(\boldsymbol{\tau})$ is a scalar and characterises the irrotational part, whereas $\mathbf{A}(\boldsymbol{\tau})$ is a vector and characterises the solenoidal part of the current. Due to the constraint of its vanishing divergence, $\mathbf{A}(\boldsymbol{\tau})$ consists of two independent scalar functions.

The basic equation expressing the relation between the primary current $\mathbf{J}^p(\boldsymbol{\tau})$ and the electric potential on the scalp, u_s was derived in [11] and corrected in [4]:

$$u_s(\mathbf{r}) = \frac{1}{4\pi} \int_{\Omega_c} \mathbf{J}^p(\boldsymbol{\tau}) \cdot \nabla_{\boldsymbol{\tau}} v_s(\mathbf{r}, \boldsymbol{\tau}) dV(\boldsymbol{\tau}), \qquad \mathbf{r} \in S_s, \tag{2}$$

where $v_s(\mathbf{r}, \boldsymbol{\tau})$ is an auxiliary function that depends on the brain-head system compartments and their conductivities, but it does not depend on $\mathbf{J}^p(\boldsymbol{\tau})$.

The divergence integral theorem implies that

$$\int_{\Omega_c} \nabla_{\boldsymbol{\tau}} \cdot (v_s(\mathbf{r}, \boldsymbol{\tau}) \mathbf{J}^p(\boldsymbol{\tau})) dV(\boldsymbol{\tau}) = \int_{S_c} v_s(\mathbf{r}, \boldsymbol{\tau}) \hat{\mathbf{n}}(\boldsymbol{\tau}) \cdot \mathbf{J}^p(\boldsymbol{\tau}) dS(\boldsymbol{\tau}). \tag{3}$$

Hence, we obtain

$$u_s(\mathbf{r}) = -\tfrac{1}{4\pi} \textstyle\int_{\Omega_c} v_s(\mathbf{r}, \boldsymbol{\tau}) (\nabla_{\boldsymbol{\tau}} \cdot \mathbf{J}^p(\boldsymbol{\tau})) dV(\boldsymbol{\tau})$$

$$+ \frac{1}{4\pi} \int_{S_c} v_s(\mathbf{r}, \boldsymbol{\tau}) (\hat{\mathbf{n}}(\boldsymbol{\tau}) \cdot \mathbf{J}^p(\boldsymbol{\tau})) dS(\boldsymbol{\tau}), \qquad \mathbf{r} \in S_s. \tag{4}$$

Next, decomposing the current J^p via the Helmholtz representation, we find

$$\nabla_{\boldsymbol{\tau}} \cdot \mathbf{J}^p(\boldsymbol{\tau}) = \Delta_{\boldsymbol{\tau}} \Psi(\boldsymbol{\tau}). \tag{5}$$

Thus, (4) becomes

$$u_s(\mathbf{r}) = -\tfrac{1}{4\pi} \textstyle\int_{\Omega_c} v_s(\mathbf{r}, \tau) (\Delta_{\boldsymbol{\tau}} \Psi(\boldsymbol{\tau})) dV(\boldsymbol{\tau})$$

$$+ \frac{1}{4\pi} \int_{S_c} v_s(\mathbf{r}, \boldsymbol{\tau}) (\hat{\mathbf{n}}(\boldsymbol{\tau}) \cdot \mathbf{J}^p(\boldsymbol{\tau})) dS(\boldsymbol{\tau}), \qquad \mathbf{r} \in S_s. \tag{6}$$

The volume integral of the right hand side of (6) can further be expressed as a surface integral S_c, using Green's second identity and the fact that the auxiliary function $v_s(\mathbf{r}, \boldsymbol{\tau})$ is harmonic [4]:

$$\int_{\Omega_c} v_s(\mathbf{r},\boldsymbol{\tau})(\Delta_{\tau}\Psi(\boldsymbol{\tau}))dV(\boldsymbol{\tau})$$

$$= \int_{S_c} \hat{\mathbf{n}}(\boldsymbol{\tau}) \cdot [v_s(\mathbf{r},\boldsymbol{\tau}) \nabla_{\boldsymbol{\tau}} \Psi(\boldsymbol{\tau}) - \Psi(\boldsymbol{\tau}) \nabla_{\boldsymbol{\tau}} v_s(\mathbf{r},\boldsymbol{\tau})]dS(\boldsymbol{\tau}). \tag{7}$$

Consequently,

$$u_s(\mathbf{r}) = \frac{1}{4\pi}\int_{S_c} \hat{\mathbf{n}}(\boldsymbol{\tau}) \cdot [\Psi(\boldsymbol{\tau}) \nabla_{\boldsymbol{\tau}} v_s(\mathbf{r},\boldsymbol{\tau}) - v_s(\mathbf{r},\boldsymbol{\tau}) \nabla_{\boldsymbol{\tau}} \Psi(\boldsymbol{\tau}) + v_s(\mathbf{r},\tau)\mathbf{J}^p(\boldsymbol{\tau})]dS(\boldsymbol{\tau}), \quad \mathbf{r} \in S_s. \tag{8}$$

In view of the decomposition (1), Eq. (8) can be written as

$$u_s(\mathbf{r}) = \frac{1}{4\pi}\int_{S_c} \hat{\mathbf{n}}(\boldsymbol{\tau}) \cdot [\Psi(\boldsymbol{\tau}) \nabla_{\boldsymbol{\tau}} v_s(\mathbf{r},\boldsymbol{\tau}) + v_s(\mathbf{r},\boldsymbol{\tau}) \nabla_{\boldsymbol{\tau}} \times\mathbf{A}(\boldsymbol{\tau})]dS(\boldsymbol{\tau}), \quad \mathbf{r} \in S_s. \tag{9}$$

This shows that the potential u_s depends on both the scalar function Ψ and the tangential components of the vector function **A**.

Following [4], in order to decouple the EEG problems we assume that **A** satisfies the condition

$$\hat{\mathbf{n}}(\boldsymbol{\tau}) \cdot \nabla_{\boldsymbol{\tau}} \times \mathbf{A}(\boldsymbol{\tau}) = 0, \qquad \boldsymbol{\tau} \in S_c. \tag{10}$$

This implies that only the scalar function Ψ affects the EEG data:

$$u_s(\mathbf{r}) = \frac{1}{4\pi}\int_{S_c} \hat{\mathbf{n}}(\boldsymbol{\tau}) \cdot [\Psi(\boldsymbol{\tau}) \nabla_{\boldsymbol{\tau}} v_s(\mathbf{r},\tau)]dS(\boldsymbol{\tau}), \qquad \mathbf{r} \in S_s. \tag{11}$$

This equation can be rewritten:

$$u_s(\mathbf{r}) = \frac{1}{4\pi}\int_{\Omega_c} [\nabla_{\boldsymbol{\tau}}\Psi(\boldsymbol{\tau}) \cdot \nabla_{\boldsymbol{\tau}} v_s(\mathbf{r},\boldsymbol{\tau})]dV(\boldsymbol{\tau}), \qquad \mathbf{r} \in S_s. \tag{12}$$

At this point it should be mentioned that the main difficulty for determining brain activity using EEG is the highly ill-posed nature of the associated spatial inverse problems: different electric currents can yield identical measurements for the electric potential. This follows from Eqs. (11) and (12).

Interestingly, using the natural principle of minimal energy, we impose that the current minimizes the L^2-and this leads to a unique current. In this connection, we define the functional E (energy) by

$$E = \int_{\Omega_c} |\mathbf{J}^p|^2 dV(\boldsymbol{\tau}). \tag{13}$$

Using again the Helmholtz decomposition (1), E can be written in the form

$$E = \int_{\Omega_c} [|\nabla_{\boldsymbol{\tau}} \Psi|^2 + |\nabla_{\boldsymbol{\tau}} \times\mathbf{A}|^2 + 2\nabla_{\boldsymbol{\tau}} \Psi \cdot (\nabla_{\boldsymbol{\tau}} \times \mathbf{A})]dV(\boldsymbol{\tau}). \tag{14}$$

However, it is shown in Proposition 9.1 and Lemma 1 of [16] that

$$\int_{\Omega_c} \nabla_\tau \Psi \cdot (\nabla_\tau \times \mathbf{A}) dV = 0. \tag{15}$$

Thus

$$E = \int_{\Omega_c} (| \nabla_\tau \Psi |^2 + | \nabla_\tau \times \mathbf{A}|^2) dV(\tau). \tag{16}$$

Moreover, considering that the electric potential on the scalp depends only on Ψ (see Eqs. (11), (12)), it follows that the minimization of E is equivalent to minimizing

$$E = \int_{\Omega_c} | \nabla_\tau \Psi|^2 dV(\tau). \tag{17}$$

In summary, the unique solution of the inverse problem of EEG is based on the solution of the following minimization problem:

$$minimize \int_{\Omega_c} | \nabla_\tau \Psi|^2 dV(\tau), \tag{18}$$

under the constraint that

$$u_s(\mathbf{r}) = \frac{1}{4\pi} \int_{S_c} \hat{\mathbf{n}}(\tau) \cdot [\Psi(\tau) \nabla_\tau v_s(\mathbf{r}, \tau)] dS(\tau), \qquad \mathbf{r} \in S_s.$$

5 Configuration of the Numerical Solution

For the numerical implementation of the constrained minimization problem (18), the first step is to consider an electrode cap with M electrodes and to discritize the cerebrum region Ω_c using N cubic voxels. For this purpose, triangulated surface meshes for the cerebrum, skull and head are needed. A visualization of the above steps is available in Fig. 2.

The next step is to compute the values of the auxiliary function v_s following [12]:

$$v_s(\mathbf{r}, \tau) - v_s(\mathbf{r}, 0) = 4\pi \int_0^T u_s(\mathbf{r}, t\hat{\tau}, \hat{\tau}) dt, \tag{19}$$

where $\hat{\tau}$ is the unit normal vector of τ. The centroid $\mathbf{c}$ is estimated from the triangulated cerebrum surface mesh, by fitting a sphere to its nodes; then $v_s(\mathbf{r}, 0) \approx 0$.

In order to compute the line integral of (19), a dataset $u_s(r, \tau, \hat{\mathbf{r}} \cdot \hat{\tau})$ is generated via OPENMEEG. Furthermore, a surrogate model is used, which is a machine learning model for regression of the function u_s. A two layers neural network was trained for the regression of u_s function, depending on the inputs $(r, \tau, \hat{\mathbf{r}} \cdot \hat{\tau})$. The variable r is the radial distance from the center of the coordinate system to the sensor position, τ is the

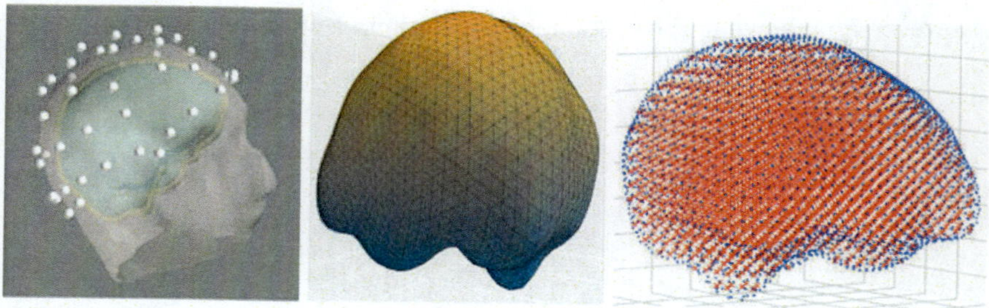

Fig. 2 **a** Visualization of the realistic head model, taken from the sample dataset in the OpenMEEG package. Electrodes are represented by white spheres. **b** Example of triangulated surface of cerebrum with 2562 nodes and 5120 triangles. **c** Example of voxelized cerebrum mesh with 3.293 voxels, marked with red crosses

Table 2 Experimental results of the surrogate model. The proposed NN provides a better approximation of u_s function than the one obtained in [12]. It is expected that this modification will be important for the full numerical solution which will be presented in a future publication

Metric	Hashemzadeh et al. (2020)	Proposed NN
RDM	0.2102	0.1781
ln(MAG)	−0.0208	−0.0159
RMSE	3.3396	2.8955

radial distance from the center of the coordinate system to the source position vector, and $(r, \tau, \hat{\mathbf{r}} \cdot \hat{\tau})$ is the cosine of the angle between the source and observation unit vectors. The relative mean distance (RDM) and natural logarithm of the magnification factor (ln(MAG)) were employed as metrics to evaluate the surrogate model. Root mean square error (RMSE) was also computed (see Table 2). The function Ψ was expanded using radial basis functions (RBFs). Using synthetic data and minimizing the energy, the function Ψ was reconstructed, achieving (RMSE) $= 0.1122$.

It is noted that a hybrid numerical-analytical solution of the inverse EEG problem was published [12], using the extra assumption that the primary current vanishes on the cortical surface. The minimization problem (18), presented in this work, represents the solution of the inverse EEG problem without the assumption that $\mathbf{J}^p$ vanishes on the cortical surface. Following the aforementioned steps, this version of EEG problem can be solved. The surrogate model's step is exactly the same in both approaches. In this connection, we are using the neural network (NN) of Fig. 3, which uses ReLU (rectified linear unit) [17] as an activation function and the Adam optimizer [18]. It is trained for a maximum of 4000 epochs with a batch size of 400, using early stopping to avoid overfitting. The sample dataset in the OpenMEEG package was adopted for comparison reasons. Standardization of inputs was also applied before the training process. Results of the proposed NN after 5-fold cross-validation can be seen in Table 2 in comparison with the previous work's results published in [12].

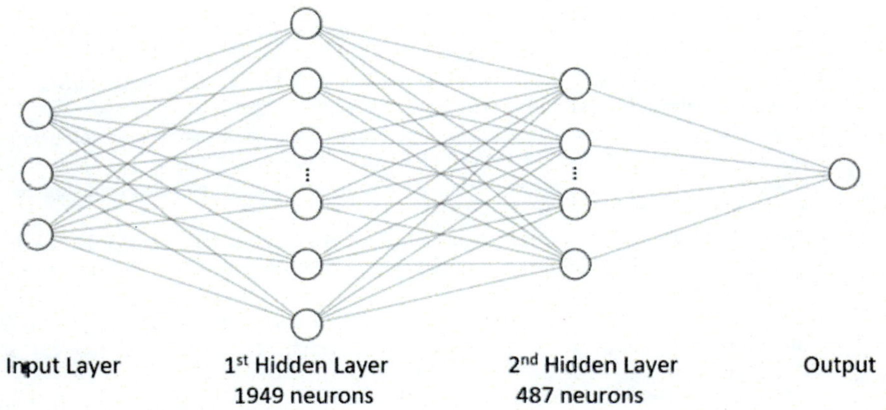

Fig. 3 Visualization of proposed neural network

6 Future Work

As stated earlier, our aim is to solve the inverse EEG problem without the assumption that $\mathbf{J}^p$ vanishes on cortical surface, expecting a more accurate reconstruction of the only function Ψ that affects the EEG data, in terms of root mean square error (RMSE). Our algorithm will be tested using real data of human EEG recordings. In a following publication an algorithm will be presented that can easily be applied for different datasets and headmodels.

References

1. Holmes, G.L., Khazipov, R.: Basic neurophysiology and the cortical basis of EEG. Clin. Neurophysiol. Primer 19–33. (2007). https://doi.org/10.1007/978-1-59745-271-7_2
2. Jackson, A.F., Bolger, D.J.: The neurophysiological bases of EEG and EEG measurement: a review for the rest of us. Psychophysiology **51**(11), 1061–71 (2014). https://doi.org/10.1111/psyp.12283
3. Britton, J.W., Frey, L.C., Hopp, J.L. et al.: Electroencephalography (EEG): an introductory text and atlas of normal and abnormal findings in adults, children, and infants. In.: St. Louis, E.K., Frey, L.C. (eds.) American Epilepsy Society (2016)
4. Dassios, G., Fokas, A.S.: Electro-Encephalography and Magneto-Encephalography: An Analytical-Numerical Approach. De Gruyter, Boston (2020)
5. Soufineyestani, M., Dowling, D., Khan, A.: Electroencephalography (EEG) technology applications and available devices. Appl. Sci. **10**(21), 1–23 (2020). https://doi.org/10.3390/app10217453
6. Carrino, F., Dumoulin, J., Mugellini, E., Khaled, O.A., Ingold, R.: A self-paced BCI system to control an electric wheelchair: evaluation of a commercial, low-cost EEG device. In: 2012 ISSNIP Biosignals and Biorobotics Conference: Biosignals and Robotics for Better and Safer Living (BRC) (2012). https://doi.org/10.1109/brc.2012.6222185
7. Frey, J., Gervais, R., Lainé, T., Duluc, M., Germain, H., Fleck, S., Lotte, F., Hachet, M.: Scientific outreach with Teegi, a tangible EEG interface to talk about neurotechnologies. In:

Proceedings of the 2017 CHI Conference Extended Abstracts on Human Factors in Computing Systems - CHI EA '17 (2017). https://doi.org/10.1145/3027063.3052971
8. Gramfort, A., Papadopoulo, T., Olivi, E., Clerc, M.: OpenMEEG opensource software for quasistatic bioelectromagnetics. BioMed. Eng. OnLine **9**(45) (2010)
9. Pascual-Marqui, R.D.: Standardized low-resolution brain electromagnetic tomography (sLORETA): technical details., Methods Find Exp Clin Pharmacol (2002)
10. Darbas M., Lohrengel S.: Review on mathematical modelling of electroencephalography. Jahresber. Dtsch. Math.-Ver. **121**, 3–39 (2019)
11. Fokas, A.: Electro-magneto-encephalography for a three-shell model: distributed current in arbitrary, spherical and ellipsoidal geometries. J. Roy. Soc. Interface **6**(34), 479–488. (2009). https://doi.org/10.1098/rsif.2008.0309
12. Hashemzadeh, P., Fokas, A.S., Schönlieb, C.B.: A hybrid analytical-numerical algorithm for determining the neuronal current via electroencephalography. J. Roy. Soc. Interface **17**(163) (2020). https://doi.org/10.1098/rsif.2019.0831
13. Hashemzadeh, P., Fokas, A.S.: Helmholtz decomposition of the neuronal current for the ellipsoidal head model. Inverse Prob. **35**(2) (2019)
14. Mosher, J.C., Leahy, R.M., Lewis, P.S.: EEG and MEG: forward solutions for inverse methods. IEEE Trans. Biomed. Eng. **46**(3) (1999)
15. Kybic, J., Clerc, M., Abboud, T., Faugeras, O., Keriven, R., Papadopoulo, T.: A common formalism for the integral formulations of the forward problem. IEEE Trans. Med. Imaging **24**(1), 12–18 (2005)
16. Cantarella, J., DeTurck, D., Gluck, H.: Vector calculus and the topology of domains in 3-space. Am. Math. Mon. **109**(5), 409–442 (2002)
17. Agarap, A.: Deep Learning using Rectified Linear Units (ReLU) (2018)
18. Kingma, D.P., Ba, J.: Adam: A Method for Stochastic Optimization (2014). CoRR, abs/1412.6980

Traveling Waves in Flowing Sand: The Dynamical Systems Approach

Ko van der Weele, Dimitrios Razis, and Giorgos Kanellopoulos

Abstract An overview is given of the various types of traveling surface waves that may be encountered in a shallow sheet of dry granular matter flowing down a chute. On the basis of the generalized Saint-Venant equations, and assuming that the material flows continuously and nowhere stagnates, we derive a dynamical system capturing all traveling waveforms that can possibly occur in the sheet. In particular, this approach enables us to elucidate the transition from a monoclinal flood wave to a periodic train of roll waves as the Froude number F_0 of the incoming flow is gradually increased through the critical value F_{cr} (which equals $2/3$ for smooth spherical particles). We show that this transition involves a series of intermediate stages, including an "undular bore" and a "solitary roll wave" that had hitherto not been reported for granular flows.

Keywords Granular chute flow · Traveling waves · Shallow water model · Dynamical systems approach

1 Introduction

Chute flow of dry granular matter is ubiquitous in agriculture, mining, and numerous other human activities where particles are transported [1–3]. In nature, the chute may be as large as a mountainside and the flow in question may take the form of a devastating landslide or rock avalanche [4–6]. One can also generate such flows in the laboratory (see Fig. 1); typically, the thickness of the granular sheet will be in the order of 10 to 15 particle diameters, and on its surface one may witness the formation

K. van der Weele (✉) · D. Razis · G. Kanellopoulos
Department of Mathematics, University of Patras, 26500 Patras, Greece
e-mail: weele@math.upatras.gr

D. Razis
Department of Civil Engineering, University of Thessaly, 38334 Volos, Greece

Department of Physics, National and Kapodistrian University of Athens, 15784 Zografou Athens, Greece

T. Bountis et al. (eds.), *Chaos, Fractals and Complexity*, Springer Proceedings in Complexity, https://doi.org/10.1007/978-3-031-37404-3_13

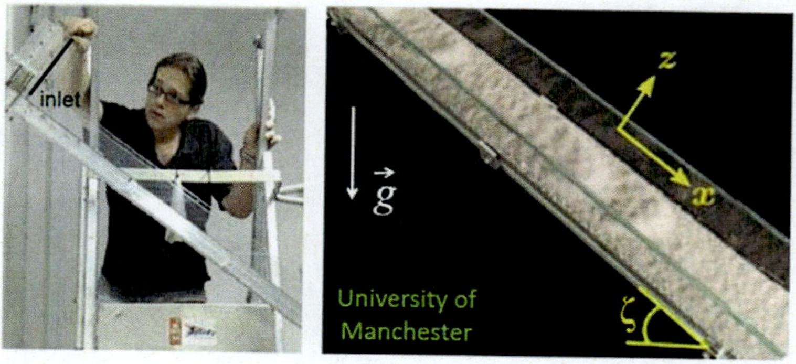

Fig. 1 Granular chute flow in the laboratory of the Manchester Centre for Nonlinear Dynamics. Left: The "sluice gate" of the reservoir at the top is lifted (by the second author) to a certain small height, upon which a thin sheet of granular particles flows downwards. The chute has an inclination angle of ζ and has intentionally been roughened by glueing a layer of particles on it. Right: Oblique view of the flow, showing three roll waves that have appeared spontaneously on the surface of the flowing sheet

of waves of relatively long wavelength and small amplitude. In the present paper we will treat the granular material as an incompressible fluid, assuming constant density throughout the flowing sheet. For the flows under consideration this is a very accurate approximation. In a more general context, however, this condition could be seriously violated, e.g. when a rapid flow of particles meets an obstacle and the grains are catapulted into ballistic flights [7] or when a fast and thin granular flow undergoes a "hydraulic jump" [8]; on such occasions, considerable density variations will be observed.

The fact that—for the flows considered here—the wavelength of the waves is considerably larger than the thickness of the sheet, which in turn is markedly larger than the wave amplitude, means that the situation is well suited for employing the so-called shallow water approximation, in which the flow is described by just two quantities: (*a*) the height of the sheet $h(x, t)$ and (*b*) the depth-averaged velocity $\bar{u}(x, t)$. These quantities are governed by two equations known as the generalized Saint-Venant equations for granular chute flow [9, 10]. This is a pair of coupled nonlinear partial differential equations for $h(x, t)$ and $\bar{u}(x, t)$, namely, the mass balance (or continuity equation):

$$\frac{\partial h}{\partial t} + \frac{\partial}{\partial x}(h\bar{u}) = 0, \tag{1}$$

and the momentum balance (or equation of motion):

$$\frac{\partial}{\partial t}(h\bar{u}) + \frac{\partial}{\partial x}(h\bar{u}^2) = gh\sin\zeta - \frac{\partial}{\partial x}\left(\frac{1}{2}gh^2\cos\zeta\right) \\ -\mu(h,\bar{u})gh\cos\zeta + \frac{\partial}{\partial x}\left(\nu h^{3/2}\frac{\partial\bar{u}}{\partial x}\right). \tag{2}$$

The mass balance expresses the fact that a net influx in any control volume within the sheet results in an increase of the local height $h(x, t)$, and vice versa. The momentum balance states that the rate of change of momentum of the granular matter in the control volume is equal to the sum of the forces acting upon it. In order of appearance these forces are (i) the gravity component in the x-direction (i.e., along the chute), (ii) the gradient of the depth-averaged pressure, (iii) the friction exerted by the chute, with $\mu(h, \bar{u})$ a variable friction coefficient introduced by Pouliquen and Forterre specifically for granular chute flow [11], and (iv) a diffusive term stemming from the in-plane stresses in the sheet, introduced by Gray and Edwards on the basis of $\mu(I)$-rheology [9]. The analytical expressions for $\mu(h, \bar{u})$ and $\nu(\zeta)$, which embody the various special properties of flowing granular materials as opposed to ordinary Newtonian fluids, are quite complicated and the reader is referred to the original literature for details [9–13].

During the past few decades, various waveforms have been found in granular chute flow. The granular counterpart of the hydraulic jump was first investigated in 1983 by Brennen et al. [8], granular roll waves were studied by Forterre and Pouliquen in 2003 [14], granular surface waves with intermediate stopping regions were examined by Edwards and Gray in 2015 [15], and the granular version of the monoclinal flood wave was predicted in 2018 by Razis et al. [12]. These waveforms were discovered separately, and described as individual case studies, until in 2019 a unified view was presented by Razis et al. [13] making use of the so-called dynamical systems approach. Apart from unifying the already discovered waveforms, this approach also brought to light several types of waves that were thus far unknown in the granular context. For reasons of brevity, we will restrict the discussion to *traveling* waves for sheets consisting of monodisperse, smooth spherical particles. We further assume that the sheet is "fully dynamic", meaning that it is everywhere in motion and does not exhibit any stopping regions.

In Sect. 2 we describe the granular monoclinal wave. In Sects. 3 and 4 we then derive the dynamical system that captures all steady traveling waveforms that can possibly occur within the framework of the generalized Saint-Venant equations (1)–(2). Then, in Sect. 5, we present the succession of different traveling waveforms as the system parameters are varied gradually. In particular, we show the transition from a monoclinal wave to periodic roll waves as the Froude number F_0 of the incoming flow is raised through the critical value 2/3. According to our model this transition involves several intermediate stages including an "undular bore" and a "solitary roll wave", which still await experimental verification. Finally, Sect. 6 contains concluding remarks.

2 Steady Uniform Flow and the Monoclinal Flood Wave

The basic solution to the above Eqs. (1)–(2) is the steady uniform flow, where the granular sheet has a constant thickness $h(x, t) = h_0$ and a corresponding constant depth-averaged velocity $\bar{u}(x, t) = \bar{u}_0$. For this solution, all derivatives vanish (both with respect to x and to t), and hence the mass balance Eq. (1) is trivially satisfied. As for the momentum balance Eq. (2), this reduces to $0 = gh\sin\zeta - \mu(h,\bar{u})gh\cos\zeta$, i.e., a simple balance between the forces of gravity and friction. This balance can also be written in the form $\mu(h,\bar{u}) = \tan\zeta$, which yields—using the explicit expression for $\mu(h,\bar{u})$ [10]—the relation between the depth-averaged velocity and the thickness under uniform flow conditions, $\bar{u}_0(h_0) \sim h_0^{3/2}$.

The uniform flow turns out to be stable as long as the Froude number, which measures the relative importance of the inertial forces versus the gravitational force in the direction of the chute,

$$F(x,t) = \frac{\bar{u}(x,t)}{\sqrt{h(x,t)g\cos\zeta}}, \tag{3}$$

lies in the interval $F_{\text{stop}} < F(x,t) < 2/3$. Below F_{stop} the flow is prone to developing stopping regions, which forms a fascinating subject [9] but will not be pursued here; our interest lies with granular sheets that are fully dynamic. We should also mention that, for the sake of simplicity we disregard a possible offset of the Froude number [14, 16–18], meaning that the results presented here concern the flow of smooth glass beads rather than the irregularly shaped particles of sand or fragmented mineral ores such as carborundum.

For the same range $F_{\text{stop}} < F(x,t) < 2/3$ we may also, starting from a uniform flow, generate a combination of *two* uniform flows of different thickness by opening the inlet a bit further, as shown in Fig. 2. The result is a traveling shock wave connecting two uniform flows of thickness h_0 and $h_+ < h_0$, respectively, known as the "granular monoclinal wave" [12]. Its wave speed is given by

$$c = \frac{h_0\bar{u}_0 - h_+\bar{u}_+}{h_0 - h_+}, \tag{4}$$

in agreement with the Rankine-Hugoniot conditions across the shock.

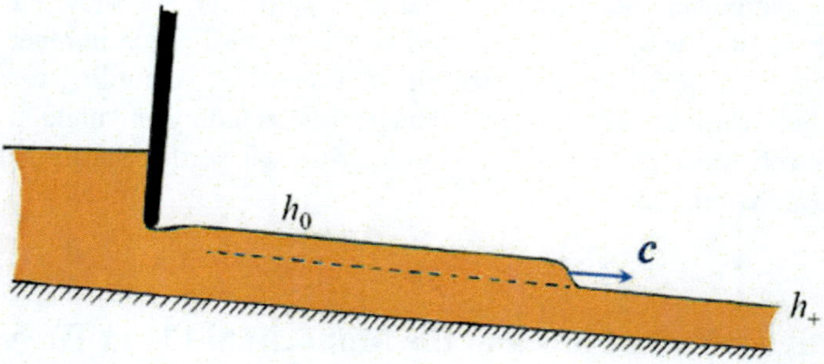

Fig. 2 Starting from a uniform flow situation, one can generate a flood wave by opening the inlet at the top of the chute a bit further. The resulting waveform is a traveling shock wave known as the granular monoclinal wave, connecting two uniform flows of different heights. It propagates along the chute with the shock speed c given by Eq. (4)

In the flat regions of the monoclinal waveform, the only forces acting on the sheet are gravity and friction, just as for the uniform flow. In the shock region, however, *all* forces of Eq. (2) contribute. Gravity and friction are still the main players, yet now also the pressure gradient, inertial forces (from the deceleration of the sheet in the shock region) and the diffusive term come into effect. The latter is by far the smallest but nonetheless quite essential, since without it, the shock wave could become infinitely steep, and even break [12].

It is interesting to note that the monoclinal wave has a mechanical analogue in the so-called antikink soliton propagating along an array of pendulums coupled via torsion springs, modeled by the sine-Gordon equation [19]. Also this antikink is a nonlinear wave connecting two plateau values (both corresponding to the downward equilibrium position of the pendulums, but differing by 2π radians) in a smooth, monotonically descending fashion.

3 One Equation to Rule All Traveling Waveforms

3.1 Focusing on Traveling Wave Solutions

In order to find possible further traveling waveforms, we introduce the traveling wave variable $\xi \equiv x - ct$ (where c stands for the wave speed) and limit our search to solutions of the form $h(x, t) = h(x - ct) = h(\xi)$, $\bar{u}(x, t) = \bar{u}(x - ct) = \bar{u}(\xi)$.

So the two independent variables x and t have now been combined and form a single variable. Accordingly, the mass conservation Eq. (1) takes the form of an ordinary differential equation (ODE):

$$-ch' + (h\bar{u})' = 0, \tag{5}$$

with the prime denoting differentiation with respect to ξ. This ODE is readily integrated to give

$$-ch + h\bar{u} = -Q, \tag{6}$$

where the integration constant $-Q = h(\bar{u} - c)$ is the constant flux of material per unit width measured in the co-moving frame.

Now, with $\bar{u} = c - Qh^{-1}$ (and hence $\bar{u}' = Qh^{-2}h'$, etc.) we can eliminate $\bar{u}$ and all its derivatives from the momentum balance Eq. (2), which then becomes

$$\frac{\nu h''}{Qh^{3/2}} - \frac{\nu\,(h')^2}{2Qh^{5/2}} + \left(\frac{1}{h^3} - \frac{g\cos\zeta}{Q^2}\right)h' + \frac{1}{Q^2}\Big(g\sin\zeta - \mu(h)g\cos\zeta\Big) = 0. \tag{7}$$

This is a second-order ODE for $h(\xi)$, which—within the model used here—governs all traveling waveforms on the chute. Interestingly, in the absence of diffusion ($\nu = 0$), the ODE (7) would only be of first order, which would severely restrict the possible variety of waveforms.

3.2 Non-dimensional Form

It is convenient and insightful to write Eq. (7) in non-dimensional form. This can be done by expressing all length scales in terms of h_0 (the height of the incoming base flow), i.e., $\xi \to \widetilde{\xi} = \xi/h_0$ and $h \to \widetilde{h} = h/h_0$, which also means that the wave speed is rescaled as $c \to \widetilde{c} = c/\bar{u}_0$, where $\bar{u}_0$ is the velocity of the base flow. We then get the following non-dimensional form of Eq. (7) [9, 13]:

$$\frac{1}{R}\left(\widetilde{h}'' - \frac{(\widetilde{h}')^2}{2\widetilde{h}}\right) - \frac{\widetilde{h}^{3/2}}{F_0^2\,(\widetilde{c}-1)}\left(\left(\frac{F_0^2\,(\widetilde{c}-1)^2}{\widetilde{h}^3} - 1\right)\widetilde{h}' + \tan\zeta - \mu(\widetilde{h})\right) = 0, \tag{8}$$

where the prime now stands for differentiation with respect to $\widetilde{\xi}$, the R denotes the "granular Reynolds number" [9], and F_0 is the Froude number of the base flow. Using the identity $R = 9F_0^2/(2\tan\zeta)$ [13], Eq. (8) can also be written as:

$$\widetilde{h}'' - \frac{(\widetilde{h}')^2}{2\widetilde{h}} - \frac{9\widetilde{h}^{3/2}}{2\,(\widetilde{c}-1)\tan\zeta}\left(\left(\frac{F_0^2\,(\widetilde{c}-1)^2}{\widetilde{h}^3} - 1\right)\widetilde{h}' + \tan\zeta - \mu(\widetilde{h})\right) = 0. \tag{9}$$

The above equation contains three dimensionless parameters: ζ (the inclination angle of the chute), F_0 (the Froude number of the uniform base flow), and $\widetilde{c}$ (the dimensionless wave speed). Instead of $\widetilde{c}$ one may just as well use $\widetilde{h}_+$ (the height of the downstream uniform part of the monoclinal waveform, see Fig. 2), since $\widetilde{c}$ is a single-valued function of $\widetilde{h}_+$ as can be seen from the non-dimensional form of Eq. (4): $\widetilde{c} = \widetilde{c}(\widetilde{h}_+) = (1 - \widetilde{h}_+^{5/2})/(1 - \widetilde{h}_+)$. In what follows we will sometimes use $\widetilde{c}$ and sometimes $\widetilde{h}_+$, depending on the occasion.

Equation (9) is a nonlinear and highly nontrivial ODE, so we will not attempt to solve it analytically but rather take the geometric approach of Dynamical Systems, where the solutions present themselves as trajectories in phase space. This has in fact been the main theme of our research programme of the past five years: to apply the methodology of dynamical systems to fluid mechanical problems [13, 17, 18, 20–22]. Since the ODE (9) is of second order and autonomous, the corresponding phase space will be two-dimensional.

4 Dynamical Systems Approach

Just like any second-order autonomous ODE, Eq. (9) can be written as a system of two coupled first-order ODEs:

$$\widetilde{h}' = \widetilde{s}, \tag{10a}$$

$$\widetilde{s}' = \frac{\widetilde{s}^2}{2\widetilde{h}} - \frac{9\widetilde{h}^{3/2}}{2(\widetilde{c}-1)\tan\zeta}\left(\left(\frac{F_0^2(\widetilde{c}-1)^2}{\widetilde{h}^3} - 1\right)\widetilde{s} + \tan\zeta - \mu(\widetilde{h})\right), \tag{10b}$$

where we have chosen the symbol $\widetilde{s}$ to represent the slope $d\widetilde{h}/d\widetilde{\xi}$ of the height profile $\widetilde{h}(\widetilde{\xi})$. Equations (10a) and (10b) constitute an autonomous dynamical system in the $(\widetilde{h}, \widetilde{s})$ phase plane.

The fixed points of this system are found by setting both $\widetilde{h}' = 0$ and $\widetilde{s}' = 0$. The first one of these conditions means that $\widetilde{s} = 0$ (for any fixed point), so the fixed points correspond to flat regions of the flow, where the slope is zero. Inserting $\widetilde{s} = 0$ into the second condition gives two fixed points: $(\widetilde{h}, \widetilde{s}) = (1, 0)$ (corresponding to the incoming base flow) and $(\widetilde{h}, \widetilde{s}) = (\widetilde{h}_+, 0)$ (representing the downstream part of the monoclinal waveform). A linear stability analysis reveals that the latter fixed point is a saddle for all parameter values of interest. The point $(1, 0)$, however, can be anything (a node, a spiral, a center, a saddle, or a hybrid borderline case) depending on the parameters ζ, F_0 and $\widetilde{h}_+$. The full range of possibilities is depicted in Fig. 3, which shows the $(F_0, \widetilde{h}_+)$ parameter diagram (at a fixed value $\zeta = 33.3°$) in the neighbourhood of the multi-critical point $(F_0, \widetilde{h}_+) = (2/3, 1)$. In the next section we will follow a typical path through this diagram, namely the vertical path at $\widetilde{h}_+ = 0.999$, for increasing values of F_0. This will illustrate how the monoclinal wave is destabilized and, via a series of intermediate stages, eventually is replaced by a periodic train of roll waves.

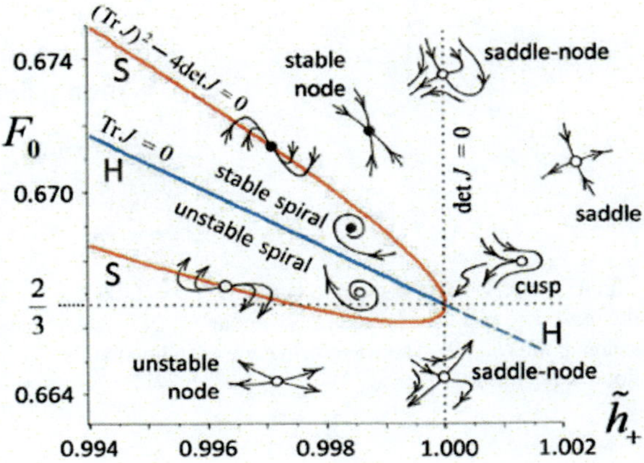

Fig. 3 The parameter diagram F_0 vs. $\widetilde{h}_+$ (at a fixed value $\zeta = 33.3°$ of the inclination angle), showing the varying nature of the fixed point $(\widetilde{h}, \widetilde{s}) = (1, 0)$ of the dynamical system (10a)-(10b). In this picture, the reader may recognize the famous Poincaré diagram for the classification of fixed points in two dimensions

5 From Monoclinal Waves to Roll Waves

5.1 *Stage 1: The Monoclinal Wave*

We start from the value $F_0 = 0.664$, just below the critical value $2/3$, and—as mentioned before—we will keep $\widetilde{h}_+ = 0.999$ and $\zeta = 33.3^\circ$ fixed along the entire path. At this first stage, the fixed point $(1, 0)$ is an *unstable node*, while $(\widetilde{h}_+, 0)$ is a saddle point as always. The $(\widetilde{h}, \widetilde{s})$ phase plane around these two fixed points is shown in Fig. 4. All trajectories in the phase plane correspond to solutions of the dynamical system (10a)-(10b), but one trajectory is of special interest, namely the heteroclinic orbit (solid red curve) connecting the fixed points $(1, 0)$ and $(\widetilde{h}_+, 0)$. The form of this orbit tells us that the incoming base flow (with height $\widetilde{h}_0 = 1$) connects to the downstream flow (with height $\widetilde{h}_+$) in such a way that the slope $\widetilde{s}$ is always negative. The waveform associated with this orbit is precisely the monoclinal wave which we already encountered in the context of Fig. 2.

In normal fluids, the monoclinal flood wave is a well-known phenomenon, being the generic waveform with which a discharge of water propagates along a channel [23]. In granular chute flow, however, it has yet to be observed in experiment.

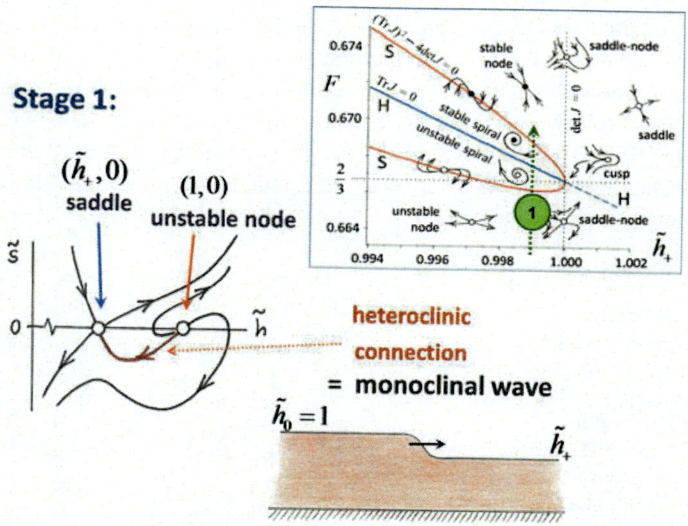

Fig. 4 Stage 1: For the parameter values $F_0 = 0.664$, $\widetilde{h}_+ = 0.999$ (indicated by a green disk in the inset) and $\zeta = 33.3^\circ$, the unstable node $(1, 0)$ is seen to have a heteroclinic connection with the saddle point $(\widetilde{h}_+, 0)$. The waveform corresponding to this heteroclinic connection is the monoclinal flood wave (cf. Fig. 2)

5.2 *Stage 2: The Undular Bore*

When we increase the Froude number of the incoming base flow to $F_0 = 0.666$, extremely close to the critical value $2/3$, we witness that the fixed point $(1, 0)$ turns into an *unstable spiral* (while $(\widetilde{h}_+, 0)$ remains a saddle point). In the $(\widetilde{h}, \widetilde{s})$ phase plane of Fig. 5 this means that the heteroclinic orbit now first spirals around $(1, 0)$ before it connects to the saddle at $(\widetilde{h}_+, 0)$. In other words, the incoming base flow develops undulations (around the height $\widetilde{h}_0 = 1$) before it permanently descends to the level of the uniform downstream flow at height $\widetilde{h}_+$. This waveform is called an undular bore and is depicted in Fig. 5.

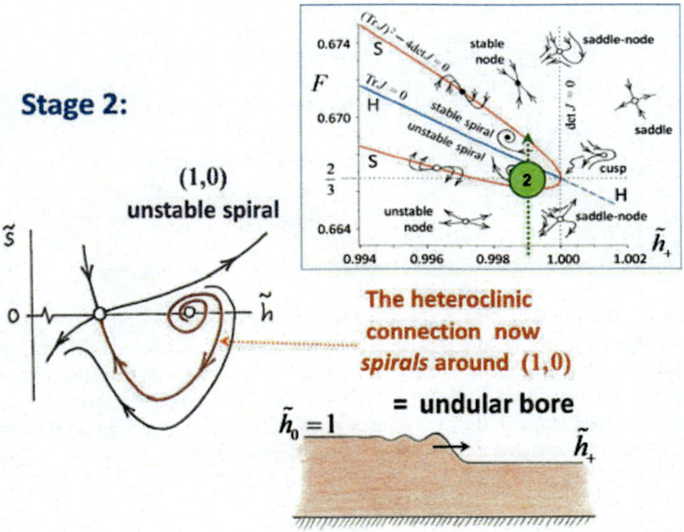

Fig. 5 Stage 2: For $F_0 = 0.666, \widetilde{h}_+ = 0.999$ (green disk in the inset) and $\zeta = 33.3°$, the fixed point $(1, 0)$ has become an unstable spiral and the heteroclinic connection first makes several turns around $(1, 0)$ before it heads towards the saddle point $(\widetilde{h}_+, 0)$. The wave corresponding to this spiraling heteroclinic connection is an undular bore

The undular bore is a common phenomenon in ordinary fluids (e.g. the tidal bores encountered in rivers and estuaries around the world [23, 24]), yet in granular chute flow this type of wave still awaits its first experimental observation.

5.3 *Stage 3: Roll Waves*

The third stage sets in with a *crisis event*: soon after F_0 crosses the critical value $2/3$, the saddle's unstable and stable manifolds precisely connect onto each other and form a homoclinic loop; see the $(\widetilde{h}, \widetilde{s})$ phase plane of Fig. 6 (green curve).

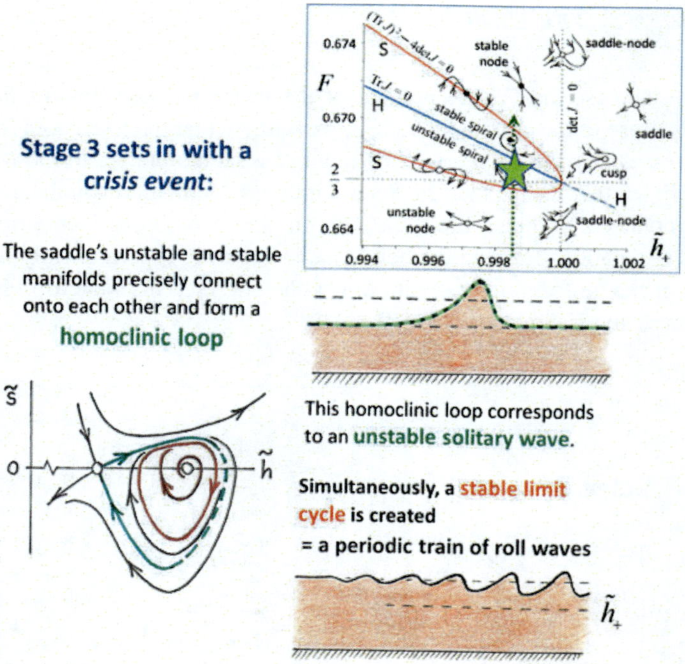

Fig. 6 Stage 3: For $F_0 = 0.667484, \tilde{h}_+ = 0.999$ (green star in the inset) and $\zeta = 33.3°$, the unstable and stable manifolds of the saddle precisely coalesce and form a homoclinic orbit (green curve, partly dashed), which corresponds to a solitary wave (dashed green-black wave in the middle plot on the right). Simultaneously, a stable limit cycle is created around the fixed point $(1, 0)$, indicated by the solid red closed curve in the $(\tilde{h}, \tilde{s})$ phase plane. The undulations around the incoming base flow $\tilde{h}_0 = 1$ build up to a stable periodic train of roll waves, as shown in the lower right plot

This homoclinic loop corresponds to the solitary waveform (dashed green-black curve) shown in the middle right plot. It starts at the level $\tilde{h}(\tilde{\xi}) = \tilde{h}_+$ for $\tilde{\xi} \to -\infty$, then forms a single hump (rising well above the level 1) and falls back to $\tilde{h}(\tilde{\xi}) = \tilde{h}_+$ for $\tilde{\xi} \to +\infty$. It is reminiscent of the celebrated solitary waves in open channel flow, such as the one described by the KdV equation [19], although it is lopsided—due to the inclination of the chute—and the forces that define its shape are different. In the KdV soliton it is dispersion that balances the nonlinear wave-steepening effects, while here it is diffusion (stemming from shear and in-plane stresses within the granular sheet) that plays this role.

Since our analysis indicates that this granular solitary wave requires a great amount of fine-tuning of the system parameters, it may be anticipated to be quite challenging to reproduce this wave in experiment.

However, simultaneous with the momentary closing of the homoclinic orbit, we witness the creation of a stable limit cycle surrounding the unstable spiral. Hence any trajectory spiraling outwards from the point $(1, 0)$ ends up on this limit cycle.

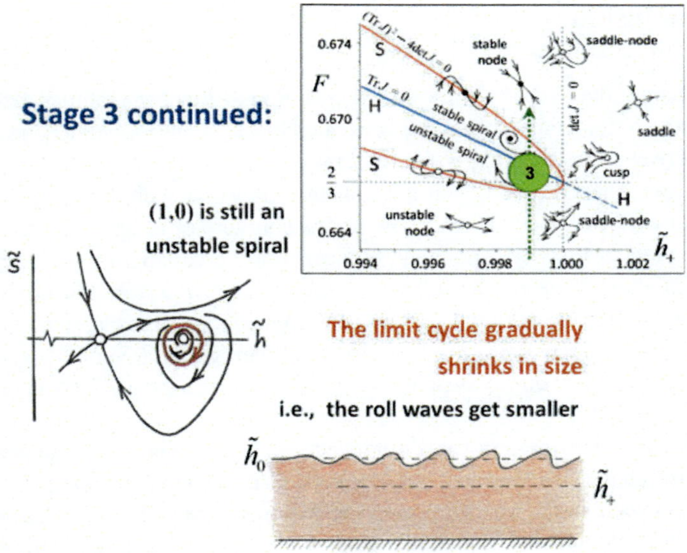

Fig. 7 Stage 3 continued: For $F_0 = 0.667487$, $\widetilde{h}_+ = 0.999$ (green disk in the inset) and $\zeta = 33.3°$, the homoclinic orbit has vanished just as suddenly as it appeared. The stable limit cycle (solid red curve) persists but has shrunk in size. Hence, the roll waves around the level of the incoming base flow ($\widetilde{h}_0 = 1$) now have a smaller amplitude

As we show in Fig. 6 (red solid line), this means that the undulations around the level $\widetilde{h}(\widetilde{\xi}) = \widetilde{h}_0 = 1$ now build up to a periodic train of roll waves.

Roll waves are a familiar phenomenon in open water channels, and they can also be seen on sloping streets on a rainy day, with the thin film of rain water typically organizing itself in the form of a series of roll waves. This same type of waveform has also been observed in the flow of granular matter on a chute; an example is seen in Fig. 1. Indeed, it is the only granular waveform from those discussed so far that has been observed experimentally [14, 15, 25].

Raising the Froude number further, the homoclinic orbit disappears again (having been in existence only fleetingly at one specific value of F_0) but the stable limit cycle persists. This situation is sketched in Fig. 7. For growing F_0, the limit cycle quickly shrinks in size, until at $F_0 = (1 - \widetilde{h}_+)/(\widetilde{h}_+(1 - \widetilde{h}_+^{3/2})) = 0.66750$ it disappears altogether: it has shrunk to a point and coincides with $(1, 0)$, causing the latter to undergo a reverse Hopf bifurcation, turning it into a *stable* spiral [13, 18].

One may of course raise the Froude number F_0 still further, but we will not pursue this here [13]. Suffice it to say that for all $F_0 > 2/3$ the flow organizes itself in such a way that a stable pattern of roll waves is established, always in such a way that the amplitude of the roll wave (or equivalently, the size of the limit cycle) is maximal. The self-organization required for this involves transient states that depend on the time t and which are not captured by the dynamical system (10a)–(10b). To study these transients one must return to the original partial differential equations expressing the balances of mass and momentum, i.e., Eqs. (1) and (2) [13, 18, 20].

6 Conclusion

In conclusion, the hydrodynamic-like description of flowing granular matter via the generalized Saint-Venant equations presented here, predicts a one-to-one correspondence between the traveling waves observed in shallow water and (fully dynamic) thin granular sheets. That is, our analysis indicates that each of the traveling waveforms known from open channel flow has its counterpart in granular chute flow.

To be precise, this correspondence concerns the so-called long-wavelength waveforms. The short-wavelength capillary waves on water, which are caused by the action of surface tension, are exempt from this correspondence. In the classical Saint-Venant equations, surface tension is simply ignored. In their granular version used here, surface tension is absent due to the lack of cohesion between the particles in dry granular matter.

As we have seen, when the Froude number F_0 of the incoming base flow is raised through the critical value $2/3$, our analysis predicts that the transition from the monoclinal wave to the roll wave pattern takes place via an intermediate waveform, namely the undular bore. The challenge now is to check this in experiment.

Observing the granular undular bore (and the monoclinal wave, for that matter) may be anticipated to require a sensitive experiment, since the amplitude of the undulations is typically very small. In order to make the undulations observable they should at least exceed the size of a granular particle, preferably by a large margin. As a roadmap for future experiments, in [13] we have outlined the desired specifications of the granular material and the chute which will optimize the chances of detecting the waveforms in question.

Let us stress that the waves studied in the present paper (traveling waves in the fully dynamic regime) by no means exhaust the possible waveforms that may be encountered in granular chute flow. For instance, for relatively small values of F_0, granular materials are able to sustain traveling waveforms that propagate over a static sublayer of the same material, with the waves crawling forward (like continuous caterpillar tracks) via a mechanism of erosion and deposition. These waves leave stopping regions behind them, i.e., the deposited material remains temporarily stagnant until it is swept into motion again by the eroding action of the next wave front. These intriguing waveforms, which in recent years have been studied by Edwards and Gray and co-workers [15, 16, 26–28], have opened up a whole new spectrum of traveling waves that have no counterpart in the flow of ordinary fluids.

Another interesting class of waveforms that we have left aside is that of standing waves, the most iconic example of which is the so-called granular jump [7, 8, 21, 22, 29–36]. The spectrum of possible waveforms becomes even wider when one considers the fact that granular materials are often multi-disperse (with particles of varying sizes and shapes, introducing segregation effects and stratification [37]) or multiphase (when the particles interact with an ambient fluid, such as in aeolian transport, mud flow or sediment migration, see e.g. [38, 39]). Indeed, granular flows can support a plethora of waveforms only matched by the multitude of wave phenomena encountered in the fields of Newtonian and non-Newtonian fluids combined.

Acknowledgements We gladly seize this opportunity to congratulate our friend and esteemed colleague Prof. Thanassis Fokas on the occasion of his 70th birthday, wishing him many happy returns of the day.

References

1. Jaeger, H., Nagel, S., Behringer, R.: Granular solids, liquids, and gases. Rev. Mod. Phys. **68**, 1259–1275 (1996). https://doi.org/10.1103/RevModPhys.68.1259
2. Aranson, I.S., Tsimring, L.S.: Patterns and collective behavior in granular media: theoretical concepts. Rev. Mod. Phys. **78**, 641–692 (2006). https://doi.org/10.1103/RevModPhys.78.641
3. Andreotti, B., Forterre, Y., Pouliquen, O.: Granular Media: Between Fluid and Solid. Cambridge University Press, New York (2013). ISBN: 978-1-107-03479-2
4. Jakob, M., Hungr, O.: Debris-flow Hazards and Related Phenomena. Praxis Publ.-Springer, Berlin (2005)3-540-20726-0
5. Pudasaini, S.P., Hutter, K.: Avalanche Dynamics. Springer, Berlin (2007). ISBN: 13 978-3-540-32686-1
6. Takahashi, T.: Debris Flow: Mechanics, Prediction and Countermeasures. 2nd edn. CRC Press, Taylor and Francis, London, UK (2014). ISBN: 9781138073678
7. Viroulet, S., Baker, J.L., Edwards, A.N., Johnson, C.G., Gjaltema, C., Clavel, P., Gray, J.M.N.T.: Multiple solutions for granular flow over a smooth two-dimensional bump. J. Fluid Mech. **815**, 77–116 (2017). https://doi.org/10.1017/jfm.2017.41
8. Brennen, C.E., Sieck, K., Paslaski, S.: Hydraulic jumps in granular material flow. Powder Technol. **35**, 31–37 (1983). https://doi.org/10.1016/0032-5910(83)85023-2
9. Gray, J.M.N.T., Edwards, A.N.: A depth-averaged $\mu(I)$-rheology for shallow granular free-surface flows. J. Fluid Mech. **755**, 503–334 (2014). https://doi.org/10.1017/jfm.2014.450
10. Razis, D., Edwards, A.N., Gray, J.M.N.T., van der Weele, K.: Arrested coarsening of granular roll waves. Phys. Fluids **26**, 123305 (2014). https://doi.org/10.1063/1.4904520
11. Pouliquen, O., Forterre, Y.: Friction law for dense granular flows: application for a mas down a rough inclined plane. J. Fluid Mech. **453**, 113–151 (2002). https://doi.org/10.1017/S0022112001006796
12. Razis, D., Kanellopoulos, G., van der Weele, K.: The granular monoclinal wave. J. Fluid Mech. **843**, 810–846 (2018). https://doi.org/10.1017/jfm.2018.149
13. Razis, D., Kanellopoulos, G., van der Weele, K.: A dynamical systems view of granular flow: from monoclinal flood waves to roll waves. J. Fluid Mech. **869**, 143–181 (2019). https://doi.org/10.1017/jfm.2019.168
14. Forterre, Y., Pouliquen, O.: Long-surface-wave instability in dense granular flows. J. Fluid Mech. **486**, 21–50 (2003). https://doi.org/10.1017/S0022112003004555
15. Edwards, A.N., Gray, J.M.N.T.: Erosion-deposition waves in shallow granular free-surface flows. J. Fluid Mech. **762**, 35–67 (2015). https://doi.org/10.1017/jfm.2014.643
16. Rocha, F.M., Johnson, C.G., Gray, J.M.N.T.: Self-channelisation and levee formation in monodisperse granular flows. J. Fluid Mech. **876**, 591–641 (2019). https://doi.org/10.1017/jfm.2019.518
17. Kanellopoulos, G.: The granular monoclinal wave: a dynamical systems survey. J. Fluid Mech. **921**, A6 (2021). https://doi.org/10.1017/jfm.2021.491
18. Kanellopoulos, G., Razis, D., van der Weele, K.: On the shape and size of granular roll waves. J. Fluid Mech. **950**, A27 (2022). https://doi.org/10.1017/jfm.2022.811
19. Drazin, P.G.: Solitons. London Mathematical Society Lecture Note Series, vol. 85. Cambridge University Press, Cambridge (1983). ISBN: 978-0-521-27422-7
20. Razis, D., Kanellopoulos, G., van der Weele, K.: Roll waves as relaxation oscillations. Phys. Fluids **35**, 063333 (2023)

21. Kanellopoulos, G., Razis, D., van der Weele, K.: On the structure of granular jumps: the dynamical systems approach. J. Fluid Mech. **912**, A54 (2021). https://doi.org/10.1017/jfm.2020.951
22. Razis, D., Kanellopoulos, G., van der Weele, K.: Continuous hydraulic jumps in laminar channel flow. J. Fluid Mech. **915**, A8 (2021). https://doi.org/10.1017/jfm.2021.31
23. Chanson, H.: Hydraulics of Open Channel Flow, An Introduction, 2nd edn. Elsevier, Amsterdam (2004). 0-7506-5978-5
24. Van Dyke, M.: An Album of Fluid Motion. The Parabolic Press, Stanford (1982). ISBN: 0-915760-02-9; see especially the photographs by D.H. Peregrine (Plates 199 and 200) on page 116, showing the famous bore on the River Severn, UK
25. Fei, J., Jie, Y., Xiong, H., Wu, Z.: Granular roll waves along a long chute: from formation to collapse. Powder Technol. **377**, 553–564 (2021). https://doi.org/10.1016/j.powtec.2020.09.007
26. Edwards, A.N., Viroulet, S., Kokelaar, B.P., Gray, J.M.N.T.: Formation of levees, troughs and elevated channels by avalanches on erodible slopes. J. Fluid Mech. **823**, 278–315 (2017). https://doi.org/10.1017/jfm.2017.309
27. Russell, A.S., Johnson, C.G., Edwards, A.N., Viroulet, S., Rocha, F.M., Gray, J.M.N.T.: Retrogressive failure of a static granular layer on an inclined plane. J. Fluid Mech. **869**, 313–340 (2019). https://doi.org/10.1017/jfm.2019.215
28. Edwards, A.N., Viroulet, S., Johnson, C.G., Gray, J.M.N.T.: Erosion-deposition dynamics and long distance propagation of granular avalanches. J. Fluid Mech. **915**, A9 (2021). https://doi.org/10.1017/jfm.2021.34
29. Hákonardóttir, K.M., Hogg, A.J.: Oblique shocks in rapid granular flows. Phys. Fluids **17**, 077101 (2005). https://doi.org/10.1063/1.1950688
30. Boudet, J.F., Amarouchene, B., Bonnier, B., Kellay, H.: The granular jump. J. Fluid Mech. **572**, 413–431 (2007). https://doi.org/10.1017/S002211200600365X
31. Gray, J.M.N.T., Cui, X.: Weak, strong and detached oblique shocks in gravity driven granular free-surface flows. J. Fluid Mech. **579**, 113–136 (2007). https://doi.org/10.1017/S0022112007004843
32. Johnson, C.G., Gray, J.M.N.T.: Granular jets and hydraulic jumps on an inclined plane. J. Fluid Mech. **675**, 87–116 (2011). https://doi.org/10.1017/jfm.2011.2
33. Faug, T.: Depth-averaged analytic solutions for free-surface granular flows impacting rigid walls down inclines. Phys. Rev. E **92**, 062310 (2015). https://doi.org/10.1103/PhysRevE.92.062310
34. Faug, T., Childs, P., Wyburn, E., Einav, I.: Standing jumps in shallow granular flows down smooth inclines. Phys. Fluids **27**, 073304 (2015). https://doi.org/10.1063/1.4927447
35. Méjean, S., Faug, T., Einav, I.: A general relation for standing normal jumps in both hydraulic and dry granular flows. J. Fluid Mech. **816**, 331–351 (2017). https://doi.org/10.1017/jfm.2017.82
36. Méjean, S., Guillard, F., Faug, T., Einav, I.: Length of standing jumps along granular flows down smooth inclines. Phys. Rev. Fluids **5**, 034303 (2020). https://doi.org/10.1103/PhysRevFluids.5.034303
37. Gray, J.M.N.T.: Particle segregation in dense granular flows. Ann. Rev. Fluid Mech. **50**, 407–433 (2018). https://doi.org/10.1146/annurev-fluid-122316-045201
38. Bruun, P.: Migrating sand waves or sand humps, with special reference to investigations carried out on the Danish North Sea Coast. Coastal Engineering 5. In: Johnson, J.W. (ed.) Proceedings of 5th Conference on Coastal Engineering, Grenoble, France, 1954, Chap. 21, pp. 269–295 (1954)
39. Tamburrino, A., Ihle, C.F.: Roll wave appearance in bentonite suspensions flowing down inclined planes. J. Hydraul. Res. **51**(3), 330–335 (2013). https://doi.org/10.1080/00221686.2013.769468

Identifying Discrete Breathers Using Convolutional Neural Networks

T. Dogkas, M. Eleftheriou, G. D. Barmparis, and G. P. Tsironis

Abstract Artificial intelligence in the form of deep learning is now very popular and directly implemented in many areas of science and technology. In the present work we study time evolution of Discrete Breathers in one-dimensional nonlinear chains using the framework of Convolutional Neural Networks. We focus on differentiating discrete breathers which are localized nonlinear modes from linearized phonon modes. The breathers are localized in space and time-periodic solutions of non-linear discrete lattices while phonons are the linear collective oscillations of interacting atoms and molecules. We show that deep learning neural networks are indeed able not only to distinguish breather from phonon modes but also determine with high accuracy the underlying nonlinear on-site potentials that generate breathers. This work can have extensions to more complex natural systems.

Keywords Discrete breathers · Convolutional neural networks

1 Introduction

Discrete breathers (DBs) or Intrinsic Localized Modes (ILMs) are time periodic and space localized modes that appear in discrete nonlinear lattices [1]. During the last over thirty year period there has been substantial amount of theoretical and experimental work that generated a body of precise knowledge regarding these modes [2]. In the present work we use modern tools of Machine Learning (ML) in order to address an entirely different question, viz. whether DBs can be recognized in some automatic form without the need of direct human intervention. We believe that this is a significant question since its precise answer may lead to much easier and direct detection of nonlinear modes in natural systems. In this preliminary work we sharpen the question to: Is it possible to recognize DBs and phonon modes in simple one-dimensional chains with nonlinear on-site potentials using Convolutional Neural

T. Dogkas · M. Eleftheriou · G. D. Barmparis · G. P. Tsironis (✉)
Department of Physics, University of Crete, P. O. Box 2208, Heraklion 71003, Greece
e-mail: gts@physics.uoc.gr

T. Bountis et al. (eds.), *Chaos, Fractals and Complexity*, Springer Proceedings in Complexity, https://doi.org/10.1007/978-3-031-37404-3_14

Networks (CNNs)? In order to construct DBs we use the numerically exact method introduced by Aubry for generation from the anticontinuous limit [3].

In this brief account we have four sections. In the first section we give some details on the generation of DBs and phonon samples to be used subsequently. In the second section a CNN model is developed in order to distinguish DBs from phonons. In the following section a CNN model is developed to identify the on-site nonlinear potential through which the linear and nonlinear modes were generated. Finally, in the last section we conclude and give some further perspectives of this work.

2 Creation of Breather and Phonon Samples

For the ML analysis in this work we create 459 samples of breathers and phonons using the anticontinuous limit method in 1D lattice with the Hamiltonian [4–6] and three different nonlinear on-site potentials as outlined in Table 1. The DBs have frequencies outside the phonon spectrum and their stability is checked through Floquet analysis. We give some details on the methods used below.

The Breather Solution
We outline very briefly the procedure we follow in order to obtain numerically exact breathers and make sure they are stable. We use a Hamiltonian in the form:

$$H = \sum_{N} \frac{p_n^2}{2} + V(x_n) + W(x_n) \tag{1}$$

where x_n the displacement at the node n of the 1D lattice, p_n is the corresponding momentum, V is one of the three nonlinear on-site potentials used, while W is the interaction potential, viz. $W(x_n) = k(x_{n-1} + x_{n+1} - 2x_n)$. The value of the parameter k is quite important since it affects the existence as well as the shape of DBs.

In the numerical procedure we first obtain the breather solution (see below) and subsequently we linearize the equations of motion around this solution. The small amplitude plane waves of the form $x_n(t) = e^{i(\omega_b t - qn)}$ propagate with frequencies:

Table 1 The three nonlinear on-site potentials used in this work, are the hard ϕ^4, the Morse and double-well potentials

Potentials	
Hard ϕ^4	$V(x) = \frac{x^2}{2} + \frac{x^4}{4}$
Morse	$V(x) = -\frac{1}{2}(1 - e^{-x})^2$
Double well	$V(x) = -\frac{(x-1)^2}{2} + \frac{(x-1)^4}{4}$

$$\omega_b = V^{''}(0) + 4W^{''}(0)\sin(\frac{q}{n}) \tag{2}$$

We solve the full equations of motion at zero coupling in order obtain a trivial breather with specific amplitude-period relationship [3, 4]. Subsequently, we create a vector $U(x_1, x_2, \cdots, x_n, p_1, p_2, \cdots, p_n)$ for the N lattice sites (we take $N = 40$) and set in the middle of the lattice the initial condition $x_n = 0.9$ and $p_n = 0$, i.e. the trivial breather. Subsequently, we increase the coupling k in small steps and solve the equations of motion; this iterative procedure is repeated until the desired coupling value is reached. Since the DBs are time periodic solutions, for periodic evolution derived from the map T we have $U_k + \Delta = T(U_k + \Delta) \Leftrightarrow U_k + \Delta = T(U_k) + \partial T x \Delta$, where $\partial T = M$ the tangent map of the system. By minimizing the square of norm $||T(U) + M\Delta - (U + \Delta)||^2$ we obtain the matrix Δ that gives the final solution for the breather and its stability. We repeat this procedure with each breather sample and thus obtain stable DB solutions as shown in Fig. 1.

Phonon Spectrum

The phonon spectrum for different couplings is given by Eq. 2. We designate the upper limit of the phonon band as ω_b, while the lower limit as ω_b'. We get acoustic-like breathers when the breather frequency, Ω_b, is less than $\omega_{b'}$ and optic-like breathers when $\Omega_b > \omega_b$. In Fig. 2, we present the phonon frequency band with upper limit $\omega_b = 1.095$, lower limit $\omega_{b'} = 1$ and coupling $k = 0.05$ for the hard ϕ^4, and with upper limit $\omega_b = 1.483$, lower limit $\omega_{b'} = 1.414$ for the Morse potential, respectively. Snapshots and the time evolution of phonons for the hard ϕ^4, Morse and double well potential with coupling $k = 0.1$, and frequencies $f = 1.042$, $f = 1.001$, and $f = 1.415$, respectively, are shown in Fig. 3.

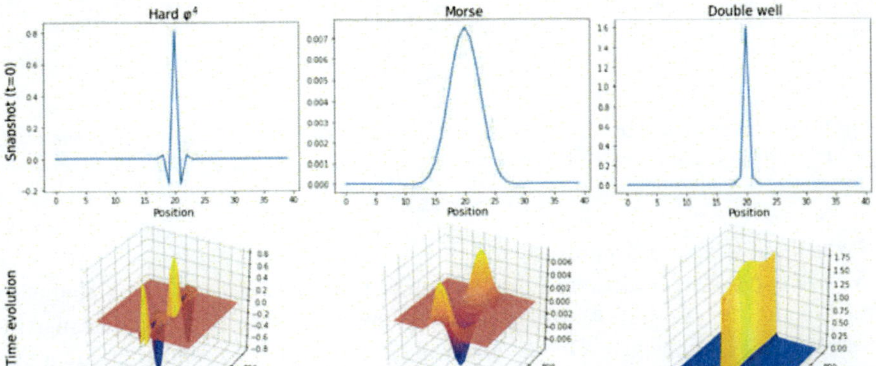

Fig. 1 A snapshot at $t = 0$ and the time evolution of an arbitrary breather for hard ϕ^4, Morse and double well potentials with coupling $k = 0.1$ and frequencies $f = 1.317$, $f = 0.967$ and $f = 0.949$ respectively

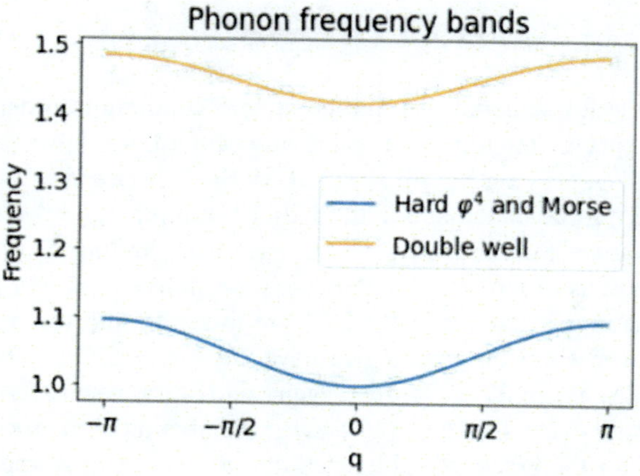

Fig. 2 Phonon frequency bands as a function of the wave-vector q, with upper limit $\omega_b = 1.095$, lower limit $\omega_{b'} = 1$ and coupling $k = 0.05$ for hard ϕ^4 and Morse potentials and with upper limit $\omega_b = 1.483$, lower limit $\omega_{b'} = 1.414$ for the double well potential

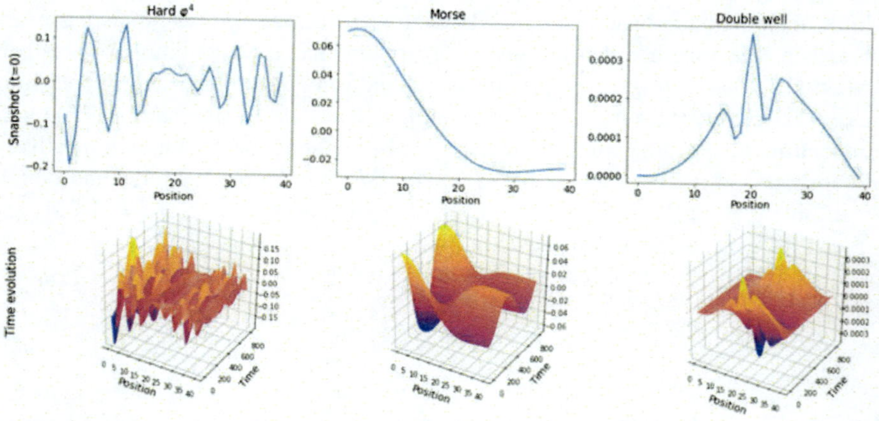

Fig. 3 A snaphot at $t = 0$ and the time evolution of phonons for the hard ϕ^4, Morse and double well potentials with coupling $k = 0.1$ and frequencies $f = 1.042$, $f = 1.001$, $f = 1.415$ respectively

Stability of Discrete Breathers

We use two different methods to examine the stability of DBs. The first method is the Floquet analysis (Fig. 4), where we obtain the eigenvalues of the tangent map M. Stable DBs solutions give Floquet eigenvalues where their imaginary and real part lie on a circle with radius r = 1. The second method is by direct observation the time evolution of DBs. In Fig. 5, we present the time evolution of a stable (left) and an unstable (right) breather for the hard ϕ^4 potential.

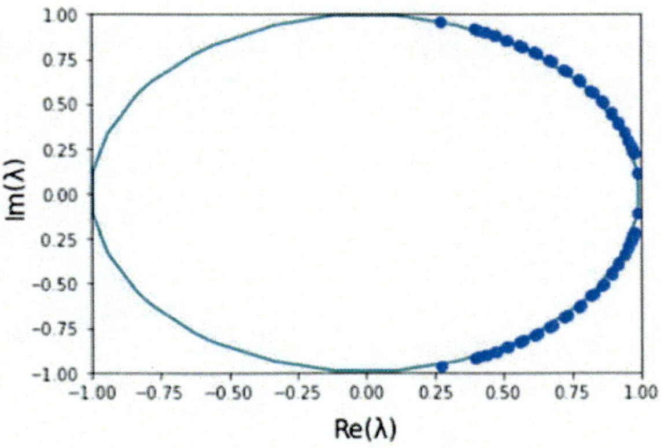

Fig. 4 The real and the imaginary part of the eigenvalues, λ, of the tangent map M matrix of a breather with potential hard ϕ^4, frequency $f = 1.227$ and coupling $k = 0.1$

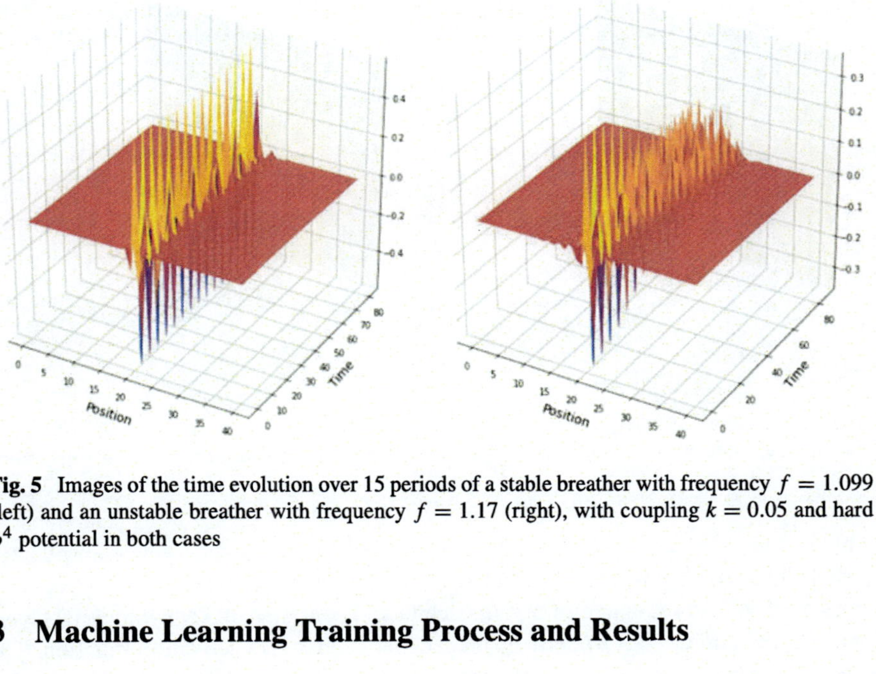

Fig. 5 Images of the time evolution over 15 periods of a stable breather with frequency $f = 1.099$ (left) and an unstable breather with frequency $f = 1.17$ (right), with coupling $k = 0.05$ and hard ϕ^4 potential in both cases

3 Machine Learning Training Process and Results

We create a dataset of 459 samples, with equally distributed phonons and DBs using several arbitrary values for both their coupling and their frequency. Each sample is a single channel gray-scaled, 2D image of size 40×822 pixels, of the time evolution of a DB or phonons. Each image was constructed in a way that at least one time period is included. In Fig. 6, we present a colorized sample of each category of DBs and phonons. The dataset was shuffled to avoid biases. Twenty percent of the samples were separated to create a hold-out set for testing. The rest 80% of the samples was

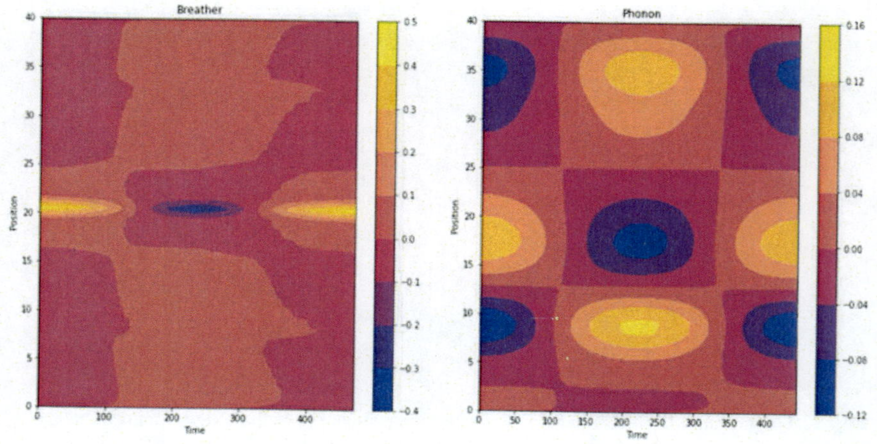

Fig. 6 Contour plots of the time evolution of a breather with frequency $f = 1.332$ and a phonon with frequency $f = 1.4161$, with coupling $k = 0.1$ for the double well potential, respectively. The x-axis represents the time, the y-axis the position of each node of the 1D chain and the colorbar the amplitude

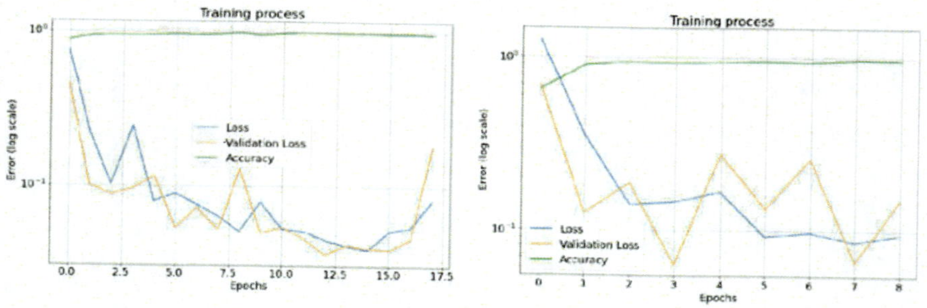

Fig. 7 The training accuracy and loss and the validation loss as a function of the number of epochs, for the breather-phonon classification (left) and the potential classification (right) model respectively

split to create a training (80%) and a validation (20%) set. The dataset was normalized using the ImageDataGenerator package [7]. Two models of identical CNNs were created. The features extractor part of each model consisted of 3 convolutional layers with 32, 64 and 64 (3×3) kernels, respectively and a *relu* activation function. The first two convolutional layers in each model were followed by a (2×2) Max-pooling layer. The classifier of each model contains two fully connected layers with 64 and 2 nodes respectively, for the case of the DB/phonons classifier, and 64 and 3 nodes layers for the three potentials classification. A *relu* activation function was used for the first layer and a *softmax* one for the output layer of each classifier. The models were allowed to train for 100 epochs, while an early stopping criterion with patience 5 epochs was monitoring the validation error in order to avoid over-fitting. The f1-score was used to evaluate the performance of the models.

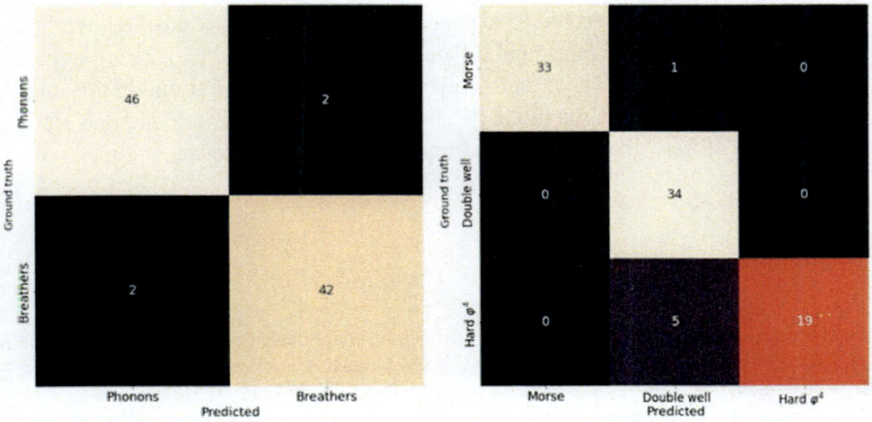

Fig. 8 The confusion matrix of the hold-out test set for the breather-phonon classification (left) and the potential classification (right) model, respectively

Table 2 Table of the accuracy and the f1-score of the test set for each model

Test set—Classification results

	Breather-Phonon	Potentials
Accuracy	95.7%	93.5%
f1-score	0.955	0.934

In Figs. 7, 8 and in Table 2, we present the training process and the results of our analysis. Both the $f1$ score is high while the confusion matrix shows very high degree of classification accuracy.

4 Conclusions

This preliminary work that applies CNNs to the breather-phonon classification problem gives promising results as a previous ML work did in chaotic systems [8]. We note that similar results were also found through the use of different ML methods in the breather problem [9]. We find here that it is indeed possible to perform an accurate classification based on the assumptions presented previously. The quantitative result of the confusion matrix on the test set shows that the classification is excellent provided the CNNs are trained appropriately. This result, however, is not unexpected since breathers and phonons as 2D images can be easily distinguished by a human eye. What is more noteworthy, however, is the possibility to find the underlying potential these modes stem from. Indeed, we see that with proper training the ML model may distinguish the specifics of the underlying dynamics. This strong feature opens up the possibility for a deeper use of Deep Learning in nonlinear physics. It is not unreasonable to expect that under specific, controlled conditions, we may be able to use experimental data to infer the precise underlying dynamics of a complex system. We note that similar indications are also found in the study of chaotic systems with ML methods [8]. We conclude that a more detailed study of the potential of machine learning in complex systems is necessary.

References

1. Sievers, A.J., Takeno, S.: Intrinsic localized modes in anharmonic crystals. Phys. Rev. Let. **61**(8), 970 (1988)
2. Flach, S., Willis, C.R.: Discrete breathers. Phys. Rep. **295**, 181 (1998)
3. Aubry, S.: Breathers in nonlinear lattices: Existence, linear stability and quantization. Phys. D **103**, 201 (1997)
4. Eleftheriou, M.: Statistical properties of classical non-linear lattices, Ph. D. thesis, University of Crete (2003)
5. Lazarides, N., Eleftheriou, M., Tsironis, G.P.: Discrete breathers in nonlinear magnetic metamaterials. Phys. Rev. Let. **95**(15). 12-10-2006
6. Lazarides, N., Tsironis, G.P.: Gain-driven discrete breathers in symmetric nonlinear metamaterials. Phys. Rev. Let. **110**(5). 30-1-2013
7. Hastie, T., Tibshirani R., Friedman, J.: The Elements Of Statistical Learning: Data Mining, Inference, and Prediction, 2nd edn. Springer (2009)
8. Almazova, N., Barmparis, G.D., Tsironis, G.P.: Analysis of chaotic dynamical systems with autoencoders. Chaos **31**(10), 103109 (2021)
9. Bajars, J., Kozirevs, F.: Data-driven intrinsic localized mode detection and classification in one-dimensional crystal lattice model. Phys. Lett. A **436**, 128071 (2022)

Subthreshold Oscillations in Multiplex Leaky Integrate-and-Fire Networks with Nonlocal Interactions

K. Anesiadis, J. Hizanidis, and A. Provata

Abstract We study the dynamics of identical Leaky Integrate-and-Fire (LIF) neurons on a multiplex composed of two identical ring networks with symmetric nonlocal coupling within each ring and one-to-one connections between rings. By varying the control parameters (intra-ring and inter-ring coupling strength) we investigate the system's behaviour and show that, under the above connectivity scheme, inter-ring coupling favors in-phase synchronization and we numerically determine the corresponding parameter region where it occurs. We also highlight the peculiar phenomenon of subthreshold oscillations occurring in one of the two rings while the elements of the other ring perform full-cycle oscillations. In the case of solitary states, the subthreshold oscillations may switch erratically between the two rings.

Keywords Chimera states · Subthreshold oscillations · Multilayered networks · Kuramoto order parameter

K. Anesiadis
School of Applied Mathematical and Physical Sciences,
National Technical University of Athens, 15772 Athens, Greece
e-mail: konanesiadis@mail.ntua.gr

K. Anesiadis · A. Provata
Institute of Nanoscience and Nanotechnology, National Center for Scientific Research "Demokritos", 15431 Athens, Greece
e-mail: a.provata@inn.demokritos.gr

J. Hizanidis (✉)
Department of Physics, University of Crete, 71003 Herakleio, Greece
e-mail: hizanidis@physics.uoc.gr

Institute of Applied and Computational Mathematics, Foundation for Research and Technology – Hellas, 70013 Herakleio, Greece

T. Bountis et al. (eds.), *Chaos, Fractals and Complexity*, Springer Proceedings in Complexity, https://doi.org/10.1007/978-3-031-37404-3_15

1 Introduction

In recent years, numerical studies on networks of coupled nonlinear oscillators have shown interesting synchronization patterns that are otherwise difficult to predict. Synchronization in large systems of interacting elements is of great importance both in physical and biological processes, including human physiology and brain functionality, as well as in technical human activities [1, 2]. A notable synchronization example is a spatiotemporal state of partial synchrony, called a chimera state, in which the system exhibits domains of coherence and incoherence at the same time. Chimera states, first introduced in networks of Kuramoto phase oscillators [3, 4], have been reported also for several other nonlinear oscillators, including the Leaky Integrate-and-Fire (LIF) [5], the FitzHugh-Nagumo (FHN) [6, 7], and the Hindmarsh-Rose (HR) [8, 9] neuronal oscillators. Especially for the former model, another peculiar phenomenon is the coexistence of oscillators that perform complete cycles and others fluctuate below the threshold potential without completing full cycles. These are called subthreshold oscillations [10] and are in the general category of partially synchronized states called bump states [11].

Regarding the network topology, various connection schemes in numerical studies have been used including global, (symmetric) nonlocal, and hierarchical/fractal [12–16]. More specifically, fractal connectivities are of particular interest as they are among the irregular connectivites for which we have observed chimeras in simulations. Furthermore, in the area of neuroscience, Diffusion Tensor Magnetic Resonance Imaging (DT-MRI) studies indicate a hierarchical/fractal geometry of the neuronal network [17, 18], encouraging the study of such connectivities for neural oscillators.

As the connectivity specifies the information flow, it may lead to full or partial synchrony or in-phase synchronization or subthreshold oscillations (bump states) and is considered essential in driving the network dynamics. Real-world networks are often composed of many interconnected subnetworks, such as neuronal networks and social networks, where each may obey a different connectivity. Since the aim of numerical investigations is to uncover the mechanisms producing nontrivial synchronization phenomena, the connectivity in the numerical simulations is kept to a considerably lower complexity than what is observed in reality.

Previous studies of LIF neuronal dynamics on a low complexity network (ring connectivity) have demonstrated the presence of solitary states which dominate for negative coupling strength σ in the parameter range $-0.5 \lesssim \sigma < 0$ [10, 13]. Solitary states are weak chimeras where there is complete in-phase synchronization except for some small clusters of incoherent oscillators. The term "weak chimeras" refers to chimera states whose incoherent domains are of infinitesimal size, while typical chimeras have both coherent and incoherent domains of finite sizes [19]. In the same studies, chimera states were reported for $\sigma < -0.5$ until they collapse in full disorder by decreasing σ towards -1. The critical value where the transition from solitary to chimera state depends on the parameters of the LIF oscillator and the network.

In this study, we consider a more general multilayered network (or multiplex) of two identical ring networks, each one of them consisting of symmetric nonlocally coupled LIF neuronal oscillators while between them there are one-to-one connections. In a previous study the LIF multiplex has been studied for low interaction between the layers [20]. Here, we examine a parameter range for the coupling strength within each ring and the coupling strength between the two rings, in which significant differences were observed in the calculated time averages of the Kuramoto order parameter of the two rings. We show how these control parameters affect the phase synchronization in the two layers of the multiplex. Although the ring networks are identical, we have identified a range of parameter values where one of the rings performs complete oscillations below the threshold while the other does not.

In Sect. 2 we introduce the single LIF model, the coupled LIF model, and the Kuramoto order parameter. In Sect. 3 we show the heatmaps of the Kuramoto order parameters for the two rings of the multiplex, independently, highlighting the area in the maps where they are distinctively different. In addition, we present the peculiar phenomenon of subthreshold oscillations. For a particular range of the control parameters, either the elements of one ring perform full-cycle oscillations while the elements of the other are not, or they switch roles irregularly in time. We conclude, in Sect. 4, by summarizing our results and presenting open problems.

2 The Model

The LIF model describing the dynamics of isolated neurons was first proposed in 1907 by Louis Lapicque [21, 22]. If we let $u(t)$ to be the time-dependent membrane potential of a nerve cell, the dynamics of the single LIF model is described by the following Eq. 1a and condition 1b:

$$\frac{du(t)}{dt} = \mu - u(t) + I(t) \,, \tag{1a}$$

$$\lim_{\delta t \to 0^+} u(t + \delta t) = u_{\text{rest}} \,, \quad \text{when} \quad u(t) > u_{\text{th}} \,. \tag{1b}$$

Equation 1a represents the integration of the membrane potential, while influx $I(t)$ may originate from external stimuli or the neighbouring neurons' collective contribution. Condition 1b represents the resetting of the potential after reaching the threshold u_{th}. Namely, the potential $u(t)$ is reset at u_{rest} immediately after its value surpasses the value of u_{th}. The parameter μ in Eq. 1a corresponds to the limiting value of the potential if resetting is not considered. Equation 1a can be analytically solved, when $I(t)$ is constant or zero. Then, the constant is incorporated in the parameter μ and the solution is $u(t) = \mu - (\mu - u_{\text{rest}})e^{-t}$, for $u_{\text{rest}} \leq u(t) \leq u_{\text{th}}$. The period T_s of oscillations of the single LIF is calculated as $T_s = \ln\left[(\mu - u_{\text{rest}}) \,/\, (\mu - u_{\text{th}})\right]$. Typically, a LIF neuron spends a period of time at the rest state after resetting, but for simplicity, we will omit this refractory period.

2.1 The Coupled LIF Dynamics in the Multiplex

There are studies of the LIF dynamics on a ring network that demonstrate a variety of synchronization patterns depending on the connectivity (nonlocal, hierarchical, reflecting, small-world, etc.), the coupling strength, and the coupling range [5, 10, 13]. In this study, we consider a two-layered multiplex, where each layer is a ring network of identical LIF oscillators. Both rings are considered identical and for convenience, we name them ring L (for left) and ring R (for right), respectively. To keep the system as simple as possible from the point of view of connectivity, we consider typical symmetric nonlocal connectivity within each ring and one-to-one connectivity across rings.

Let us denote by $\sigma^{\rm L}_{jk}$ the intra-ring connectivity between nodes (j, k) in ring L; similarly for ring R. To avoid having many different parameters, we assume that the connections within each layer are tantamount. Since both rings are considered identical, the general form of the nonlocal intra-ring connectivity with coupling range K around node j is:

$$\sigma^{\rm L}_{jk} \equiv \sigma^{\rm R}_{jk} \equiv \sigma_{jk} = \begin{cases} \sigma, \ \forall k : [j - K \leq k \leq j + K] \\ 0, \ \text{elsewhere} \end{cases} \tag{2}$$

Regarding the inter-ring connections, let us denote by $\sigma^{\rm L\to R}_j$ the connectivity between the j-th nodes of rings L and R, and, similarly, for the opposite direction (note that these are the only connections between the two rings if any). As before, for the sake of simplicity, we assume common values for all connections, $\sigma^{\rm L\to R}_j \equiv \sigma^{\rm R\to L}_j \equiv s$.

Let $u^{\rm L}_j(t)$, $j = 1, \ldots, N$ represent the membrane potential of the j-th neuron in the left ring, where N is the ring size. Then, the dynamics of the j-th coupled LIF neuron of the ring L in the multiplex is described as follows:

$$\frac{du^{\rm L}_j(t)}{dt} = \mu - u^{\rm L}_j(t) + \frac{\sigma}{2K} \sum_{k=j-K}^{j+K} \left[u^{\rm L}_k(t) - u^{\rm L}_j(t)\right] + s\left[u^{\rm R}_j(t) - u^{\rm L}_j(t)\right], \tag{3a}$$

$$\lim_{\delta t \to 0^+} u^{\rm L}_j(t + \delta t) = u_{\rm rest}, \quad \text{when} \quad u^{\rm L}_j(t) > u_{\rm th}. \tag{3b}$$

The notation and definitions are similar for the ring R. In Eq. 3, we consider nonlocal diffusive-like connectivity with coupling range K, common in both rings. In the above expressions all the indices in the rings L and R are taken mod N. Other common parameters of all nodes are the limiting membrane potential value μ, the rest state potential $u_{\rm rest}$ and the threshold potential $u_{\rm th}$.

In this study we use as working parameter set: $\mu = 1$, $u_{\rm rest} = 0$, $u_{\rm th} = 0.98$, $N = 500$ and $K = 120$. For these parameters, the single (uncoupled) rings present chimera states when coupling strengths take negative values and subthreshold oscillations for positive ones. The inter-ring coupling s in the multiplex connectivity varies in the range $0 \leq s \leq 1$, while the intra-ring coupling σ varies in the range

$-1 \leq \sigma \leq 0$. All simulations start from random initial conditions, while periodic boundary conditions are considered for all indices.

2.2 *Kuramoto Order Parameter*

For quantifying the synchronization within each ring the Kuramoto order parameter Z is employed [3, 6], denoted by Z^{L} and Z^{R} for the rings L and R respectively. To define Z we first need to define the phase of every oscillator. Then, the instantaneous Kuramoto order parameter which measures synchronization in ring L is defined as:

$$Z^{\mathrm{L}}(t) = \frac{1}{N^{\mathrm{L}}} \left| \sum_{k=1}^{N} e^{i\phi_k^{\mathrm{L}}(t)} \right| \tag{4}$$

where $|\cdot|$ stands for the magnitude of the complex number in the argument. Similarly, the Kuramoto order parameter $Z^{\mathrm{R}}(t)$ is defined for ring R. The order parameter generally takes values in the range $0 \leq Z(t) \leq 1$. When $Z \simeq 0$ then the ring elements are asynchronous and when $Z \simeq 1$ they are in-phase synchronous. Intermediate values of Z indicate partial network synchronization. Typically, solitary states exhibit almost absolute coherence with the incoherent oscillators consisting only of a small fraction of the network and thus Z is very close to 1. On the other hand, typical chimeras have finite domain of incoherence, reflected in the Kuramoto order parameter taking values considerably less than 1.

A reasonable choice for the phase definition of a LIF oscillator is by setting a Poincaré section at the threshold potential u_{th}, the point that signifies the completion of a full oscillation cycle, as seen in [5]. However, phases of oscillators that perform subthreshold oscillations for long times in the numerical simulations are ill-defined. Instead, in this study the instantaneous phase $\phi_j^{\mathrm{L}}(t)$ of the j-th oscillator in ring L is defined as:

$$\phi_j^{\mathrm{L}}(t) = \frac{2\pi u_j^{\mathrm{L}}}{u_{\mathrm{th}}}. \tag{5}$$

Similarly, are defined the phases in the ring R. Both definitions were found to produce consistent results.

3 Results

We investigate the multiplex for negative (repulsive) intra-ring couplings and positive (attractive) inter-ring couplings. Our numerical results for $(\sigma, s) \in [-1, 0] \times [0, 1]$ indicate that the multiplex hosts complete synchrony and disorder, solitary states, chimera states, and traveling waves. The Kuramoto order parameter can demonstrate

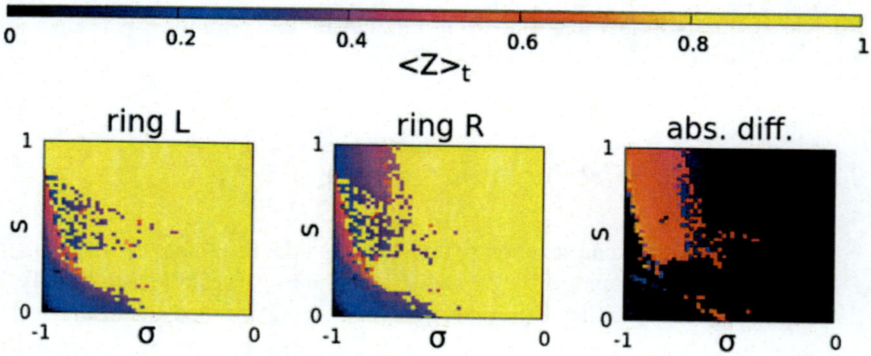

Fig. 1 Heat maps showing the magnitude of the time-averaged Kuramoto order parameters of the rings L and R for varying coupling strength values $-1 \leq \sigma \leq 0, 0 \leq s \leq 1$, as well as their absolute difference. For each value in the heat maps, the evolution time of the system was 3000 Time Units (TUs). The initial 600 TUs were discarded as transient. Other parameters are: $N = 500$, $K = 120$, $\mu = 1$, $u_{\text{rest}} = 0$ and $u_{\text{th}} = 0.98$. All simulations were performed starting from the same random initial conditions

the synchronization state. Figure 1 shows the calculated Kuramoto order parameter for the rings L and R in the multiplex, as well as their absolute difference, for varying σ (horizontal axis) and s (vertical axis). The calculations were performed for a step of 0.02 for σ and s, resulting in a heat map of 51×51 color boxes. The heat map indicates where the phase transitions are and what synchronization patterns to expect.

The bottom areas in each of the first two maps, corresponding to a low coupling between rings, are consistent with what we know about the uncoupled network. In panels Fig. 1a, b (from left to right) Z values ascend from 0 to 1 as σ increases from -1 to just below 0, with a discontinuity at the critical point. The critical point is shifted to the left (it becomes more negative) as s increases from 0 to 0.30, together with the purple area that hosts chimera states. This is shown in Fig. 2, where the Kuramoto order parameter versus the intra-ring coupling σ is plotted for four different levels of s. Together with the critical point, the purple area that hosts chimera states is also moved. For instance, both layers of the multiplex for $\sigma = -1.00$ and $s = +0.10$ are disordered within. However, for $s = +0.20$ a drifting chimera state is developed; see also last row of Fig. 2. Overall, the degree of in-phase synchronization of the network appears to be enhanced by the synergy of the two rings, expressed by s.

3.1 *Subthreshold Oscillations*

Another critical region seen in Fig. 1 is the boundary where the absolute difference of the Kuramoto order parameters goes from zero to finite values; see Fig. 3 for $\sigma = -1.00$. We notice in the whole range that there are some scattered tuples (σ, s) for which the two rings have large variations in synchronization, although it is more

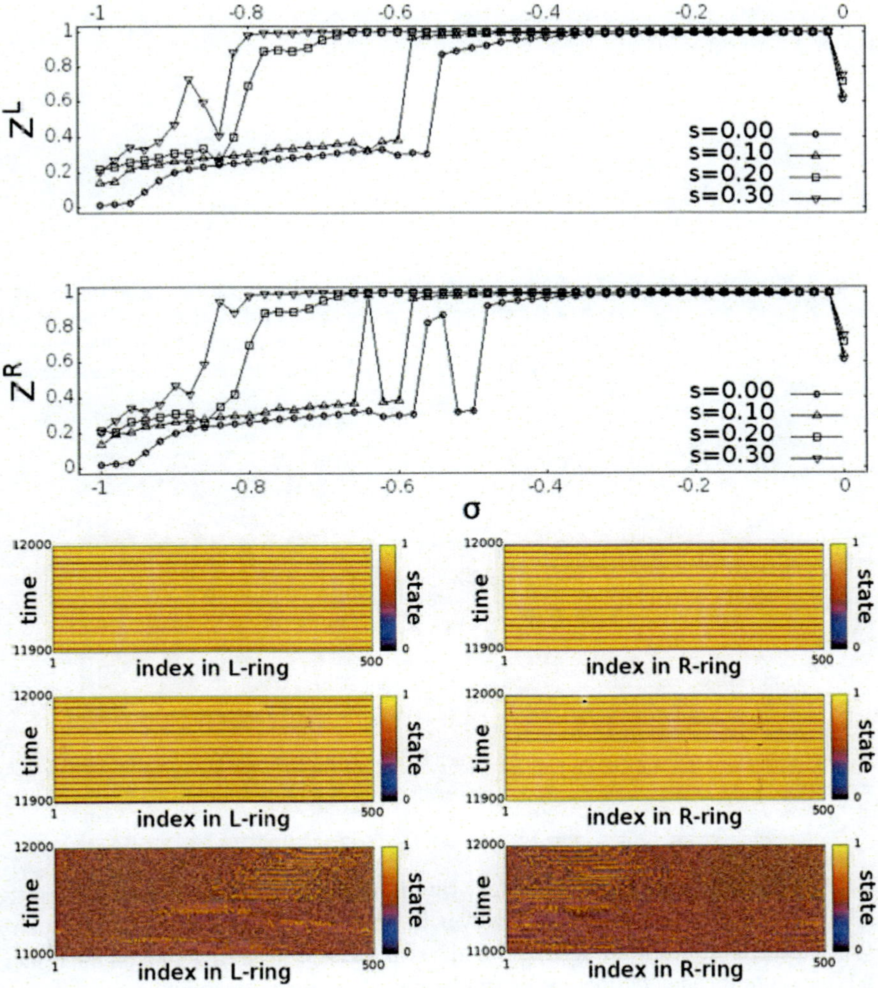

Fig. 2 In the first and second rows, we show the average Kuramoto order parameters for the two rings versus the intra-ring coupling strength σ, for $s = 0.00, 0.10, 0.20$ and 0.30. Temporal averages are taken over $\Delta T = 3000$ TUs, after excluding the first 600 TUs as transients. The rest of the rows (from top to bottom) show the spacetime plots of the multiplex for $\sigma = -0.10$, $s = 0.20$ (in-phase sync), for $\sigma = -0.40$, $s = 0.30$ (solitary states), and for $\sigma = -1.00$, $s = -0.20$ (chimera states). Other parameters are as in Fig. 1

pronounced for $\sigma < -0.5$ and $s > 0.3$. To investigate this discrepancy we plot in Fig. 4 the spacetime plots of the multiplex for $\sigma = -0.90$ and $s = 0.90$. For these values, $Z^{\mathrm{L}} \simeq 0.95$ and $Z^{\mathrm{R}} \simeq 0.35$.

In this case, the ring with the high-order Kuramoto parameter exhibits throughout its range non-complete cycles of the nominal oscillation called subthreshold oscillations, in the sense that none of its oscillators reaches the threshold potential. The

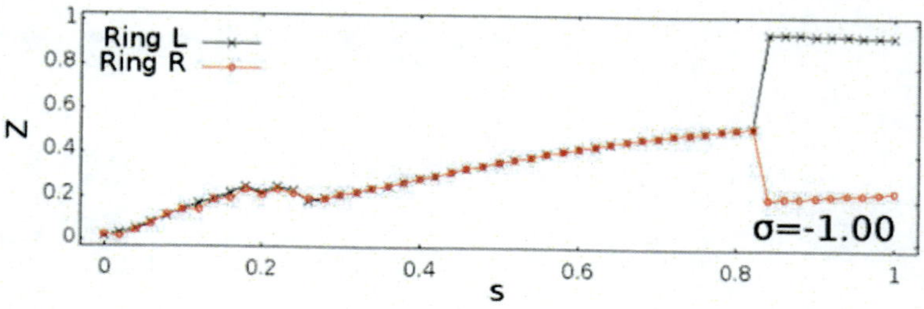

Fig. 3 The average Kuramoto order parameters for the two rings versus the inter-ring coupling strength s, for $\sigma = -1.00$. Both rings exhibit the same degree of coherence, but after a critical value of s, the Kuramoto order parameter in rings L and R diverges significantly. Temporal averages are taken over $\Delta T = 3000$ TUs, after excluding the first 600 TUs as transients. Other parameters are as in Fig. 1

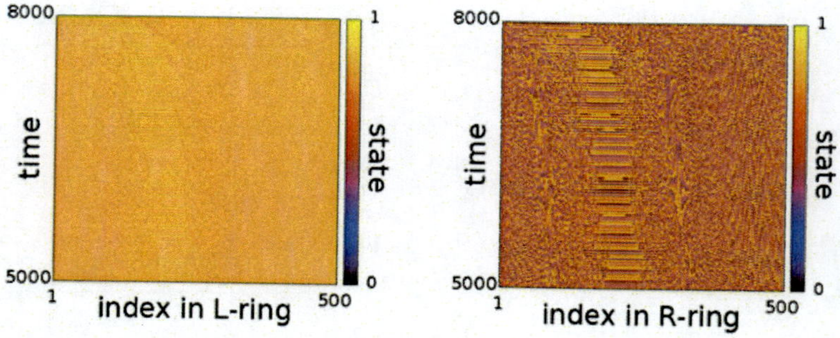

Fig. 4 Spacetime plots of the multiplex (rings L and R) for $\sigma = -0.90$ and $s = 0.90$. Ring R exhibits a chimera state, while ring L exhibits exclusively subthreshold oscillations

Z value of the subthreshold oscillatory ring is high due to the small deviation of the oscillators' states as the network evolves in time. Indicatively, Fig. 4 shows such a case of the multiplex for $\sigma = -0.90$ and $s = 0.90$. The spacetime plot on the right is a typical one-headed chimera state (with one coherent and one incoherent domain), whereas the spacetime plot on the left follows in pattern the other plot but without any of ring L oscillators completing a full cycle (the coherent domain continues drifting for the next 4000 TUs). In fact, while in the ring R all oscillator potentials take values $u^{\mathrm{R}} \in (0, 0.98)$, in the ring L $u^{\mathrm{L}} \in (0.70, 0.98)$. This is not the usual subthreshold oscillations, where the elements' potential stay and fluctuate slightly just below the threshold. They now perform oscillations of finite range but never reaching the rest state potential. This is a result of multiplexing and has not been observed in other connectivities, so far.

Subthreshold oscillations, however, are also observed outside this candidate critical area but in a switching manner; that is, as the multiplex evolves in time, the state switches from one ring to the other erratically. Indicatively, Figs. 5, 6 show such a case of the multiplex for $\sigma = -0.40$ and $s = +0.80$. Both spacetime plots exhibit either solitary state and subthreshold oscillations or vice versa. The switch is random

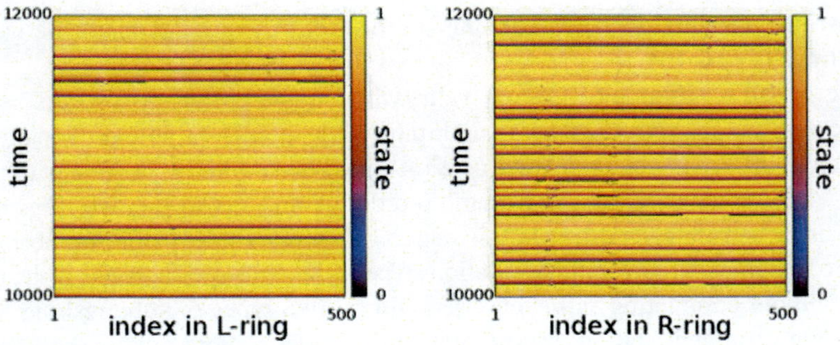

Fig. 5 Spacetime plots of ring L and ring R, where switching between solitary state and subthreshold oscillation occurs. The coupling strengths are set $\sigma = -0.40$ and $s = +0.80$

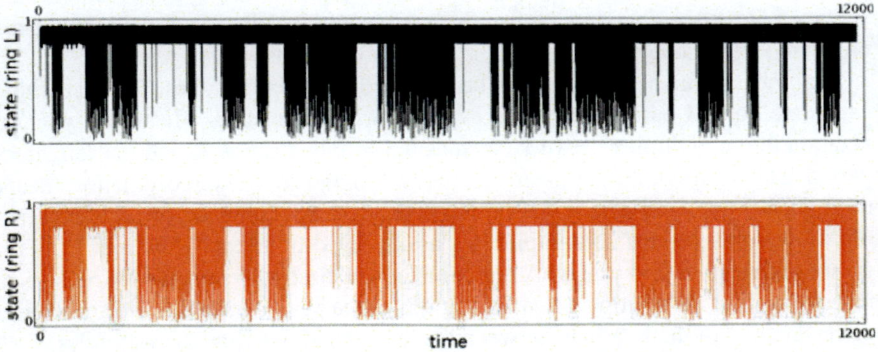

Fig. 6 Time-series of an oscillator in ring L and its homologous in ring R (red) where switching between solitary state and subthreshold oscillation occurs. The coupling strengths are set $\sigma = -0.40$ and $s = +0.80$

but takes some time to happen, the time depending on the magnitude of the coupling strengths (for low magnitude the switching goes unnoticed). Note that this kind of behaviour is not observed for regular (non-weak) chimera states, although we cannot rule out the possibility.

4 Conclusions and Open Problems

In the presented study we discuss the interplay of negative intra-ring σ and positive inter-ring s coupling strength in a multiplex of two identical layers for $(\sigma, s) \in [-1, 0] \times [0, 1]$. Each layer consists of a ring network of symmetric nonlocally coupled LIF oscillators. The two layers are coupled together via one-to-one connections. The single-ring network case is known in the bibliography, as well the case of weak inter-ring connectivity ($s = 0.10$), thus we focus on stronger interaction between the two rings ($s > 0.10$).

For small magnitude of s, the σ value where a phase transition occurs becomes more negative for both rings of the multiplex while their Kuramoto order parameters coincide; $Z^L = Z^R$ (Z is defined in Eq. 4). However, when s is greater than 0.30 and $-1.00 < \sigma < -0.50$, the Kuramoto order parameters of the two rings deviate significantly from each other. In this region of the parameter plane, we observe the phenomenon of subthreshold oscillations where elements oscillate below the threshold, without completing full-cycle oscillations. Two types of subthreshold oscillations are observed: stationary and switching. In the stationary type, subthreshold oscillations occur only in the elements of one of the two rings (presumably for long periods of time); see Fig. 4. In the switching type, solitary states and subthreshold oscillations alternate erratically in the two rings (the switching type is seen only for solitary states); see Figs. 5, 6. Notice that the boundaries of criticality might depend on the model parameters as well as the network configuration.

A number of open problems may be proposed for further studies. There are other candidate control parameters to be explored with interesting effects on synchronization in the multiplex network, such as the coupling range K and the refractory period. The nature of the stationary versus switching subtheshold oscillations needs to be further explored. Other questions relate to the phenomenon of the subthreshold oscillations itself; for example, is it a long-lived transient? How robust it is if noise is induced in the system? A more immediate direction concerns the numerical investigation of the multiplex in other regions of the parameter plane to gain a better understanding of the interplay between the inter- and intra-ring connectivities.

References

1. Pikovsky, A., Rosenblum, M., Kurths, J.: Synchronization - A Universal Concept in Nonlinear Sciences. Cambridge University Press, Cambridge (2001)
2. Boccaletti, S., Pisarchik, A.N., Del Genio, C.I., Amann, A.: Synchronization: From Coupled Systems to Complex Networks. Cambridge University Press, Cambridge (2018)
3. Kuramoto, Y., Battogtokh, D. Coexistence of coherence and incoherence in nonlocally coupled phase oscillators. Nonlinear Phenomena in Complex Systems. vol. 5, pp. 380–385 (2002). https://doi.org/10.48550/arXiv.cond-mat/0210694
4. Abrams, D.M., Strogatz, S.H.: Chimera states for coupled oscillators. Phys. Rev. Lett. **93**, 174102 (2004). https://doi.org/10.1103/PhysRevLett.93.174102
5. Olmi, S., Politi, A., Torcini, A.: Collective chaos in pulse-coupled neural networks. EPL **92**, 60007 (2010). https://doi.org/10.1209/0295-5075/92/60007
6. Omelchenko, I., Omel'chenko, O.E., Hövel, P., Schöll, E.: When nonlocal coupling between oscillators becomes stronger: patched synchrony or multichimera states. Phys. Rev. Lett. **110**, 224101 (2013). https://doi.org/10.1103/PhysRevLett.110.224101
7. Omelchenko, I., Provata, A., Hizanidis, J., Schöll, E., Hövel, P.: Robustness of chimera states for coupled FitzHugh-Nagumo oscillators. Phys. Rev. E **91**, 022917 (2015). https://doi.org/10.1103/PhysRevE.91.022917

8. Hizanidis, J., Kanas, V., Bezerianos, A., Bountis, T.: Chimera states in networks of nonlocally coupled Hindmarsh-Rose neuron models. Int. J. Bif. Chaos **24**, 1450030 (2014). https://doi.org/10.1142/S0218127414500308
9. Hizanidis, J., Kouvaris, N.E., Zamora-López, G., Díaz-Guilera, A., Antonopoulos, C.G.: Chimera-like states in modular neural networks. Sci. Rep. **6**, 19845 (2016). https://doi.org/10.1038/srep19845
10. Tsigkri-DeSmedt, N.D., Hizanidis, J., Schöll, E., Hövel, P., Provata, A.: Chimeras in leaky integrate-and-fire neural networks: effects of reflecting connectivities. Eur. Phys. J. B. **90**, 139 (2017). https://doi.org/10.1140/epjb/e2017-80162-0
11. Laing, C.R., Omel'chenko, O.: Moving bumps in theta neuron networks. Chaos **30**, 043117 (2020). https://doi.org/10.1063/1.5143261
12. Ulonska, S., Omelchenko, I., Zakharova, A., Schöll, E.: Chimera states in networks of Van der Pol oscillators with hierarchical connectivities. Chaos **26**, 094825 (2016). https://doi.org/10.1063/1.4962913
13. Tsigkri-DeSmedt, N.D., Hizanidis, J., Hövel, P., Provata, A.: Multi-chimera states and transitions in the leaky integrate-and- fire model with nonlocal and hierarchical connectivity. Eur. Phys. J. Special Topics **225**, 1149–1164 (2016). https://doi.org/10.1140/epjst/e2016-02661-4
14. Chouzouris, T., Omelchenko, I., Zakharova, A., Hlinka, J., Jiruska, P., Schöll, E.: Chimera states in brain networks: empirical neural vs modular fractal connectivity. Chaos **28**, 045112 (2018). https://doi.org/10.1063/1.5009812
15. zur Bonsen, A., Omelchenko, I., Zakharova, A., Schöll, E. Chimera states in networks of logistic maps with hierarchical connectivities. Eur. Phys. J. B. **91**, 65 (2018). https://doi.org/10.1140/epjb/e2018-80630-y
16. Andrzejak, R.G.: Chimeras confined by fractal boundaries in the complex plane. Chaos **31**, 053104 (2021). https://doi.org/10.1063/5.0049631
17. Katsaloulis, P., Verganelakis, D.A., Provata, A.: Fractal dimension and lacunarity of tractography images of the human brain. Fractals **17**, 181–189 (2009). https://doi.org/10.1142/S0218348X09004284
18. Katsaloulis, P., Ghosh, A., Philippe, A.C., Provata, A., Deriche, R.: Fractality in the neuron axonal topography of the human brain based on 3-D diffusion MRI. Eur. Phys. J. B **85**, 150 (2012). https://doi.org/10.1140/epjb/e2012-30045-y
19. Ashwin, P., Burylko, O.: Weak chimeras in minimal networks of coupled phase oscillators. Chaos **25**, 013106 (2015). https://doi.org/10.1063/1.4905197
20. Anesiadis, K., Provata, A.: Synchronization in multiplex leaky integrate-and-fire networks with nonlocal interactions. Front. Netw. Physiol. **2**, 910862 (2022). https://doi.org/10.3389/fnetp.2022.910862
21. Lapicque, L.: Recherches quantitatives sur l'excitation èlectrique des nerfs traitèe comme une polarization. J. Physiol. Pathol. Gènèrale. **9**, 567–578 (1907)
22. Brunel, N., van Rossum, M.C.W.: Quantitative investigations of electrical nerve excitation treated as polarization (translation of "Recherches quantitatives sur l'excitation èlectrique des nerfs traitèe comme une polarization"). Biol. Cybern. **97**, 341–349 (2007). https://doi.org/10.1007/s00422-007-0189-6

Networks' Modulation: How Different Structural Network Properties Affect the Global Synchronization of Coupled Kuramoto Oscillators

Juliette Courson, Thanos Manos, and Mathias Quoy

Abstract In a large variety of systems (biological, physical, social etc.), synchronization occurs when different oscillating objects tune their rhythm when they interact with each other. The different underlying network defining the connectivity properties among these objects drives the global dynamics in a complex fashion and affects the global degree of synchrony of the system. Here we study the impact of such types of different network architectures, such as Fully-Connected, Random, Regular ring lattice graph, Small-World and Scale-Free in the global dynamical activity of a system of coupled Kuramoto phase oscillators. By fixing the external stimulation parameters, we choose different fractions of nodes from the system first randomly and then informed by their respective strong/weak connectivity properties (centrality, shortest path length and clustering coefficient) and we measure the global degree of synchrony. Our main finding is, that in Scale-Free and Random networks a sophisticated choice of nodes based on graph connectivity properties exhibits a systematic trend in achieving higher degree of synchrony. For the other types of graphs considered, the choice of the stimulated nodes (randomly vs selectively using the aforementioned criteria) seems to not have a noticeable effect.

J. Courson (✉)
Laboratoire de Physique Théorique et Modélisation (LPTM), CNRS, UMR 8089,
CY Cergy Paris Université, Cergy-Pontoise Cedex, France
e-mail: juliette.courson@cyu.fr

J. Courson · T. Manos · M. Quoy
Equipes Traitement de l'Information et Systèmes (ETIS), CNRS, UMR 8051, ENSEA,
CY Cergy Paris Université, Cergy-Pontoise Cedex, France

J. Courson
Department of Computer Science, University of Warwick, Coventry, UK

M. Quoy
IPAL CNRS Singapore, Singapore, Singapore

T. Bountis et al. (eds.), *Chaos, Fractals and Complexity*, Springer Proceedings in Complexity, https://doi.org/10.1007/978-3-031-37404-3_16

1 Introduction

Complex networks is a powerful tool in various fields that allow us to investigate and understand the real world [4]. For example, different ensembles of neurons connected by synapses coordinate their activity to perform certain tasks (in biology), infrastructures like the Internet are formed by routers and computer cables and optical fibers (in hardware communication) and the human personal or professional relationships (in social sciences) to name a few [10].

Nonlinearity is a very important feature in complex systems giving a rich repertoire of different activity patterns, such as stable, unstable, periodic etc. A modification of some parameter might also produce a change in their stability, and therefore in the dynamics of the system. Furthermore, such systems may have a high sensitivity to initial conditions, or to any external input, that could completely change their dynamics [30].

Such dynamics often yield to a self-organized coherent activity, i.e. to synchronization. The latter can be loosely defined as the capacity of different oscillating objects to adjust their rhythm due to their interaction and plays a key role in a large variety of systems, whether biological, physical, or even social (see e.g. [23]). In a more formal way, synchronization emerges from the interaction of several autonomous oscillators, also called self-sustained oscillators. That is, nonlinear dynamical systems that produce oscillations without any need of external source. Their dynamics is given by a nonlinear differential equation or, in the case of multiple coupled oscillators, by several coupled differential equations.

The relative way that autonomous oscillators are connected within a given network can affect their global activity and synchronization properties. Neural networks can be represented as a graph of connections between the different neurons. Since the introduction of small-world networks and scale-free networks (see e.g. [2, 31]), the field of network graph analysis has attracted the attention of many studies aimed to better understand complex systems (see e.g. [5, 6, 14, 17, 29]). Furthermore, modern network connectivity techniques allow us to capture various aspects of their topological organization, as well as to quantify the local contributions of individual nodes and edges to network's functionality (see e.g. [27]).

In neuroscience, synchronization plays a very important role. The human brain is a very large and complex system whose activity comprises the rapid and precise integration of a gigantic amount of signals and stimulus to perform multiple tasks (see e.g. [8, 12, 27]). One example occurs in epileptic seizures, where periods of abnormal synchronization in the neural activity can spread within different regions of the brain, and cause an attack in the affected person (see e.g. [32]). More examples are found in other brain diseases such as Parkinson disease, where an excessively synchronized activity in a brain region correlates with motor deficit (see e.g. [7, 18] and references therein) or tinnitus (see e.g. [11, 19, 20] and references therein).

In this study, we focus at a rather theoretical framework. We set out to investigate the impact of different network architectures, such as Fully-Connected, Random, Regular ring lattice graph, Small-World and Scale-Free in the global dynamical activity of a system of coupled Kuramoto phase oscillators [16]. The Kuramoto model has been broadly used to study various types of oscillatory complex activity, see e.g. [1, 24, 26] (to name only a few) and references therein. Our goal is to investigate the impact of the network (graph) structure in the system's global degree of synchronization when applying identical and fixed external stimulus to different subsets of nodes which are chosen according to various network connectivity criteria. We find that, in scale-free and random networks, a sophisticated choice of nodes based on graph connectivity properties exhibits a systematic trend in achieving higher degree of synchrony. For the other types of graphs considered, the choice of the stimulated nodes (randomly vs selectively using the aforementioned criteria) seems to not have a noticeable effect.

2 Methods and Materials

2.1 Connectivity Measurements

We here study the dynamics of phase oscillators coupled via binary, undirected graphs $G = (V, E)$, containing a set of N vertices $V = \{v \in [\![1 : N]\!]\}$ and a set $E = \{(v, w) \in [\![1 : N]\!]^2\}$ of edges. Let A be the corresponding adjacency matrix, with $A_{vw} = 1$ if there is a connection between node v and node w, 0 otherwise. Self-connections are excluded, so $A_{vv} = 0$ for any vertex v. For our analysis later on, we will use the following graph connectivity measurements [22]:

- **Shortest path length**. The shortest path length $L_{v,w}$ between any two nodes v and w is the number of connections on the shortest path going from one to another, computed following Dijkstra's algorithm. We define the shortest path length of a node v as the average shortest path between v and any other node of the network:

$$< L_v >= \sum_{w \in V} \frac{L_{v,w}}{N}. \tag{1}$$

  Note that $L_{v,w}$ might not be defined if there is no way connecting node v to node w. The lower the shortest path length, the fastest the information goes from one node to another. For example, when building a subway network (that is, a graph where different stations are interconnected), one might want to minimize the stations' average shortest path length so the users can easily navigate across the city.
- **Centrality**. The eigenvector centrality is used to quantify the importance of a node in the network. Let λ be the highest eigenvalue for A, so that all the corresponding eigenvector's components are non null. The eigenvector centrality x_v of vertex v is defined as the vth component the eigenvector, namely:

$$x_v = \frac{1}{\lambda} \sum_{w \in V} A_{wv} x_w. \quad (2)$$

Keeping in mind the subway network example, a station with a high centrality would be densely connected to other stations, in particular to other central ones.
- **Clustering**. Let $k_v = \sum_w A_{vw}$ be the degree of node v. In the case of a undirected graph, $\frac{k_v(k_v-1)}{2}$ edges can exist in the direct neighborhood of v. With n_v the number of edges that actually exist in this neighborhood, the local clustering coefficient is defined as [12]:

$$C_v = \frac{2n_v}{k_v(k_v - 1)}. \quad (3)$$

That is, in a subway network where stations B and C are the next stop after station A on their line, clustering would give the probability that there exists a line directly connecting B and C.

2.2 *Neural Networks as Graphs*

We investigate synchronization properties in various network configurations that exhibit in general different characteristics. In more detail we here employ the neural networks described in the following list (see e.g. [31]):

- **Fully-Connected networks**. They contains $(N - 1)^2$ edges connecting every node in one layer to every node in the other layer.
- **Regular networks**. They consist of a lattice of N nodes, each being connected to their k nearest neighbors.
- **Small-World networks**. A Small-World network is constructed from a Regular one after multiple random rewiring phases: going clockwise over the lattice, a vertex and the edge to its nearest neighbor are selected. The edge is removed, and the vertex reconnected to a random node with probability p, without duplicating any existing edge. This random rewiring is repeated, considering the following nearest neighbour, until having rewired with probability p all edges of the network. Choosing p in an adequate range of values, the built network exhibits both high mean clustering and low mean characteristic path length, that is small-worldness.
- **Random networks**. Using the same procedure as for Small-World networks, setting $p = 1$ produces a Random graph, with all edges being systematically randomly rewired.
- **Scale-Free networks**. They are networks whose degree distribution follows a power-law:

$$P(k) \propto k^{-\gamma}. \quad (4)$$

with k the node degrees, γ a real constant. The Barabási-Albert model gives a procedure for the construction of scale-free networks [3], by starting with a small

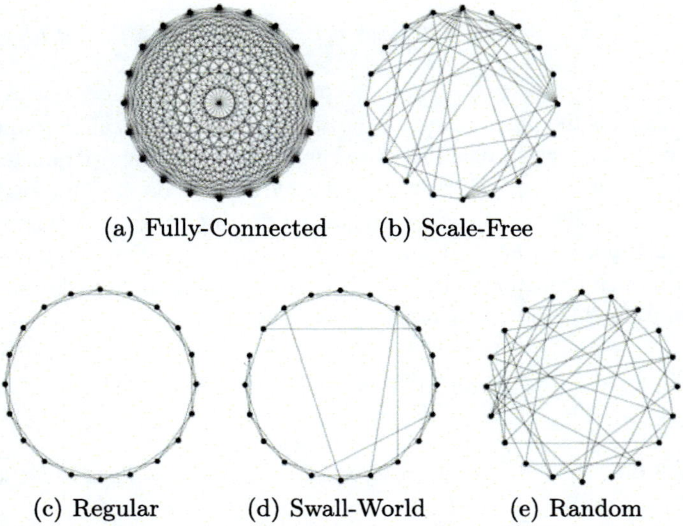

Fig. 1 **Network graphs**. Small graphs of size $N = 20$ showing the different network structures: **a** Fully-Connected graph, **b** Scale-Free graph with initial size $m_0 = 5$, **c** Regular graph with node degree $k = 4$, **d** Small-World graph with initial node degree $k = 4$ and rewiring probability $p = 0.2$ and **(c)** Random graph with initial node degree $k = 4$. See text for more details

Fully-Connected network of m_0 nodes then adding one by one the $N - m_0$ remaining nodes, connecting them to the m already present nodes with probability

$$p_v = \frac{k_v}{\sum_w k_w}, \tag{5}$$

$w \in [\![0 : m - 1]\!]$. Scale-free networks exhibit nodes with degrees that are several standard deviations away from the average degree of the network. These highly connected nodes are called *hubs*. Note that, however, for smaller networks with size $N < 100$, the scale-free property might not be properly observable.

Figure 1 provides a visual representation of the above mentioned graphs. Note that for visualization purposes we here show only a small fraction of the actual networks that we use later in our simulations where the number of nodes is set be $N = 500$.

2.3 *The Kuramoto Model*

We use of the Kuramoto model to study the neural activity of the coupled system. To this end we consider a population of N phase oscillators [16]:

$$\dot{\theta}_i = \omega_i + F\delta_{i,C}\sin(\Omega t + \theta_i) + \frac{K}{k_i}\sum_j A_{ij}\sin(\theta_j - \theta_i), \tag{6}$$

where θ_i denotes the phase of the $i-$th oscillator, ω_i its respective frequency (Hz) drawn from a Lorentz probability distribution $g(\omega)$ of scale parameter $\gamma = 0.5$, centered in $x_0 = 1$. A is the binary adjacency matrix coupling the oscillators, k_i the degree of oscillator i and K is the global coupling constant. We apply external stimulus in a subset of the oscillators with fixed amplitude F and frequency Ω. The term δ_i is a binary function indicating this subset of nodes where the stimulation is applied in different realization in our simulations,

$$\delta_i = \begin{cases} 1 \text{ if node } i \text{ is in the stimulated subset} \\ 0 \qquad\qquad\qquad \text{else.} \end{cases} \tag{7}$$

We set the time-step at 0.01s and we integrated the system with an Euler scheme (no noise is considered).

The system's degree of synchrony is measured using its order parameter r [16]:

$$re^{i\psi} = \frac{1}{N}\sum_{j=1}^{N} e^{i\theta_j}, \tag{8}$$

where Ψ denotes the population's mean phase. The order parameter r tends to 1 for a perfectly synchronized population and to 0 in the absence of synchronization respectively. Due to the presence of strong fluctuations, all r time-series shown in this paper are determined using a moving average on r, on time windows of length 2s sliding each 0.1s. The final states of a population, r_f, are computed by averaging these moving-averaged r time-series over a 15s time-window where the system has reached its stable state. The system's degree of synchrony depends on the coupling strength K's relative position to a critical coupling strength K_c, whose value depends on the network configuration (see e.g. [9, 21]). Here, we set this value at $K = 0.2$ so that all considered networks are desynchronized in the absence of any external stimulation.

3 Results

Network modulation. In order to adequately tune the stimulus' amplitude and frequency values such that they can lead the system into a synchronous state, we first perform a systematic analysis in the parameter space (F, Ω). Hence, we begin by applying external stimulus to all the nodes for each pair of parameters and measure the final order parameter r_f. Such a parameter map reveals the presence of the well-known Arnold tongues [23], namely regions in the plane (F, Ω) for which the system gets synchronized.

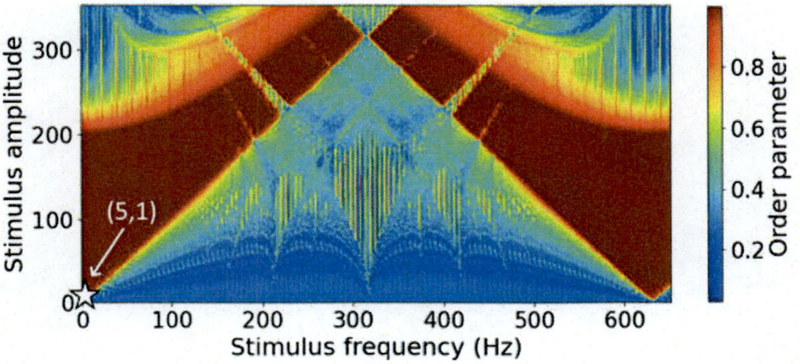

Fig. 2 **Synchronization regions in the stimulation frequency-amplitude parameter space for a Regular network**. Final order parameter reached for a Regular network of size $N = 20$, and degree $k = 4$ where all nodes are stimulated, for different pairs of stimulus intensity (F) and frequency (Ω) values in Eq. (6). Each data point corresponds to a single simulation over 30s, the final order parameter being averaged over the last 15s. The color map shows large main synchronization regions, as well as small higher-order synchronization areas. The white star symbol at the bottom-left part of the figure indicates the chosen parameters $(F, \Omega) = (5, 1)$ for the forthcoming simulations

In Fig. 2, we present the different synchronization regions (presence of several Arnold tongues) for a Regular network of a relatively small size $N = 20$ and mean neighborhood $k = 4$. For every other studied network, the maps depicts similar features with large synchronization regions at relatively small amplitudes of the external current, $F < 200$. These tongues get thinner with higher values of F. Inside the main Arnold tongues, the oscillators are phase-locked at the forcing frequency Ω and $r_f \approx 1$. Inside zones of weaker degree of synchrony $r_f < 1$, some oscillators are phase-locked, while the oscillators of higher natural frequencies keep rotating independently. The white star symbol in the bottom-left part of the figure indicates the chosen parameters $(F, \Omega) = (5, 1)$ for our forthcoming simulations, resulting in a partial phase-locking of the network. Note that we have performed similar analysis with larger sizes but smaller parameter grid size and the overall picture turns out to be consistent. We have also prepared similar plots for all considered network configurations (figures not shown here).

Simulation protocol. We set the values $F = 5$, $\Omega = 1$Hz for the stimulus intensity and frequency in Eq. (6), so that the network is weakly entrained without being completely phase-locked. We then measure the degree of synchronization (with the order parameter) in different networks described in Sect. 2.2. The system starts evolving for 4s without any external input before we start applying the stimulation to a subset of nodes until 30s. More precisely, we stimulate different fractions of nodes, i.e., 25, 50 and 75% in each given network. These nodes can be either chosen randomly, or depending on particular connectivity properties (as described in Sect. 2.1). For the latter case, we first sort the nodes according to their connectivity relative measurements (from higher to lower), i.e. the eigenvector centrality, average shortest path length and clustering coefficient. The resulting time series are smoothed with a

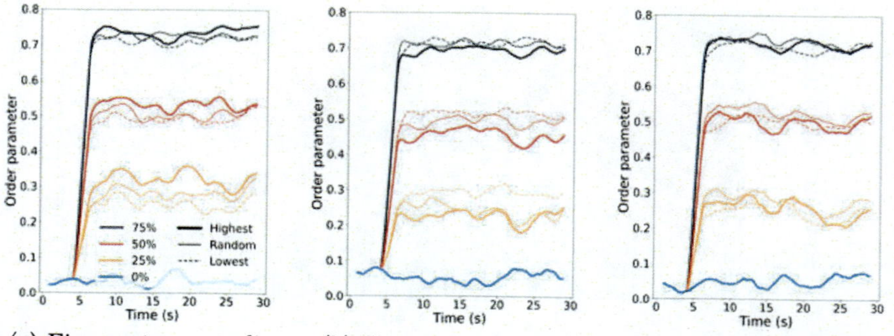

(a) Eigenvector centrality (b) Shortest path length (c) Clustering coefficient

Fig. 3 Representative r time-series for different stimulation setups of Scale-Free networks. The order parameter as a function of time, for $N = 500$ oscillators, when stimulating 0% (blue), 25% (orange), 50% (red) and 75% (black) of the nodes. The stimulated nodes are chosen randomly, then based on their **a** eigenvector centrality **b** average shortest path length and **c** clustering coefficient values. Bold solid lines (resp. dashed lines) correspond to the stimulation of nodes with the highest (resp. lowest) values, while thin solid lines correspond to the stimulation of randomly chosen nodes

moving average, and their r_f is averaged over the last 15s. In order to obtain a statistically relevant value of the final value of the order parameter r_f, we performed 20 simulations for each different network-setup (randomizing the initialization/generation of the networks, the natural frequencies and the initial conditions for each simulation).

In Fig. 3, we show representative time-series for various stimulation setups for Scale-Free networks of size $N = 500$ and initial size $m_0 = 5$. Each panel corresponds to a different initialization, from which the order parameter evolution is computed depending on the amount of stimulated nodes and the way they are selected. Thin solid lines correspond to randomly selected nodes. In that case, a first subset containing 25% of the nodes is created. Then for the stimulation of larger in size subsets, another 25% of nodes is successively added, so that larger subsets of random nodes always contains the smaller one. The bold solid lines (resp. dashed lines) show the time-series when the stimulated nodes are the ones with the highest (resp. lowest) eigenvector centrality (panel (a)), average shortest path length (panel (b)) and clustering coefficient (panel (c)). Note that higher values of the connectivity measurement do not necessarily lead to stronger synchrony. In particular, lower values for the nodes' average shortest path length depicts shorter connections to the rest of the network, and therefore a more efficient synchronization.

Optimization of global synchronization. In Fig. 4, we present the main finding of our work, namely a systematic comparison of the synchronization efficiency when applying identical stimulus in different types of graph networks. Each panel is split into 3 columns showing the statistical summary for the ensembles of different realizations and choosing to stimulate different subsets of nodes, i.e. randomly (middle-orange boxplots in the legends) or with highest (upper-red boxplots in legends) or lower (lower-blue boxplots in the legends) connectivity measurement. Panel (a)

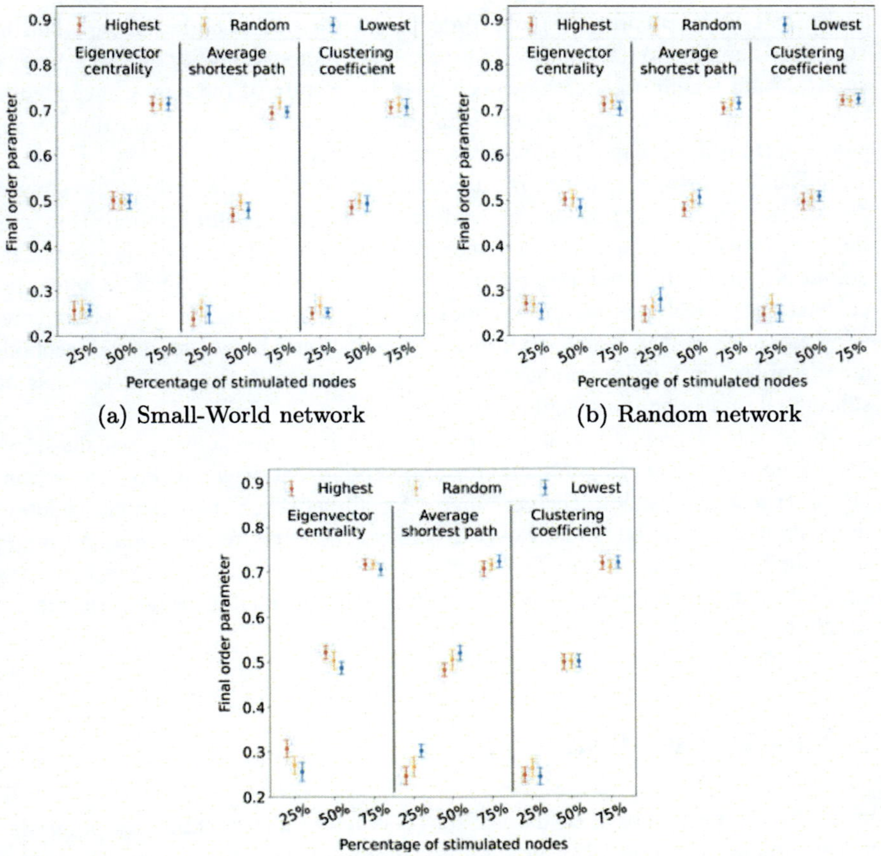

(a) Small-World network

(b) Random network

(c) Scale-Free network

Fig. 4 **Synchronization efficiency comparison for different types of graph networks**. Final order parameter obtained, for **a** Small-World networks **b** Random networks **c** Scale-Free networks of size $N = 500$. The final value of the order parameter r_f is computed for different stimulation subset sizes, composed of randomly chosen nodes (middle-orange boxplots in the legends), nodes with the highest connectivity measurement (left-red boxplots) and nodes with the lowest connectivity measurement (right-blue boxplots). r_f shown are an average over 20 simulations. The analysis is performed with three different connectivity measurements: eigenvector centrality (left column in each panel), average shortest path length (central column in each panel) and clustering (right column in each panel)

refers to a Small-World network of size $N = 500$, initial degree $k = 4$ and rewiring probability $p = 0.2$, (b) to a Random network of size $N = 500$, initial degree $k = 4$ and rewiring probability $p = 1$ and (c) to a Scale-Free network of size $N = 500$ and initial size $m_0 = 5$ respectively. Note this analysis is not performed on Regular and Fully-Connected networks, since all nodes of such networks have identical connectivity properties and does not allow any connectivity-based selection.

In Scale-Free networks, the global order parameter reaches higher values when the stimulus is applied to the nodes with higher eigenvector centrality or lower average shortest path length. In such networks, a small fraction of the nodes have significantly higher connectivity, and stimulating preferentially these nodes enables strong synchronization compared to stimulating random nodes.

In Random networks, selecting stimulated nodes according to their lowest average shortest path length instead of randomly enhances synchronization. However, there is no benefit in choosing more central nodes. Indeed, the high degree of randomness (induced by a rewiring probability $p = 1$, see Sect. 2.2) in these networks' connections causes disparity in the nodes' average shortest path length, without creating any node of way higher centrality. In Small-World networks, the connections are distributed in a more homogeneous way, and hence the node selection has no substantial impact on the system's final synchrony.

For all three aforementioned networks and all stimulation subset sizes, the selection of the nodes according to their clustering coefficient does not show any advantage over a simple random choice. Finally, for all networks and stimulation subset selection methods, although they overall achieve higher degree of synchrony, larger stimulated subsets containing 75% of the population do not allow to observe any clear advantage in particular selection of the nodes, since all three possible subsets largely overlap.

4 Summary and Discussion

In this study, we investigated the impact of structure and connectivity properties in a modulated network. We sought out to identify efficient ways to synchronize a population of Kuramoto phase oscillators using nodes' stimulation with fixed small amplitude and frequency. To this end, we first performed a parameter sweep exploration for stimulus amplitude and frequency parameters to identify settings that allow the system to synchronize. Then, we computed the evolution of characteristic networks of Kuramoto oscillators, where external stimulation is applied to different subpopulations with identical fixed low amplitude and frequency. In order to measure the system's synchrony of each network-type and stimulation configuration, we calculated the global order parameter for ensembles of different random system initializations.

We showed that by choosing this subpopulations based on their respective network connectivity properties (i.e. high eigenvector centrality and lower short-path length), we were able to enhance the networks' global degree of synchronization in comparison to the one achieved by randomly choosing them. However, this is not the case when using the clustering coefficient as a selection criteria.

From a neuroscience point of view Scale-Free networks play an important role in the structure and function of mammal brains, see for example [25] (a study on the scale-free dynamics and the emergence of collective organisation occurs in rodents) or [13] (investigating the fractal structure of the human brain and its dynamics).

Furthermore, in Alzheimer patients' brain, the functional connectivity structure is found to exhibit properties similar to Random network graphs (see e.g. [15, 28] and references therein). Thus, understanding how to optimally synchronize systems with similar network structures can improve the overall expected performance of a given external simulation protocol.

Acknowledgements JC was supported by the LABEX MME DII (ANR-16-IDEX-0008) Ph.D. grant and CY cognition grant of the CY Cergy Paris University. TM was supported by the Labex MME DII (ANR-11-LBX-0023-01) French national funding program. MQ was supported by a CNRS financing for a research semester in the IPAL Lab in Singapore. We also acknowledge the use of the Osaka compute-cluster resources of CY Cergy Paris University.

References

1. Acebrón, J.A., Bonilla, L.L., Vicente, C.J.P., Ritort, F., Spigler, R.: The Kuramoto model: a simple paradigm for synchronization phenomena. Rev. Mod. Phys. **77**(1), 137 (2005)
2. Barabási, A., Réka, A.: Emergence of scaling in random networks. Science **286**(5439), 509–512 (1999)
3. Barabási, A.L.: The barabási-albert model. In: Network Science. Cambridge University Press, Cambridge (2016)
4. Barrat, A., Barthélemy, M., Vespignani, A.: Dynamical Processes on Complex Networks. Cambridge University Press, Cambridge (2008)
5. Bassett, D.S., Bullmore, E.: Small-world brain networks. Neuroscientist **12**(6), 512–523 (2006)
6. Berry, H., Quoy, M.: Structure and dynamics of random recurrent neural networks. Adapt. Behav. **14**, 129–137 (2006)
7. Brown, P.: Oscillatory nature of human basal ganglia activity: relationship to the pathophysiology of Parkinson's disease. Mov. Disord. **18**(4), 357–363 (2003)
8. Chialvo, D.: Emergent complex neural dynamics. Nat. Phys. **6**, 744–750 (2010)
9. Chiba, H., Medvedev, G.S., Mizuhara, M.S.: Bifurcations in the Kuramoto model on graphs. Chaos (Woodbury, N.Y.) **28**(7), 073109 (2018)
10. Dorogovtsev, S.N., Mendes, J.F.F.: Evolution of Networks. From biological nets to the internet and WWW. Oxford. Oxford University Press (2003)
11. Eggermont, J.J., Tass, P.A.: Maladaptive neural synchrony in tinnitus: origin and restoration. Front. Neurol. **6** (2015)
12. Fornito, A., Zalesky, A., Bullmore, E.: Fundamentals of Brain Network Analysis. Elsevier/Academic Press (2016)
13. Grosu, G.F., Hopp, A.V., Moca, V.V., Bârzan, H., Ciuparu, A., Ercsey-Ravasz, M., Winkel, M., Linde, H., Mureşan, R.C.: The fractal brain: scale-invariance in structure and dynamics. Cerebral Cortex (2022)
14. Jeong, H., Tombor, B., Albert, R., Oltvai, Z.N., Barabási, A.L.: The large-scale organization of metabolic networks. Nature **407**(6804), 651–654 (2000)
15. Kang, M., Petrasek, Z.: Random graphs: theory and applications from nature to society to the brain. Internationale Mathematische Nachrichten **227**, 1–24 (2014)
16. Kuramoto, Y.: Chemical Oscillations, Waves, and Turbulence. Dover Books on Chemistry Series, Dover Publications (2003)
17. Li, W., Cai, X.: Statistical analysis of airport network of china. Phys. Rev. E **69**, 046106 (2004)
18. Manos, T., Diaz-Pier, S., Tass, P.A.: Long-term desynchronization by coordinated reset stimulation in a neural network model with synaptic and structural plasticity. Front. Physiol. **12** (2021)

19. Manos, T., Zeitler, M., Tass, P.A.: How stimulation frequency and intensity impact on the long-lasting effects of coordinated reset stimulation. PLoS Comput. Biol. **14**(5), 1–31 (2018)
20. Manos, T., Zeitler, M., Tass, P.A.: Short-term dosage regimen for stimulation-induced long-lasting desynchronization. Front. Physiol. **9** (2018)
21. Mirollo, R., Strogatz, S.: The spectrum of the partially locked state for the Kuramoto model. J. Nonlinear Sci. **17**, 309–347 (2007)
22. Newman, M.E.J.: Networks: An Introduction. Oxford University Press, Oxford, New York (2010)
23. Pikovsky, A., Rosenblum, M.G., Kurths, J.: Synchronization, a universal concept in nonlinear sciences. Cambridge University Press, Cambridge (2001)
24. Popovych, O.V., Jung, K., Manos, T., Diaz-Pier, S., Hoffstaedter, F., Schreiber, J., Yeo, B.T., Eickhoff, S.B.: Inter-subject and inter-parcellation variability of resting-state whole-brain dynamical modeling. Neuroimage **236**, 118201 (2021)
25. Ribeiro, T.L., Chialvo, D.R., Plenz, D.: Scale-free dynamics in animal groups and brain networks. Front. Syst. Neurosci. **14** (2021)
26. Rodrigues, F.A., Peron, T.K.D., Ji, P., Kurths, J.: The Kuramoto model in complex networks. Phys. Rep. **610**, 1–98 (2016)
27. Sporns, O.: Networks of the Brain. The MIT Press (2010)
28. Stam, C.J., De Haan, W., Daffertshofer, A., Jones, B.F., Manshanden, I., Van Cappellen van Walsum, A.N., Montez, T., Verbunt, J.P.A., De Munck, J.C., Vn Dijk, B.W., Berendse, H.W., Scheltens, P.: Graph theoretical analysis of magnetoencephalographic functional connectivity in Alzheimer's disease. Brain **132**, 213–224 (2009)
29. Strogatz, S.H.: Exploring complex networks. Nature **410**(6825), 268–276 (2001)
30. Strogatz, S.H.: Nonlinear Dynamics and Chaos: With Applications to Physics, Biology, Chemistry, and Engineering, 2nd edn. Westview Press, a member of the Perseus Books Group, Boulder, CO (2015)
31. Watts, D.J., Strogatz, S.H.: Collective dynamics of 'small-world' networks. Nature **393**(6684), 440–442 (1998)
32. Wong, R.K., Traub, R.D., Miles, R.: Cellular basis of neuronal synchrony in epilepsy. Adv. Neurol. **44**, 583–592 (1986)

Neural Correlates of Human-Machine Trust in Autonomous Vehicles Context

Andrei Dragomir, Ioulietta Lazarou, Manuel S. Seet, Spiros Nikolopoulos, Ioannis Kompatsiaris, and Anastasios Bezerianos

Abstract Poor mental states-such as fatigue, low vigilance and low trust-in-automation-have been known to interfere with the appropriate use and interaction with vehicular automation. This has spurred strong interest in driver state monitoring systems (DSMS) that support adaptive interfacing between human drivers and automated driving system to enhance road safety and driver experience. While there have been thriving developments in fatigue and vigilance monitoring, research on trust monitoring is still in its infancy. Trust-in-automation has predominantly been measured subjectively via self-report measures, with fewer studies attempting to measure trust objectively owing to the difficulties in capturing this relatively abstract mental state. Nevertheless, recent progress has unveiled promising potential for objective trust monitoring that can be implemented in future intelligent vehicles. This review presents a framework for understanding the cognitive, affective and behavioural components of driver trust, and surveys current approaches and developments in objective trust measurement in autonomous vehicle contexts using behavioural and brain-based techniques. Approaches are evaluated for strengths and limitations in both their conceptual validity in capturing trust-relevant information, measure reliability, and their practical value in real-world driving settings. Future directions for improving trust monitoring towards practical implementation are also discussed.

Keywords Human-machine interaction · Trust · Brain · Neuroimaging · Electroencephalography · Autonomous vehicles · Driving

A. Dragomir (✉) · M. S. Seet
The N.1 Institute for Health, National University of Singapore, 28 Medical Dr. 05-COR, Singapore 117456, Singapore
e-mail: andreid@nus.edu.sg

I. Lazarou · S. Nikolopoulos · I. Kompatsiaris · A. Bezerianos
Information Technologies Institute, Centre for Research and Technology Hellas (CERTH-ITI), Thessaloniki, Greece

T. Bountis et al. (eds.), *Chaos, Fractals and Complexity*, Springer Proceedings in Complexity, https://doi.org/10.1007/978-3-031-37404-3_17

1 Introduction

Major advances in automated driving technology are making autonomous vehicles (AV) increasingly feasible for widespread public use in the foreseeable future. Large-scale implementation of fully automated driving can precipitate many potential societal benefits, chief among which is significant improvement to transportation safety [1]. Currently, a substantial proportion of traffic accidents and injuries are caused by human errors related to poor driver vigilance [2], fatigue [3], stress [4, 5] etc. With the introduction of AVs, the driving task can be delegated to vehicular automation which can drive more safely than human drivers [6].

Ironically, concerns have been raised about how the safety improvements associated with driving automation might be negated by deteriorated driver states induced by automated driving. Fully automated driving (SAE Level 5) is still far from feasible given the current state of technology, and so the first generations of intelligent vehicles for public road use would be equipped with partially automated functions (SAE Levels 3–4). Under partial automated driving, human drivers are required to standby to take over vehicular control whenever a take-over request (TOR) is issued due to system limits [7], or to intervene during system malfunctions [8]. The process of taking over control requires situational awareness, action planning and execution [9]; all of which place acute demands on drivers' perceptual, cognitive and motor functions. To complicate this issue, long periods of automated driving will likely see human drivers being visually and cognitively disengaged from the driving task [10] or experiencing stress and mental fatigue due to prolonged monitoring [11]. These driver states have been shown to compromise the ability to regain situational awareness and manoeuvre the vehicle towards safety on short notice [9]. Another driver mental state that will be relevant in the era of automated driving is trust-in-automation, defined as the attitude concerning whether the vehicular automation can perform the driving task without compromising the safety of the driver and other road user. Over-trust or "complacency" promotes vigilance reduction and cognitive disengagement [8], which subsequently affects take-over performance as described above. Meanwhile, under-trust encourages drivers to revert to lower levels of automated driving or fully manual driving (SAE Level 0), re-introducing the risk of human errors into the driving task [7].

These issues have motivated strong interest in adaptive automation that flexibly adjusts its functions based on the current state of the driver. For example, when the human driver is losing attentiveness during automated driving, the vehicle may issue an alert to remind the driver to pay attention [12]. If the driver is showing signs of fatigue, the vehicle may activate advanced driver assistance systems (ADAS) and/or recommend the driver to take a break. These systems rely on driver state monitoring systems (DSMS) that record vehicle behaviour patterns or driver bio-behavioural signals, use them to estimate driver state information which is then relayed to the vehicle control/interface system. Over the last few decades, there have been thriving developments in monitoring systems for many driver states (including vigilance

[2, 13], fatigue [14, 15], workload [16]). Some forms of DSMS (e.g. driver drowsiness and attentiveness) have already been implemented in commercially available vehicles [17].

In contrast, the frontier on monitoring trust-in-automation is in its infancy. Interest in trust monitoring emerged relatively recently when driver trust-in-automation becomes increasingly pertinent as partially automated driving becomes increasingly achievable. It is highly desirable to track driver trust online during automated driving, so that the vehicle can adapt role allocation or interactions with the driver in real-time to calibrate appropriate trust-in-automation [18]. For example, if a driver is losing trust-in-automation, an interface can communicate to make system operations more transparent, issue a trust-recovering message that addresses a system error [9], or issue an invitation for the driver to take over vehicle control [19]. The aim is to foster optimal interactions between human drivers and autonomous vehicles that mitigates the safety concerns of over-trust or under-trust.

However, quantifying trust can be difficult because it is a relatively abstract psychological construct. Most research thus far measured trust-in-automation via self-report questionnaires. However, subjective measures of trust are based on introspection and thus have limited accuracy. Moreover, continually probing trust with questionnaires is too disruptive for the driver, and so is not feasible for a trust monitoring system. This has spurred researchers to seek for objective measures of trust, but this quest has proven difficult because trust is a high-level cognitive construct that does not have intuitively identifiable manifestations in behaviour or physiology. This stands in contrast with other mental states like stress and vigilance that have more direct physiological proxies, namely heart rate variability and eye behaviour respectively. Nonetheless, recent studies have experimented with and have found promise in objective measures that exploit indirect effects of trust. These developments unveil potential for driver trust monitoring that may be practically implemented in future intelligent vehicles.

The present focused review will survey the state-of-the-art advances in trust monitoring in AV contexts via brain measurements. In this paper, a unified theoretical framework is presented to help understand the cognitive, affective and behavioural components of driver trust. Then, current approaches in objective trust measurement are examined and evaluated. Lastly, future directions are recommended for advancing the practical potential of driver trust monitoring.

2 Understanding Trust-in-Automation

Driver trust-in-automation can be defined as an attitude concerning whether the vehicular automation can perform the driving task without compromising the safety of the driver and other road users [20]. When there is high trust, the driver accepts being made vulnerable to the actions of the vehicular automation while being confident that doing so will not result in any negative outcome. Conversely, when there is low trust, the driver becomes unwilling to accept that vulnerability due to lack of confidence of the automation's capabilities.

Trust-in-automation has focused on two broad aspects: factors of trust and effects of trust. Concerning the first aspect, a large body of research have identified the many factors influencing trust-in-automation [21], which have been organised in proposed frameworks. These factors include (1) individual factors such as age, personality, trust propensity, beliefs and expectations about driving automation; (2) automation factors such as the level of automation, vehicle appearance/anthropomorphism, system reliability and system transparency; and (3) environment factors such as situational uncertainty and risk. This article will only cover briefly how these factors modulate trust, in the next section (see section Psychological and Cognitive Factors of Trust Between Drivers and AVs); for an in-depth review, please refer to [21, 22].

The second aspect–the effects of driver trust–will be the main interest of the present review, as it holds the key to discovering objective trust measures. To discern the best approaches to operationalise and measure driver trust, it is useful to organise the effects of trust in a theoretical framework. No known comprehensive framework has been proposed before. Therefore, to meet this need, we present a unified framework of how trust-in-automation manifests in driver cognition, affect and behavioural. Such an approach is commonly used to systematically analyse abstract psychological constructs in terms of more concretely defined and observable components.

2.1 Cognitive Component

The cognitive effects of trust-in-automation include risk/reward evaluation, confidence, expectancy, mental engagement and situational awareness. High-trusting drivers evaluate automation use as more beneficial/rewarding than risky. They tend to have more confidence in automation performance and lower expectancy for errors, and so are less mentally engaged with the driving task and have reduced situational awareness. On the other hand, low-trusting drivers evaluate automation use as more risky than beneficial/rewarding. They tend to have lower confidence in automation performance, higher error expectancies and so are more mentally engaged and more acutely aware of the situation-these mental states heighten alertness and predispose the driver toward intervening. These cognitive constituents would be reflected in brain activity, with some, such as reward/risk evaluation, rooted in subcortical structures located deep in the brain (such as the amygdala and the ventral striatum) [23], and others, such as expectancy and mental engagement, based in cortical regions found on the outer layers of the brain (such as the orbitofrontal cortex, the anterior and posterior cortex) [24].

2.2 *Affective Component*

Trust-in-automation influences drivers' affective experience, which encompasses stress, arousal, and comfort [25]. Low trust is associated with high stress/arousal and low comfort: these experiences are underpinned by uncertainty of automation performance, and/or a sense of vulnerability or anxiety for any negative outcomes of automated driving errors (e.g. accidents, near-misses, injuries). On the other hand, high trust tends to be associated with low stress/arousal and high comfort, as the driver becomes relaxed while accepting the reliability of the automated vehicle. These affective components of trust bring about changes in periphery physiological activity such as heart rate, electrodermal activity and muscular tension [25].

2.3 *Behavioural Component*

Trust-in-automation impacts drivers' behaviour in two respects: Interventional behaviour and Anticipatory behavior. (1) Interventional behaviour refers to drivers' actions that reduces the level of automation or takes control away from it entirely. The general assumption is that a driver with lower trust would be more likely to take over vehicle control when the vehicular automation is functioning normally. This behavioural component would be reflected in the duration and frequency of automation disengagements [25]. (2) Anticipatory behaviours refer to actions that indicate driver preparation to intervene during automating driving [26]. Reduced trust would compel the driver to be increasingly observant of the driving task and be in a physically ready state to re-control the vehicle, despite no explicit instruction to do so from the vehicle (e.g. no TOR). These behavioural components would be reflected in eye-gaze patterns and body motion of the driver. Under this framework, driver trust is most directly reflected in cognition, with affect and behaviour being the consequent, downstream effects [27]. In other words, drivers' affective experience and behavioural patterns are primarily driven by the driver's cognitive stances regarding the automation. This premise should be considered when evaluating the extent to which different measures are truly capturing driver trust.

3 Psychological and Cognitive Factors of Trust Between Drivers and AVs

Several studies have indicated that people are reluctant with regards to vehicle automation due to safety reasons [28, 29]. In particular, American Automobile Association in 2018 reported that over 70% of U.S. drivers are afraid of using a fully AV, while approximately 75% of participants in a recent study [28] have moderate to high skepticism and apprehensions about AVs [29]. To improve the safety and acceptance of AVs, many studies have underlined the necessity of understanding driver's behavior and expectation in different situations according to each driving behavior [30]. Therefore, interacting with the vehicle technology, drivers usually exhibit specific cognitive processes that are regarded as behavioural adaptation (BA). In particular, several studies have demonstrated that the movement and type of surrounding vehicles can influence driving behavior, including driver reaction time and decision-making [31], drivers' expectations, perceived risk influences, trust in automation the driver's obstacles [32], poorer lane discipline, sudden reactions to safety-critical events, increased speed, or decreased time headway [33].

A previous study of vehicle technology indicated that BA should be considered in three stages: immediate, short term, and long term [34], which refers to adaptation that may occur soon after a driver experiences a change in a safety system or to a significant but lower rate of behavioural change for drivers, where the adaptation may be characterized by gradual changes in drivers [34]. Previous research has demonstrated that drivers exhibit decreased vigilance, such as increased mind wandering and less frequent eye blinks [35]. In this common vein, other studies have demonstrated that pre-existing knowledge and experience may influence trust of the drivers [32]. In particular, aspects of the driver's personality such as confidence and locus of control can also influence trust that rely on AVs [36]. Moreover, driving experience usually increases self-reported trust in automation through learning the reliability and predictability of the system [32]. Recent studies have suggested that people have a strong desire to be able to take control back from automated systems [37]), likely because they feel safer when they are in control of the vehicle rather than riding as a passive passenger. For example, a recent study [38] modeled risk acceptability for self-driving vehicles in a sample of responses from Chinese participants. On the other hand, another concern with AVs is overtrust in the automation as the driver is still responsible for certain aspects of driving such as monitoring the roadway. Drivers may over-rely on the automation and disengage from driving [39]. One on-road study showed that the interaction between AV maneuvers (e.g., lane change, immediate acceleration) and their parameters (e.g., acceleration, jerk, and quickness) influenced driver comfort [40]. For example, participants with a high level of trust tended to monitor the road less [41]; and longer fixation duration and higher fixation count on the driving environment were associated with greater situation awareness [42].

Recent advances have shown that drivers have reported greater intentions to behave more aggressively towards AVs compared to other human drivers [38]. For instance, survey-based work has revealed trust in AVs to be influenced by factors such as gender [28, 43], age [44, 45], personality [46], cultural differences [47], daily driving behaviours [43] and experience [44]. Regarding age and in particular, older adults, research has indicated that they raise concerns using AVs due to issues related to trust and confidence, such as not having an operator nearby during failures [48]. Nonetheless, the results of recent surveys have suggested that trust in self-driving technology is extremely low in the general population but especially older adults do not yet perceive the benefits or question the usability. In general, older adults tend to suffer the negative performance effects of imperfect automation more than younger age groups. Situations such as night-time driving [49], prolonged driving [50], and extreme temperatures [51] can induce fatigue; whereas mobile phones [52], in-vehicle systems [53], and other non-driving linked behaviors can induce problems with regards to sustained attention. Cognitive dysfunction and attentional deficits during driving with AV, have been demonstrated by measures of visual attention indexed by ocular behaviors. On the other hand, some may indicate overtrust and this over-reliance may come from older adults' inability to properly identify and diagnose automation errors due to age related limitations in working memory. Lower working memory may also make it more difficult for older adults to generate alternative courses of action, a working memory-intensive activity, if they are conscious of an automation failure.

4 Current Approaches to Objective Trust Measures Using Brain Signals

Researchers have increasingly explored neuro-monitoring of trust-in-automation. This interest has been fuelled by recent work suggesting that driver behavioural cues may not necessarily reflect driver trust [27]; instead, trust is rooted in driver cognition which is best probed with brain-based measures.

Consequently, this section will examine how neuroimaging techniques have been used to measure trust-in-automation. Current techniques which offer high spatial resolution when assessing brain activity, such as functional magnetic resonance imaging (fMRI) or magnetoencephalography (MEG), have limited practical application when investigating human trust in automation due to their lack of portability, as well as reduced temporal dynamic resolution. On the other hand, techniques such as electroencephalography (EEG) and functional near-infrared spectroscopy (fNIRS) couple their higher temporal dynamics resolution with reduced equipment size and wearability making them more suitable to inherently out-of-the-lab context needed to objectively assess trust.

4.1 Approaches Based on EEG

Given that trust is predominantly a cognitive construct, EEG probably stands as the most direct and accurate method of assessing driver trust with the highest signal-to-noise ratio and in dynamic contexts [25]. However, EEG-based measures are often less practical than other measures. Gel-based EEG may give the best signal quality, but they are time-consuming and laborious to set up and have limited portability. Nevertheless, there have been promising developments in wireless wearable EEG devices using dry sensors that significantly improve ease and speed of setup and allow for extended experimental times [54].

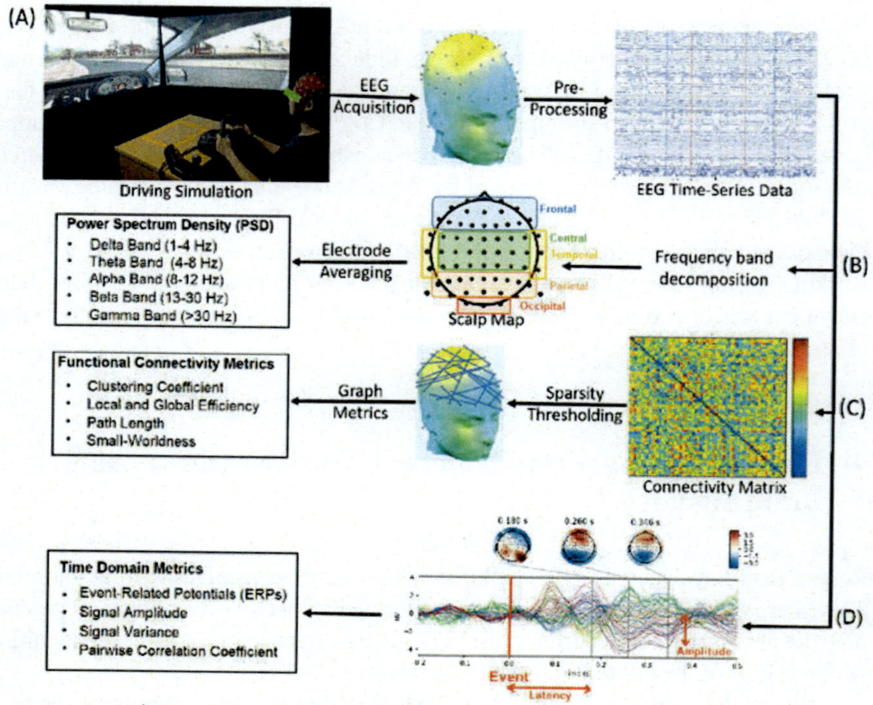

Fig. 1 Overview of most popular EEG measures used to assess neural correlates of trust in AV contexts. **A** EEG signals are recorded during driving task; **B** Recorded signals are decomposed into frequency bands and power spectral density (PSD) is estimated to quantify trust-related changes in brain activity; **C** functional connectivity measures are estimated to assess signal interdependencies and their changes correlated with changes in trust; **D** Time domain measures (such as peak-to-peak amplitude, signal variance, event related potentials) are estimated directly from the recorded EEG signals. (Adapted with permission from [55])

Depending on the device type and sensors coverage across the scalp, EEG provides adequate coverage to assess trust-related brain activity. The main regions where trust-relevant information can be accurately detected include the frontal, temporal and occipital areas [55, 56]. From recorded EEG signals a number of measures can be estimated that can subsequently be used in quantitative analyses. The most popular measures include: event related potentials (ERPs) and other time-domain measures, power spectral density (PSD) and functional connectivity (FC) (Fig. 1).

4.1.1 Time-Domain Measures

ERPs have limited practicality, as they can only be quantified with respect to a precisely defined events that are difficult to specify outside of controlled laboratory conditions. As such, few studies used ERPs to directly investigate trust in AVs.

Relevant examples are the works of [57, 58], which used passenger-driver trust as a paradigm for investigating trust in AV, arguing that passenger comfort is an important factor in the acceptance of AVs on a wide scale. Authors investigated whether certain driving events, such as braking, lane changing and aggressive driving, are correlated with brain measures. Specifically, in [58] authors showed that the neural response of passengers in the interval 200–500 ms after the event (also called the P300 ERP) is sensibly stronger in the case of low trust, compared to high trust condition (Table 1). This was noted for ERPs recorded in frontal and parietal regions. In [57] authors demonstrate that it is possible to predict with high accuracy (>80%) the driving events based on ERPs occurring prior to the actual event, and that ERPs timing may reflect the level of passenger trust.

Other time-domain measures, such as EEG signal mean amplitude, variance or pairwise correlations between signals are noise prone and thus less reliable. Nevertheless, studies showed that they can be effectively used to objectively classify the level of trust of a driver in a vehicle equipped with an automated obstacle detection system [59]. Authors showed that time domain measures recorded over frontal, central and parieto-occipital regions, in conjunction with frequency domain features and electrodermal activity measures can be used in a binary classification context (trust–distrust), with an accuracy of about 70%.

4.1.2 Power Spectral Measures

Power-based metrics are more flexible when compared to ERPs, as EEG power is sustained over a longer period. They also allow for a mechanistic interpretation of the results, when decomposing the signals into frequency bands which can subsequently be linked to known cognitive functions and the related brain areas. The typical frequency bands in which EEG signal activity is investigated are: delta ([1–4]Hz), theta ([4–8]Hz, alpha ([8–13]Hz), beta ([13–30]Hz) and gamma ([30–40]Hz).

An example of a study which investigated trust and its relation to related cognitive functions (e.g. motivational state and action planning) is that of Seet el at. [55] (Table 1). In this study, the authors investigate the differential impact of AV malfunction on human trust and the relevant neural correlates. The authors found that human trust in AV is specifically affected in vehicles operating in full autonomous mode (SAE level 5), when drivers are unable to take over control from the vehicle in situations of malfunctions. They conclude that the deterioration in trust does not originate from the malfunctions but is rather due to the inability of human operators to react and avoid the risk of negative outcomes (such as crashing) resulting from the AV malfunctions. The cognitive origins of the trust alterations in the are underlined by the selective decrease of right frontal power in the alpha band. This has as consequence an increase in left-lateralised frontal activity, which according to the framework of activation versus withdrawal motivation [60], points to an increased motivational preference to be actively engaged in the task.

Power spectral measures, estimated using the discrete wavelet transform, have been used also in [59] to develop a multimodal (EEG and electrodermal signals)

Table 1 Overview of studies investigating neural correlates of trust in AV contexts. ERP = event related potentials; PSD = power spectral density; wPLI = weighted phase lag index; PLI = phase lag index; VLPC = ventrolateral prefrontal cortex; DLPC = dorso-lateral prefrontal cortex; VMPC = ventromedial prefrontal cortex; AV = autonomous vehicle; HV = human-driven vehicle

Paper	Trust manipulation approach	Technique	Relevant brain locations
DSouza et al. [58]	Passenger-driver trust, changes with driving events (braking aggressive acceleration, lane changes)	EEG, ERP	Frontal and parietal regions, 200–500 ms after event (P3)
Belcher et al. [57]	Passenger-driver trust, changes with driving events (braking aggressive acceleration, lane changes)	EEG, ERP	Across cortex (20 channels); prediction prior to the driving event, 250–500 ms
Akash et al. [59]	Driving with an automated obstacle detection sensor (random faults in detecting obstacles)	EEG, time domain (signal amplitude, variance)	Frontal, central and parieto-occipital regions
Seet et al. [55]	AV malfunction (fails to stop at traffic lights), SAE levels 3 and 5	EEG, PSD	Frontal asymmetry in the alpha frequency band (8–13 Hz)
Akash et al. [59]	Driving with an automated obstacle detection sensor (random faults in detecting obstacles)	EEG, frequency domain (discrete wavelet transform)	Parietal and central regions in the theta (4–8 Hz) and beta frequency bands (16–32 Hz)
Seet et al. [55]	AV malfunction (fails to stop at traffic lights), SAE levels 3 and 5	EEG, functional connectivity (wPLI)	Frontal regions, clustering coefficient metric in the alpha frequency band (8–13 Hz)
Xu et al. [64]	AV malfunction (fails to stop at traffic lights), SAE levels 4 and 5	EEG, functional connectivity (PLI)	Across cortex, path length, local efficiency and clustering coefficient metrics in the theta band (4-8Hz); global efficiency and small-worldness metrics in the beta band (13–30 Hz)
Perello-March et al. [68]	Manipulation of participants' expectations regarding AV credibility (SAE levels 3 and 4)	fNIRS	Increased hemodynamic activity in the VLPC; lateralized activation in the DLPC for the low trust group
Unni et al. [69]	Safety critical situations in which drivers interact with other traffic participants (AVs and HVs)	fNIRS	Increased activation in the left and right DLPC and left VLPC, as well as VMPC in high trust compared to low trust

classifier model for determining human trust in an automated driving system. Power spectral measures were estimated from a 9 channel EEG system covering frontal, central and parieto-occipital regions. The most relevant spectral measures for discriminating trust versus distrust conditions were found in the theta, alpha and, especially, beta bands. Authors speculated that the relevance of the beta band in trust classification might indicate the interplay between trust and cognitive task demands, as well as emotion, given the known role of beta oscillations in modulating cognitive workload and emotional states. Furthermore, the relevance of the theta band activity was linked to the involvement of this oscillatory rhythm in decision-making related mechanisms as previously reported by other studies [23].

The relevance of the alpha and beta bands in characterizing trust in automation was highlighted also by the study of Oh et al. [61], who investigated trust level and its neural correlates during the decision of choosing between manual and automated control in a driving task. Similarly to the observations of other studies, authors reported significantly higher alpha and beta band power in high trust (vs. low trust) experimental trials in the majority of study participants. Furthermore, authors investigated also gamma band power and reported decreased power in high versis low trust trials for the majority of study participants.

4.1.3 Functional Connectivity Measures

Approaches based on FC measures are relatively newer and they assess correlations or interdependencies (in time and/or in phase) among signals recorded in different areas of the brain. This allows for a natural modeling of brain function which accounts for the information transfer across the brain that occurs during cognitive function [62]. Specifically, after estimating interdependencies, pairwise connections (or edges) are established between brain regions represented by EEG sensors (or underlying cortical regions), giving rise to a representation typically know as functional connectivity networks (or brain networks). These functional networks change configuration as a result of cognitive manipulations, tasks performance or after exposure to stimuli. The changes in networks' configuration across different conditions can then be quantified using graph theoretic metrics.

Seet et al. [55] employed graph theoretic measures estimated from functional connectivity networks to investigate the differential impact of AV malfunction on human trust, in addition to the power spectral measures described above (Fig. 1). As functional connectivity measure they used the weighted phase-lag index, a measure which indexes phase synchonization in electrophysiological signals and is less sensitive to volume conduction effects [63]. They found that the clustering coefficient (a measure of how much brain regions tend to cluster, or interact, together) in the frontal right region significantly decreases in the alpha band, in cases when AV malfunctions occur in full automation mode. The authors interpret this as a decrease in frontal right hemisphere's functional segregation that underpins the disruption of cortical function supporting executive cognition.

In a similar experimental paradigm, Xu et al. [64] investigated the changes in the networks' functional integration (characterized by graph theoretic measures such as path length and global efficiency), as well as local segregation (characterized using graph theoretic measures such as clustering coefficient and local efficiency). The balance between optimal functional network integration and segregation is a hallmark of higher cognition function. The authors investigated changes in these measures in the theta and beta bands between normal and malfunction trials, separately for conditional automation (SAE level 4) and full automation (SAE level 5) modes. They found significantly higher path length and decreased network efficiency in the theta band in malfunction trials in the conditional automation mode, specifically for the low trust condition. Moreover, there was increased local efficiency and clustering coefficient in the theta band in the case of malfunction trials (when compared to normal function trials) in the high trust condition. In the full automation mode results indicated a decrease in functional integration in the case of malfunction trials, characterized by global efficiency, in the high trust condition. Furthermore, a disruption of regular functional network architecture was observed in the beta band, quantified by a decreased small-worldness measure in malfunction trials, in the low trust condition. Collectively, these results suggest the significant changes, both in local-specialized and in global-integrative brain function, that are associated to trust in AV, in the context of vehicle malfunction.

4.2 Approaches Based on fNIRS

fNIRS is a relatively newer neuroimaging technique, which combines the advantage of fMRI in terms of spatial resolution and that of EEG in terms of portability, making is suitable for out-of-the-lab testing. It assesses brain activity by measuring the hemodynamic response to neural activity relying on the different absorption properties of biological chromophores. This makes it different from fMRI, which relies on the paramagnetic properties of hemoglobin [65]. fNIRS passes near-infrared light (650–950 nm) between pairs of source and detector sensors, located on the scalp at a fixed separation distance, typically between 2 and 5 cm. However, it must be noted that fNIRS is limited to measuring only the neural activity of superficial layers of the cortex, in contrast to fMRI which allows whole brain coverage.

fNIRS has been increasingly used recently in cognitive assessment studies in AV contexts, particularly in areas such as monitoring cognitive workload, attention and fatigue [66] (please see [67] for a review of fNIRS applications in driving research). However, while cognitive assessment studies based on fNIRS abund, there are only a handful of works investigating trust in AV contexts. Recently, Perello-March et al. [68] investigated trust by manipulating drivers expectations on AV (SAE Levels 3–4) credibility. Their results provide support for the hypothesis suggesting two distinct but interrelated cortical mechanisms for trust and distrust (low trust). High trust is suggested by the authors to be behaviorally linked to decreased attentional monitoring and working memory, while low trust is rather tied to affective (or emotional) mechanisms. Authors reported increased hemodynamic activity in the ventrolateral prefrontal cortex (VLPC), as well as a lateralized activation in the dorso-lateral prefrontal cortex (DLPC) for the low trust group, when compared to the high trust group. These findings are supported also by previous literature on trust in automation in other areas (e.g. human-robot collaboration [24].

In another recent work, Unni et al. [69] used fNIRS to investigate trust related decision-making in AV driving contexts in which drivers were confronted with safety critical situations in which drivers interact with other traffic participants (AVs and human drivers–HV). Specifically, human drivers were asked to find traffic gaps to turn left in front of incoming rightwards traffic (AV and HV) in an intersection. Their behavioral results showed that participants were more certain in decision-making when confronted with AV incoming traffic, observation underscored by smaller traffic gap size, when compared to HV incoming traffic. When comparing cortical activity, largest activation changes were observed in the left and right DLPC and left VLPC, as well as ventromedial prefrontal areas (VMPC), with increased activation differences being noted when turning in front of AV as compared to HV traffic.

5 Conclusions and Future Directions

Trust is a crucial element which pervades the realm of human-human interaction. Human-machine trust plays a significant role in the acceptance and adoption of new technologies, as well as in optimizing human performance in environments which require collaboration or ergonomic interaction between humans and machines [70]. Studying trust and particularly trust-related changes, in human cognition, affect and behaviour is essential for designing human-machine interfacing to mitigate trust-eroding effects.

While current trust assessment approaches are dominated by self-reports, behavioral and physiological measures, brain-based techniques have a significant potential for objective assessment of human-machine trust, particularly in AV related applications. Self-report measures, while very well established, have the disadvantage of being obtrusive to the task, unable to capture sufficient dynamic variations, and being often to subjective, or lacking accuracy [25]. Behavioral measures, while less obtrusive, are often more difficult to capture and interpret, as trust is not the sole factor influencing behavior. In addition, there is significant individual behavioral variability and strong internal determinants influencing trust. Physiological measures (such as electrodermal activity, eye gazing, heart rate, while being easier to capture/record have significant disadvantages as well. Specifically, they reflect physiological activity downstream from the brain, where the main mechanisms of trust are modulated. As such, peripheral physiological measures are often less reliable, due to confounding cognitive and behavioral factors.

Approaches for assessing human-machine trust in AV based on neural signals are still in their infancy. Most existing studies utilize EEG and fNIRS due to their portability and ease of applying them in ecological experiments. AV contexts, and vehicular transportation research generally, require mobility and the capability to capture dynamic changes in human-machine trust and in underlying neural correlates. However, research in this area can still make use of the vast body of research in the trust in automation area, as well as human–machine interaction, both in terms of rigorous theoretical models of trust, as well as in terms of neuroimaging findings and related neural mechanisms [24, 25]. So far, findings of AV studies confirmed previously known neural mechanisms of trust, as well as behavioral aspects from cognitive neuropsychology research, such as: (i) the relevance of the approach versus withdrawal motivation and decision-making mechanisms [55]; (ii) the role played by cognitive load and affect on trust [59]; and (iii) attentional monitoring and working memory [68]. With respect to these, neural activity in the theta, alpha, beta and gamma bands was reported to be highly correlated with different levels of trust in AV, as overviewed in Sect. 4.1.

The promise of AV of improving public road safety and driving experience depends on public acceptance and and uptake of this technology and human-machine trust plays a key role in this process. Current research challenges stand in translating the findings and the brain signals measurement techniques from lab-based, simulated driving conditions to actual road driving conditions. For this to happen, streamlined approaches using reduced sensor layouts for unobtrusive monitoring are needed. Seet et al. [55] showed that it is possible to capture trust dynamics using only few frontal EEG sensors. Future studies can use dry sensor, wireless devices to expand these findings to actual driving conditions. Additionally, it would be beneficial if future work will focus on contextualizing existing theoretical models of trust with actual brain and physiological measurements to create rigorous, unifying theoretical frameworks for investigating trust in AV.

References

1. Teoh, E.R., Kidd, D.G.: Rage against the machine? Google's self-driving cars versus human drivers. J. Saf. Res. **63**, 57–60 (2017)
2. Seet, M., Bezerianos, A., Panou, M., Bekiaris, E., Thakor, N.V., Dragomir, A.: Individual susceptibility to vigilance decrement in prolonged assisted driving revealed by alert-state wearable EEG Assessmen. IEEE Trans. Cognit. Dev. Syst. (2022). In Press
3. Wang, H., Dragomir, A., Abbasi, N.I., Li, J., Thakor, N.V., Bezerianos, A.: A novel real-time driving fatigue detection system based on wireless dry EEG. Cognit. Neurodyn. **12**(4), 365–376 (2018)
4. Chung, W.Y., Chong, T.W., Lee, B.G.: Methods to detect and reduce driver stress: a review. Int. J. Automot. Technol. **20**(5), 1051–1063 (2019)
5. Sciaraffa, N., Di Flumeri, G., Germano, D., Giorgi, A., Di Florio, A., Borghini, G., Vozzi, A., Ronca, V., Varga, R., van Gasteren, M., Babiloni, F.: Validation of a light EEG-based measure for real-time stress monitoring during realistic driving. Brain Sci. **12**(3), 304 (2022)
6. Hergeth, S., Lorenz, L., Krems, J.F.: Prior familiarization with takeover requests affects drivers' takeover performance and automation trust. Hum. Factors **59**(3), 457–470 (2017)

7. Gold, C., Körber, M., Lechner, D., Bengler, K.: Taking over control from highly automated vehicles in complex traffic situations: the role of traffic density. Hum. Factors **58**(4), 642–652 (2016)
8. Miller, D., Johns, M., Mok, B., Gowda, N., Sirkin, D., Lee, K., Ju, W.: Behavioral measurement of trust in automation: the trust fall. In: Proceedings of the Human Factors and Ergonomics Society Annual Meeting, pp. 1849–1853. SAGE (2016)
9. Collet, C., Musicant, O.: Associating vehicles automation with drivers functional state assessment systems: a challenge for road safety in the future. Front. Hum. Neurosci. **13**, 131 (2019)
10. Larue, G.S., Rakotonirainy, A., Pettitt, A.N.: Driving performance impairments due to hypovigilance on monotonous roads. Accid. Anal. Prev. **46**(3), 2037–2046 (2011)
11. Schmidt, E.A., Schrauf, M., Simon, M., Fritzsche, M., Buchner, A., Kincses, W.E.: Drivers' misjudgement of vigilance state during prolonged monotonous daytime driving. Accid. Anal. Prev. **41**(5), 1087–1093 (2009)
12. Cummings, M.L., Bauchwitz, B.: Safety implications of variability in autonomous driving assist alerting. IEEE Trans. Intell. Transp. Syst. **23**(8), 12039–12049 (2022)
13. Kong, W., Lin, W., Babiloni, F., Hu, S., Borghini, G.: Investigating driver fatigue versus alertness using the granger causality network. Sensors **15**(8), 19181–19198 (2015)
14. Di Flumeri, G., Ronca, V., Giorgi, A., Vozzi, A., Aricò, P., Sciaraffa, N., Zeng, H., Dai, G., Kong, W., Babiloni, F.: EEG-based index for timely detecting user's Drowsiness occurrence in automotive applications. Front. Hum. Neurosci. **16** (2022)
15. Bose, R., Wang, H., Dragomir, A., Thakor, N.V., Bezerianos, A., Li, J.: Safety implications of variability in autonomous driving assist alerting. IEEE Trans. Cogn. Dev. Syst. **12**(2), 323–331 (2019)
16. Lei, S., Roetting, M.: Influence of task combination on EEG spectrum modulation for driver workload estimation. Hum. Factors **53**(2), 68–179 (2011)
17. Mabry, J.E., Glenn, T.L., Hickman, J.S.: Commercial motor vehicle operator fatigue detection technology catalog and review. Technical Report (2019)
18. Metcalfe, J.S., Marathe, A.R., Haynes, B., Paul, V.J., Gremillion, G.M., Drnec, K., Atwater, C., Estepp, J.R., Lukos, J.R., Carter, E.C., Nothwang, W.D.: Building a framework to manage trust in automation. In: Micro-and Nanotechnology Sensors, Systems, and Applications IX, pp. 351–361. SPIE (2017)
19. Molnar, L.J., Ryan, L.H., Pradhan, A.K., Eby, D.W., Louis, R.M.S., Zakrajsek, J.S.: Understanding trust and acceptance of automated vehicles: an exploratory simulator study of transfer of control between automated and manual driving. Transp. Res. F: Traffic Psychol. Behav. **58**, 319–328 (2018)
20. Kyriakidis, M., de Winter, J.C., Stanton, N., Bellet, T., van Arem, B., Brookhuis, K., Martens, M.H., Bengler, K., Andersson, J., Merat, N., Reed, N.: A human factors perspective on automated driving. Theor. Issues Ergon. Sci. **20**(3), 223–249 (2019)
21. Hoff, K.A., Bashir, M.: Trust in automation: integrating empirical evidence on factors that influence trust. Hum. Factors **57**(3), 407–434 (2015)
22. Schaefer, K.E., Chen, J.Y., Szalma, J.L., Hancock, P.A.: A meta-analysis of factors influencing the development of trust in automation: implications for understanding autonomy in future systems. Hum. Factors **58**(3), 377–400 (2016)
23. Drnec, K., Marathe, A.R., Lukos, J.R., Metcalfe, J.S.: From trust in automation to decision neuroscience: applying cognitive neuroscience methods to understand and improve interaction decisions involved in human automation interaction. Front. Hum. Neurosci. **10**, 290 (2016)
24. Hopko, S.K., Mehta, R.K.: Trust in shared-space collaborative robots: shedding light on the human brain. Hum. Factors 00187208221109039 (2022)
25. Kohn, S.C., De Visser, E.J., Wiese, E., Lee, Y.C., Shaw, T.H.: Measurement of trust in automation: a narrative review and reference guide. Front. Psychol. **12** (2021)
26. He, D., DeGuzman, C.A., Donmez, B.: Anticipatory driving in automated vehicles: the effects of driving experience and distraction. Hum. Factors 00187208211026133 (2021)
27. Victor, T.W., Tivesten, E., Gustavsson, P., Johansson, J., Sangberg, F., Ljung Aust, M.: Automation expectation mismatch: incorrect prediction despite eyes on threat and hands on wheel. Hum. Factors **60**(8), 1095–1116 (2018)

28. Schoettle, B., Sivak, M.: A survey of public opinion about autonomous and selfdriving vehicles in the U.S., the U.K., and Australia (2014). http://deepblue.lib.umich.edu/handle/2027.42/108384
29. Piao, J., McDonald, M., Hounsell, N., Graindorge, M., Graindorge, T., Malhene, N.: Public views towards implementation of automated Vehicles in urban areas. Transp. Res. Proc. **14**, 2168–2177 (2016)
30. Lee, J.G., Kim, K.J., Lee, S., Shin, D.H.: Can autonomous vehicles be safe and trustworthy? Effects of appearance and autonomy of unmanned driving systems. Int. J. Hum. Comput. Interact. **31**(10), 682–691 (2015)
31. Jurecki, R., Poliak, M., Jaśkiewicz, M.: Young adult drivers: simulated behaviour in a car-following situation. Promet-Traffic Transp. **29**(4), 381–390 (2017)
32. Hoff, K.A., Bashir, M.: Trust in automation: integrating empirical evidence on factors that influence trust. Hum. Factors **57**(3), 407–434 (2015)
33. Young, M.S., Stanton, N.A.: Back to the future: brake reaction times for manual and automated vehicles. Ergonomics **50**, 46–58 (2007)
34. Manser, M.P., Creaser, J., Boyle, L.N.,: Behavioural adaptation: methodological and behavrioal issues. In Behavioural Adaptation and Road Safety: Theory, Evidence and Action, pp. 339–359 (2013)
35. Körber, M., Cingel, A., Zimmermann, M., Bengler, K.: Vigilance decrement and passive fatigue caused by monotony in automated driving. Proc. Manuf. **3**, 2403–2409 (2015)
36. Walker, G.H., Stanton, N.A., Salmon, P.: Trust in vehicle technology. Int. J. Veh. Des. **70**, 157 (2016)
37. König, M., Neumayr, L.: Users' resistance towards radical innovations: the case of the self-driving car. Transp. Res. F: Traffic Psychol. Behav. **44**(3), 42–52 (2017)
38. Liu, P., Yang, R., Xu, Z.: How safe is safe enough for self-driving vehicles? Risk Anal. **39**(2), 315–325 (2019)
39. Price, M.A., Venkatraman, V., Gibson, M.C., Lee, J.D., Mutlu, B.: Psychophysics of trust in vehicle control algorithms. (SAE Technical Paper) (2016). https://doi.org/10.4271/2016-01-0144
40. Bellem, H., Schonenberg, T., Krems, J.F., Schrauf, M.: Objective metrics of comfort: Developing a driving style for highly automated vehicles. Transp. Res. F: Traffic Psychol. Behav. **41**, 45–54 (2016)
41. Hergeth, S., Lorenz, L., Vilimek, R., Krems, J.F.: Keep your scanners peeled: gaze behavior as a measure of automation trust during highly automated driving. Hum. Factors **58**, 509–519 (2016)
42. Shinohara, Y., Currano, R., Ju, W., Nishizaki, Y.: Visual attention during simulated autonomous driving in the US and Japan. In: Proceedings of the 9th International Conference on Automotive User Interfaces and Interactive Vehicular Applications, Oldenburg, pp. 144–153 (2017)
43. Howard, D., Dai, D.: Public perceptions of self-driving cars: the case of Berkeley, California. In: Proceedings of the 93rd Transportation Research Board Annual Meeting, Washington, D.C., pp. 1–21 (2014)
44. Bansal, P., Kockelman, K.M.: Are we ready to embrace connected and self-driving vehicles? A case study of Texans. Transportation **45**(2), 641–675 (2018)
45. Haghzare, S., Campos, J.L., Bak, K., Mihailidis, A.: Older adults' acceptance of fully automated vehicles: effect of exposure, driving style, age, and driving conditions. Accid. Anal. Prev. **150**, 105919 (2021)
46. Jing, P., Du, L., Chen, Y., Shi, Y., Zhan, F., Xie, J.: Factors that influence parents' intentions of using autonomous vehicles to transport children to and from school. Accid. Anal. Prev. **152**, 105991 (2021)
47. Kaye, S., Lewis, I., Forward, S., Delhomme, P.: A priori acceptance of highly automated cars in Australia, France, and Sweden: a theoretically-informed investigation guided by the TPB and UTUAT. Accid. Anal. Prev. **137**, 105441 (2020)
48. Faber, K., van Lierop, D.: How will older adults use automated vehicles? Assessing the role of AVs in overcoming perceived mobility barriers. Transp. Res. Part A Policy Pract. **133**, 353–363 (2020)

49. Phipps-Nelson, J.O., Redman, J.R., Rajaratnam, S.M.: Temporal profile of prolonged, nighttime driving performance: breaks from driving temporarily reduce time on task fatigue but not sleepiness. J. Sleep Res. **20**, 404–415 (2011)
50. Finkleman, J.M.: A large database study of the factors associated with workinduced fatigue. Hum. Factors **36**, 232–243 (1994)
51. Xianglong, S., Hu, Z., Shumin, F., Zhenning, L.: Bus drivers' mood states and reaction abilities at high temperatures. Transp. Res. Part F Traffic Psychol. Behav. **59**, 436–444 (2018)
52. Strayer, D.L., Drews, F.A.: Cell-phone-induced driver distraction. Curr. Dir. Psychol. Sci. **16**, 128–131 (2007)
53. Arexis, M., Maquestiaux, F., Gaspelin, N., Ruthruff, E., Didierjean, A.: Attentional capture in driving displays. Br. J. Psychol. **108**, 259–275 (2017)
54. Chi, Y.M., Wang, Y.T., Wang, Y., Maier, C., Jung, T.P., Cauwenberghs, G.: Dry and noncontact EEG sensors for mobile brain-computer interfaces. IEEE Trans. Neural Syst. Rehabil. Eng. **20**(2), 228–235 (2011)
55. Seet, M., Harvy, J., Bose, R., Dragomir, A., Bezerianos, A., Thakor, N.: Differential impact of autonomous vehicle malfunctions on human trust. IEEE Trans. Intell. Transp. Syst. **23**(1), 548–557 (2022)
56. Park, C., Shahrdar, S., Nojoumian, M.: EEG-based classification of emotional state using an autonomous vehicle simulator. In: 2018 IEEE 10th Sensor Array and Multichannel Signal Processing Workshop (SAM), pp. 297–300. IEEE (2018)
57. Belcher, M.A., Huang, I., Battacharya, S., Hairston, D.W., Metcalfe, J.S.: EEG-based prediction of driving events from passenger cognitive state using Morlet Wavelet and Evoked Responses. Transp. Eng. **8**, 100107 (2022)
58. DSouza, K., Dang, T., Metcalfe, J.S., Battacharya, S.: Brain-based indicators of passenger trust during open-road driving. In: 2021 IEEE 94th Vehicular Technology Conference (VTC2021-Fall), pp. 1–6. IEEE (2021)
59. Akash, K., Hu, W.L., Jain, N., Reid, T.: A classification model for sensing human trust in machines using EEG and GSR. ACM Trans. Interact. Intell. Syst. (TiiS) **8**(4), 1–20 (2018)
60. Harmon-Jones, E., Gable, P.A.: On the role of asymmetric frontal cortical activity in approach and withdrawal motivation: an updated review of the evidence. Psychophysiology **55**(1), e12879 (2018)
61. Oh, S., Seong, Y., Yi, S., Park, S.: Neurological measurement of human trust in automation using electroencephalogram. Int. J. Fuzzy Logic Intell. Syst. **20**(4), 261–271 (2020)
62. Dragomir, A., Omurtag, A. : Brain's networks and their functional significance in cognition. In: Handbook of Neuroengineering, pp. 1–30 (2021)
63. Vinck, M., Oostenveld, R., Van Wingerden, M., Battaglia, F., Pennartz, C.M.: An improved index of phase-synchronization for electrophysiological data in the presence of volume-conduction, noise and sample-size bias. Neuroimage **55**(4), 1548–1565 (2011)
64. Xu, T., Dragomir, A., Liu, X., Yin, H., Wan, F., Wang, H.: An EEG study of human trust in autonomous vehicle basing on graphic theoretical analysis. Front. Neuroinform. **16**, 907942 (2022)
65. Pinti, P., Tachtsidis, I., Hamilton, A., Hirsch, J., Aichelburg, C., Gilbert, S., Burgess, P.W.: The present and future use of functional near-infrared spectroscopy (fNIRS) for cognitive neuroscience. Ann. N. Y. Acad. Sci. **1464**(1), 5–29 (2020)
66. Sibi, S., Baiters, S., Mok, B., Steiner, M., Ju, W.: Assessing driver cortical activity under varying levels of automation with functional near infrared spectroscopy. In: 2017 IEEE Intelligent Vehicles Symposium (IV), pp. 1509–1516. IEEE (2017)
67. Balters, S., Baker, J.M., Geeseman, J.W., Reiss, A.L.: A methodological review of fNIRS in driving research: relevance to the future of autonomous vehicles. Front. Hum. Neurosci. **15**, 637589 (2021)
68. Perello-March, J.R., Burns, C.G., Woodman, R., Elliott, M.T., Birrell, S.A.: Using fNIRS to verify trust in highly automated driving. IEEE Trans. Intell. Transp. Syst. (2022)

69. Unni, A., Trende, A., Pauley, C., Weber, L., Biebl, B., Kacianka, S., Lüdtke, A., Bengler, K., Pretschner, A., Fränzle, M., Rieger, J.W.: Investigating differences in behavior and brain in human-human and human-autonomous vehicle interactions in time-critical situations. Front. Neuroergonomics **3** (2022)
70. Parasuraman, R., de Visser, E., Wiese, E. and Madhavan, P.: Human trust in other humans, automation, robots, and cognitive agents: neural correlates and design implications. In: Proceedings of the Human Factors and Ergonomics Society Annual Meeting, vol. 58, no. 1, pp. 340–344. ISAGE (2014)

Fokas and Mathematics

High-Order Localized Wave Solutions of the New (3+1)-Dimensional Kadomtsev-Petviashvili Equation

Yulei Cao, Athanassios S. Fokas, and Jingsong He

Abstract The Korteweg-de Vries equation (KdV) is a classic representative of one-dimensional integrable systems, while the Kadomtsev-Petviashvili (KP) equation is a representative of two-dimensional integrable systems, which is an extension of the KdV equation in two dimensions. However, constructing three-dimensional integrable nonlinear equations has always been the most vital open problems in integrability. In this paper, a new three-dimensional KP equation is investigated. By applying Hirota bilinear method and long wave limit method, the multi-soliton, rational and semi-rational solutions are presented.

Keywords (3+1)-dimensional KP equation · Bilinear method · Long wave limit method · High-order localized wave

1 Introduction

There are a range of integrable nonlinear evolution equations in (1+1)-dimensional and (2+1)-dimensional systems [1, 2]. The most celebrated models in (1+1)-dimensional systems are the KdV equation and the nonlinear Schrodinger (NLS) equation [3–6]. As is well known, every one-dimensional integrable nonlinear equation has several integrable extensions in (2+1)-dimensions. The two physically

Y. Cao
School of Mathematics and Science, Nanyang Institute of Technology, Nanyang 473004, Henan, P. R. China

A. S. Fokas
DAMTP, University of Cambridge, Cambridge CB3 0WA, UK

Viterbi School of Engineering, USC, Los Angeles, CA 90089, USA

Centre of Mathematics, Academy of Athens, 11527 Athens, Greece

J. He (✉)
Institute for Advanced Study, Shenzhen University, Shenzhen 518060, Guangdong, P. R. China
e-mail: hejingsong@szu.edu.cn

T. Bountis et al. (eds.), *Chaos, Fractals and Complexity*, Springer Proceedings in Complexity, https://doi.org/10.1007/978-3-031-37404-3_18

essential extensions of KdV are the KPI and KPII equations. A two-dimensional analogue of the NLS equation is the DS equation. Integrable nonlinear evolution equations are widely used in physical systems. KP and DS equations have significant applications in weak dispersive media [7–10], optics and fluid dynamics [11–13].

One of the most essential tasks in soliton theory is to construct the $(3+1)$-dimensional [$(3+1)$-d] integrable evolution equations [14]. Substantial progress in this direction was made in [15], which introduced a $4+2$ dimensional generalization of KP and DS equations. Furthermore, Refs. [15–17] provide the solution of the Cauchy problem for the generalization of KP and DS equations in 4+2 dimensions. Additionally, Ref. [18] indicates the existence of integrable nonlinear evolution equations in any dimension, which involve a nonlocal commutator. Moreover, the issue of degenerating the nonlinear equation from the $(4+2)$ to a $(3+1)$-d equation has also been discussed [15, 16, 19, 20]. Recently, we introduced a new (3+1)-d KP equation [21]

$$u_{xt} + \alpha u_{xxxx} + \beta(uu_x)_x + \frac{\gamma}{4}u_{yy} - \frac{\gamma}{4}u_{zz} + i\frac{\gamma}{2}u_{yz} = 0, \ u \in \mathbb{C}, \ x, y, z, t \in \mathbb{R}, \tag{1}$$

the α, β, γ are complex constants. In this paper, various nonlinear wave solutions of Eq. (1) are proposed, and their dynamics are also discussed.

2 Soliton Solutions of the $(3+1)$-Dimensional KP Equation

The $(3+1)$-d KP equation admits Lax pairs and is completely integrable [21]. In this section, based on the Hirota bilinear method the multi-soliton solutions of the $(3+1)$-d KP Eq. (1) are presented [22]. Through the variable transformation

$$u = 12\frac{\alpha}{\beta}(\ln f)_{xx}, \tag{2}$$

the $(3+1)$-dimensional KP Eq. (1) possess the following bilinear form

$$\left(D_xD_t + \alpha D_x^4 + \frac{\gamma}{4}D_y^2 - \frac{\gamma}{4}D_z^2 + \frac{\gamma}{2}i\, D_yD_z\right) f \cdot f = 0, \tag{3}$$

the function f is a complex, and D is the Hirota's bilinear differential operator [22], defined as

$$D_x^m D_t^n f(x,t)\cdot g(x,t) = \left(\frac{\partial}{\partial_x} - \frac{\partial}{\partial_{x'}}\right)^m\left(\frac{\partial}{\partial_t} - \frac{\partial}{\partial_{t'}}\right)^n f(x,t)g(x',t')\bigg|_{x'=x,t'=t}.$$

Based on this bilinear form, the $(3+1)$-d KP Eq. (1) has the following N-soliton solutions:

$$u = 12\frac{\alpha}{\beta}(\ln f)_{xx}, \quad f = \sum_{\mu=0,1} \exp\left(\sum_{j<k}^{(N)} \mu_j \mu_k A_{jk} + \sum_{j=1}^{N} \mu_j \eta_j\right), \tag{4}$$

with

$$\begin{aligned} \eta_j &= k_j\left(x + p_j y + q_j z - \left[\frac{\gamma}{4}(p_j + i\,q_j)^2 + \alpha k_j^2\right] t\right) + \eta_j^0, \\ e^{A_{jk}} &= \frac{(p_j - p_k + i\,q_j - i\,q_k)^2 \gamma - 12\alpha\,(k_j - k_k)^2}{(p_j - p_k + i\,q_j - i\,q_k)^2 \gamma - 12\alpha\,(k_j + k_k)^2}, \end{aligned} \tag{5}$$

the k_j, p_j, q_j, η_j^0 are real constants, and the subscript j represents an integer. The $\sum_{j<k}^{(N)}$ summation is over all possible combinations of the N elements with the specific condition $j < k$. The notation $\sum_{\mu=0}$ indicates summation over all possible combinations of $\mu_1 = 0, 1$, $\mu_2 = 0, 1, \cdots, \mu_n = 0, 1$.

The above parameters must satisfy the condition

$$\gamma_I q_j^2 - 2\gamma_R p_j q_j - 4\alpha_I k_j^2 - \gamma_I p_j^2 = 0, \tag{6}$$

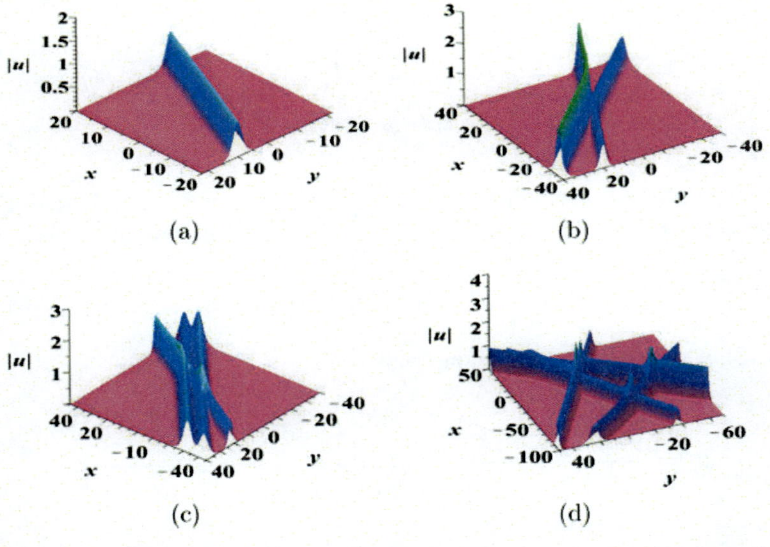

Fig. 1 The three-dimensional plots of multi-soliton solutions of the (3 + 1)-d KP Eq. (1); **a** One-soliton solution with parameters $N = 1, \alpha = 2, \beta = 2, \gamma = 2 + \frac{i}{2}, k_1 = \frac{1}{2}, p_1 = 2, q_1 = 1$; **b** Two-soliton solution with parameters $N = 2, \alpha = 2, \beta = 2, \gamma = 2 + \frac{i}{2}, k_1 = \frac{1}{2}, p_1 = 2, q_1 = 1, k_2 = \frac{1}{2}, p_2 = 1, q_2 = 2$; **c** Three-soliton solution with parameters $N = 3, \alpha = 2, \beta = 2, \gamma = 2 + \frac{i}{2}, k_1 = \frac{1}{2}, p_1 = 2, q_1 = -1, k_2 = \frac{1}{2}, p_2 = 1, q_2 = 2, k_3 = \frac{1}{2}, p_3 = 1, q_3 = \frac{1}{2}$; **d** Four-soliton solution with parameters $N = 4, \alpha = 2, \beta = 2, \gamma = 2 + \frac{i}{2}, k_1 = \frac{1}{2}, p_1 = 2, q_1 = -1, k_2 = \frac{1}{2}, p_2 = 1, q_2 = 3, k_3 = \frac{1}{2}, p_3 = -2, q_3 = 2; k_4 = \frac{1}{2}, p_4 = 3, q_4 = 0$

here we have assumed that $\gamma = \gamma_R + i\gamma_I$ and $\alpha = \alpha_R + i\alpha_I$. The dynamics of these soliton solutions are revealed in Fig. 1.

3 Rational Solutions of the (3 + 1)-Dimensional KP Equation

In this section, we investigate the rational solutions of the (3 + 1)-d KP Eq. (1). Apply the following parameter restrictions to the Eq. (4)

$$N = 2n, \quad \eta_j^0 = i\pi \quad (1 \le j \le N), \tag{7}$$

and letting $k_j \to 0$, then the function f in (4) consists of polynomial functions [23, 24]. Further utilizing (2), we obtain the N-th order rational solutions of the (3 + 1)-d KP equation, where the function f is defined as

$$f = f^{[n]} = \prod_{k=1}^{N} \theta_k + \frac{1}{2}\sum_{k,j}^{(N)} \alpha_{kj} \prod_{l \neq k,j}^{N} \theta_l + \cdots + \frac{1}{M!2^M} \sum_{i,j,\ldots,m,n}^{(N)} \overbrace{\alpha_{kj}\alpha_{kl}\cdots\alpha_{mn}}^{M} \prod_{p \neq k,j,\ldots m,n}^{N} \theta_p + \cdots, \tag{8}$$

with

$$\begin{aligned} \alpha_{jk} &= \frac{24\alpha}{\frac{1}{2}\left[(p_j - p_k)^2 - (q_j - q_k)^2\right] + i\,(p_j - p_k)(q_j - q_k)}, \\ \theta_j &= x + p_j y + q_j z - \frac{\gamma}{4}\left(p_j + i\,q_j\right)^2 t. \end{aligned} \tag{9}$$

In what follows, we will discuss the dynamics of these rational solutions. Firstly, the simplest rational solution of the (3+1)-d KP equation is considered. By taking

$$N = 2, \ \alpha = \alpha_R + i\,\alpha_I, \ p_1 = p_2^* = p_R + i\,p_I, \ \gamma = \gamma_R + i\,\gamma_I, \ q_1 = q_2^* = q_R + i\,q_I, \tag{10}$$

in Eq. (8), where we impose the constraints

$$p_I = \frac{\sqrt{\gamma_R^2 + \gamma_I^2} - \gamma_R}{\gamma_I} q_I, \quad p_R = \frac{\gamma_R^2 + \gamma_I^2 - \gamma_R\sqrt{\gamma_R^2 + \gamma_I^2}}{\gamma_I\sqrt{\gamma_R^2 + \gamma_I^2}} q_R. \tag{11}$$

The analytical expression of the firs-order rational solution of the (3 + 1)-d KP Eq. (1) is written as

$$
\begin{aligned}
&u = 12\frac{\alpha}{\beta}(\ln f)_{xx}, \quad f = \mathscr{R}_1^2 + \mathscr{R}_2^2 + \kappa_{12}, \quad \kappa_{12} = \frac{6\gamma_I^2\,(\alpha_R + i\,\alpha_I)}{\left(\sqrt{\gamma_R^2+\gamma_I^2} - \gamma_R\right)(\gamma_R^2+\gamma_I^2)\,q_I^2}, \\
&\mathscr{R}_1 = \frac{\gamma_R^2+\gamma_I^2 - \gamma_R\sqrt{\gamma_R^2+\gamma_I^2}}{\gamma_I\sqrt{\gamma_R^2+\gamma_I^2}}\,q_I y + q_I z + \frac{\left(\gamma_R^2+\gamma_I^2 - \gamma_R\sqrt{\gamma_R^2+\gamma_I^2}\right)\sqrt{\gamma_R^2+\gamma_I^2}}{\gamma_I^2}\,q_R q_I t, \\
&\mathscr{R}_2 = x + \frac{\gamma_R^2+\gamma_I^2 - \gamma_R\sqrt{\gamma_R^2+\gamma_I^2}}{\gamma_I\sqrt{\gamma_R^2+\gamma_I^2}}\,q_R y + q_R z + \frac{(q_R^2 - q_I^2)\left(\gamma_R^2+\gamma_I^2 - \gamma_R\sqrt{\gamma_R^2+\gamma_I^2}\right)\sqrt{\gamma_R^2+\gamma_I^2}}{2\gamma_I^2}\,t.
\end{aligned} \tag{12}
$$

If

$$
\alpha_R\left(\sqrt{\gamma_R^2+\gamma_I^2} - \gamma_R\right) > 0 \quad or \quad \alpha_I \neq 0, \tag{13}
$$

this first-order rational solution is smooth. Taking

$$
\alpha = 1,\ p_1 = \sqrt{2} - 1 + (\sqrt{2} - 1)i,\ q_1 = 1 + i,\ \gamma = 1 + i,
$$

in (12), the first order rational solution of (3 + 1)-d KP Eq. (1) is a real function, rewritten as

$$
\begin{aligned}
&u = 12(\ln f)_{xx}, \\
&f = [x + (\sqrt{2} - 1)y + z]^2 + [(\sqrt{2} - 1)y + z + (2\sqrt{2} - 2)t]^2 + 3(2 + \sqrt{2}).
\end{aligned} \tag{14}
$$

As seen in Fig. 2, this rational solution is a lump wave in the (x, y)-plane.

Furthermore, we also give the second-order lumps by taking $N = 4$ in Eq. (8). The expression of function f is as follows

$$
\begin{aligned}
f = {}&\theta_1\,\theta_2\,\theta_3\,\theta_4 + a_{12}\,\theta_3\,\theta_4 + a_{13}\,\theta_2\,\theta_4 + a_{14}\,\theta_2\,\theta_3 + a_{23}\,\theta_1\,\theta_4 \\
&+ a_{24}\,\theta_1\,\theta_3 + a_{34}\,\theta_1\,\theta_2 + a_{12}\,a_{34} + a_{13}\,a_{24} + a_{14}\,a_{23},
\end{aligned} \tag{15}
$$

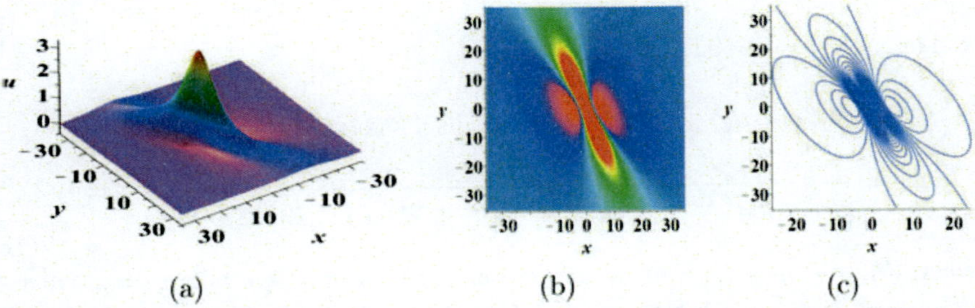

Fig. 2 The first-order lump solution (14) of the (3 + 1)-d KP equation with parameters $\alpha = 1, p_1 = p_2^* = \sqrt{2} - 1 + (\sqrt{2} - 1)i, \gamma = 1 + i, q_1 = q_2^* = 1 + i, \beta = 1, t = 0, z = 0$; **a** a three dimensional plot; **b** a density plot; **c** a contour plot

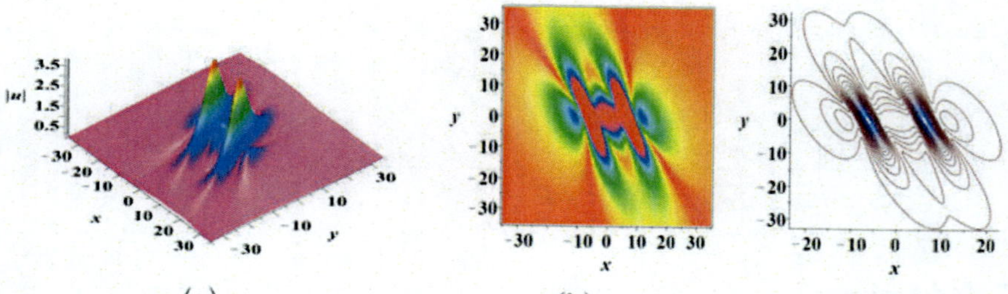

Fig. 3 The second-order lumps of the (3 + 1)-d KP equation with parameters $\alpha = 1+i, q_1 = q_2^* = 1+i, \beta = 1, p_1 = p_2^* = (\sqrt{2}-1)+(\sqrt{2}-1)\,i, q_3 = q_4^* = 2+i, p_3 = p_4^* = (\sqrt{2}-1)+(\sqrt{2}-1)\,i, \gamma = 1+i, t = 0, z = 0$; **a** a three dimensional plot; **b** a density plot; **c** a contour plot

where θ_j and a_{jk} are defined in (9). Figure 3a represents a three dimensional plot of corresponding lumps, Fig. 3b represents the density plot, Fig. 3c is the contour plot.

4 Semi-rational Solutions of the (3 + 1)-Dimensional KP Equation

In this section, by using the long-wave limit method semi-rational solutions of the (3 + 1)-dimensional KP equation are also revealed. Taking

$$1 < 2j < N, \quad 1 \le k \le 2j, \quad \eta_k^0 = i\pi, \tag{16}$$

and letting $k_j \to 0$ for all k_j, then the functions f in (4) consists of polynomial function and exponential function. The case of $N = 3$ is first considered. Taking

$$N = 3, \ \alpha = 1+i, \ \eta_1^0 = \eta_2^0 = i\pi, \ \gamma = 1+i, \ \beta = 1, \tag{17}$$

and taking $k_1, \ k_2 \to 0$ in Eq. (4), we obtain

$$\begin{aligned} f &= (\theta_1\theta_2 + a_{12}) + (\theta_1\theta_2 + a_{12} + a_{13}\theta_2 + a_{23}\theta_1 + a_{12}a_{23})e^{\eta_3}, \\ a_{j3} &= \frac{48\alpha k_3}{(p_j - p_3)^2 - (q_j - q_3)^2 + 2i(q_j - q_3)(p_j - p_3) - 12\alpha k_3^2}, \ j = 1, 2, \end{aligned} \tag{18}$$

where θ_j, a_{12}, η_3 are given by Eqs. (9) and (5). Further, we take $q_1 = q_2^* = 1 + i, \ p_1 = p_2^* = (\sqrt{2}-1)+(\sqrt{2}-1)\,i, k_3 = 1, \ p_3 = \frac{1}{3}, q_3 = 1,$ and $\eta_3^0 = 0$ in Eq. (18), as shown in Fig. 4, this semi-rational solution is composed of a lump and a soliton.

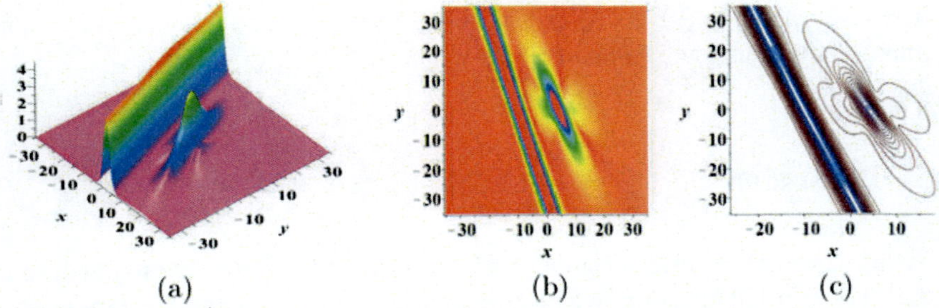

Fig. 4 The semi-rational solution composed of a lump and a soliton of the (3 + 1)-d KP equation with parameters $\beta = 1, q_1 = q_2^* = 1 + i, \gamma = 1 + i, p_1 = p_2^* = (\sqrt{2} - 1) + (\sqrt{2} - 1)\,i, \alpha = 1 + i, k_3 = 1, p_3 = \frac{1}{3}, q_3 = 1, t = 0, z = 0$; **a** a three dimensional plot; **b** a density plot; **c** a contour plot

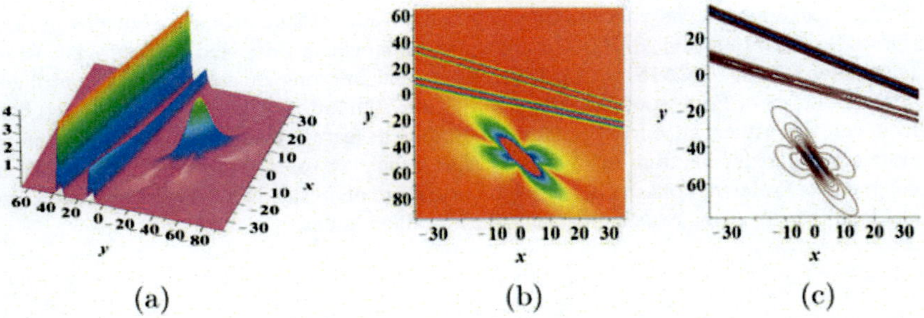

Fig. 5 The semi-rational solution composed of a lump and two solitons of the (3 + 1)-d KP equation with parameters $\alpha = 1 + i, q_1 = q_2^* = 1 + i, \beta = 1, p_1 = p_2^* = (\sqrt{2} - 1) + (\sqrt{2} - 1)\,i, \gamma = 1 + i, k_3 = \frac{1}{2}, k_4 = 1, q_3 = -2 + \sqrt{7}, q4 = \frac{-3+\sqrt{2}}{2}, p_3 = 2, p_4 = \frac{3}{2}, t = 0, z = 0$; **a** a three dimensional plot; **b** a density plot; **c** a contour plot

Additionally, the semi-rational solutions consisting of more solitons and more lumps are also proposed for larger N. Taking

$$N = 4,\ \alpha = 1 + i,\ \eta_1^0 = \eta_2^0 = i\pi,\ \gamma = 1 + i,\ \beta = 1, \tag{19}$$

and taking $k_1, k_2 \to 0$ in Eq. (4), we obtain

$$\begin{aligned} f = {} & e^{A_{34}}(a_{13}a_{23} + a_{13}a_{24} + a_{13}\theta_2 + a_{14}a_{23} + a_{14}a_{24} + a_{14}\theta_2 \\ & + a_{23}\theta_1 + a_{24}\theta_1 + \theta_1\theta_2 + a_{12})e^{\eta_3+\eta_4} + (a_{13}a_{23} + a_{13}\theta_2 \\ & + a_{23}\theta_1 + \theta_1\theta_2 + a_{12})e^{\eta_3} + (a_{14}a_{24} + a_{14}\theta_2 + a_{24}\theta_1 + \theta_1\theta_2 \\ & + a_{12})e^{\eta_4} + \theta_1\theta_2 + a_{12}\,, \\ a_{js} = {} & \frac{48\alpha k_s}{(p_j - p_s)^2 - (q_j - q_s)^2 + 2i(q_j - q_s)(p_j - p_s) - 12\alpha k_s^2},\ j = (1, 2),\ s = (3, 4), \end{aligned} \tag{20}$$

here θ_j is shown in (9). As described in Fig. 5, this semi-rational solution is superimposed by a lump and two solitons.

5 Discussion

In this paper, the multidimensional analogue of KP equation is investigated, namely the $(3+1)$-d KP equation, which is completely integrable and has Lax pairs. We constructed nonsingular multi-soliton solutions of the $(3+1)$-d KP equation using the Hirota bilinear method and the perturbation expansion technique. Additionally, high-order lumps and semi-rational solutions of the $(3+1)$-d KP equation are also revealed by using the long-wave limit method.

Acknowledgements The work was supported by the National Natural Science Foundation of China (Grant No. 12071304 and No. 11871446), and the Guangdong Basic and Applied Basic Research Foundation (Grant 2022A1515012554). Yulei Cao and Jingsong He are pleased to contribute this article in honor of Professor Athanassios S. Fokas's 70th birthday. One of us (Jingsong) has collaborated with Athanassios for over 10 years. Both of us learned the higher dimensional integrable system and its reduction from Athanassios, which is the main topic of this joint paper with him. We congratulate Professor Fokas on his extraordinary record of outstanding mathematical achievement, and look forward to many more years of fruitful collaborations.

References

1. Faddeev, L.D., Takhtadjan, L.A.: Hamiltonian Methods in the Soliton Theory. Springer Series in Soviet Mathematics. Springer, Berlin/Heidelberg, Germany (1987)
2. Ablowitz, M.J., Clarkson, P.A.: Solitons, Nonlinear Evolution Equations and Inverse Scattering. Cambridge University Press, Cambridge, UK (1991)
3. Russell, J.S.: Report of the committee on waves. In: Proceedings of the 7th Meeting of the British Association for the Advancement of Science, Liverpool, UK, 1838, pp. 417–496 (1838)
4. Korteweg, D.J., de Vries, G.: On the change of form of long waves advancing in a rectangular canal, and a new type of long stationary waves. Philos. Mag. Ser. **39**, 422–443 (1895)
5. Chiao, R., Garmire, E.C., Townes, C.: Self-trapping of optical beams. Phys. Rev. Lett. **13**, 479–482 (1964)
6. Zakharov, V.: Stability of periodic waves of finite amplitude on the surface of a deep fluid. J. Appl. Mech. Tech. Phys. **9**, 190–194 (1968)
7. Ablowitz, M.J., Fokas, A.S., Musslimani, Z.H.: On a new nonlocal formulation of water waves. J. Fluid Mech. **562**, 313–344 (2006)
8. Kadmotsev, B.B., Petviashvili, V.I.: On the stability of solitary waves in weakly dispersing media. Sov. Phys. Doklady **15**, 539–541 (1970)
9. Infeld, E., Rowlands, G.: Nonlinear Waves, Solitons and Chaos. Cambridge University Press, Cambridge, UK (2000)
10. Ablowitz, M.J.: Nonlinear Dispersive Waves: Asymptotic Analysis and Solitons. Cambridge University Press, Cambridge, UK (2011)
11. Davey, A., Stewartson, K.: On three-dimensional packets of surface waves. Proc. R. Soc. Lond. A **338**, 101–110 (1974)
12. Ablowitz, M.J., Segur, H.: On the evolution of packets of water waves. J. Fluid Mech. **92**, 691–715 (1979)

13. Ioannou-Sougleridis, I., Frantzeskakis, D.J., Horikis, T.P.: A Davey-Stewartson description of two-dimensional solitons in nonlocal media. Stud. Appl. Math. **144**, 3–17 (2020)
14. Fokas, A.S.: Integrable nonlinear evolution equations in three spatial dimensions. Proc. R. Soc. A **478**, 20220074 (2022)
15. Fokas, A.S.: Integrable nonlinear evolution partial differential equations in $4 + 2$ and $3 + 1$ dimensions. Phys. Rev. Lett. **96**, 190201 (2006)
16. Fokas, A.S.: Kadomtsev-Petviashvili equation revisited and integrability in $4 + 2$ and $3 + 1$. Stud. Appl. Math. **122**, 347–359 (2009)
17. Fokas, A.S., Weele, M.C.V.: Complexification and integrability in multidimensions. J. Math. Phys. *59*, 091413 (2018)
18. Fokas, A.S.: Nonlinear Fourier transforms and integrability in multidimensions. Contemp. Math. **458**, 71–80 (2008)
19. Yang, Z.Z., Yan, Z.Y.: Symmetry groups and exact solutions of new (4+1)-dimensional Fokas equation. Commun. Theor. Phys. **51**, 876–880 (2009)
20. Wang, X.B., Tian, S.F., Feng, L.L., Zhang, T.T.: On quasi-periodic waves and rogue waves to the (4+1)-dimensional nonlinear Fokas equation. J. Math. Phys. **59**, 073505 (2018)
21. Fokas, A.S., Cao, Y.L., He, J.S.: Multi-solitons, multi-breathers and multi-rational solutions of integrable extensions of the Kadomtsev-Petviashvili equation in three dimensions. Fractal Fract. **6**, 425 (2022)
22. Hirota, R.: The Direct Method in Soliton Theory. Cambridge University Press, Cambridge, UK (2004)
23. Ablowitz, M.J., Satsuma, J.: Solitons and Rational solutions of Nonlinear Evolution Equations. J. Math. Phys. **19**, 2180–2186 (1978)
24. Satsuma, J., Ablowitz, M.J.: Two-dimensional lumps in nonlinear dispersive systems. J. Math. Phys. **20**, 1496–1503 (1979)

Progress in Initial-Boundary Value Problems for Nonlinear Evolution Equations and the Fokas Method

A. Alexandrou Himonas

Dedicated to Professor Athanassios S. Fokas

Abstract The unified transform method (UTM), also known as the Fokas method [13], provides a novel approach for solving initial-boundary value problems (ibvp) for linear and integrable nonlinear partial differential equations. In particular, it gives solution formulas for forced linear ibvp. This motivated the initiation of a new program (by Fokas and collaborators) for studying the well-posedness in Sobolev spaces of ibvp for nonlinear evolution equations by employing this method, which is analogous to the way well-posedness of initial value problems (ivp) are studied. Using the Fokas formula we are able to derive linear estimates in Sobolev, Hadamard and Bourgain spaces. Then, using as an iteration map the one defined by the UTM formula when the forcing is replaced by the nonlinearity, and utilizing the linear estimates in combination with the multilinear estimates suggested by the nonlinearity we show that this map is a contraction in appropriate solution spaces. For the Korteweg-de Vries (KdV) and Nonlinear Schrödinger (NLS) equations this program has been implemented for ibvp in one-space dimension [16, 17, 29] and significant progress has been made in higher dimensions [24]. In this review we will try to present key points of this remarkable story. It is based on collaborative work with A. Fokas, D. Mantzavinos and F. Yan.

Keywords Korteweg-de Vries equation · Integrability · Initial-boundary value problem · Fokas unified transform method · Well-posedness in Sobolev spaces · Linear space-time estimates · Bilinear estimates in Bourgain spaces

2020 Mathematics Subject Classification Primary: 35Q55 · 35G31 · 35G16

A. A. Himonas (✉)
University of Notre Dame, Notre Dame, IN 46556, USA
e-mail: himonas.1@nd.edu
URL: https://math.nd.edu/people/faculty/alex-himonas/

T. Bountis et al. (eds.), *Chaos, Fractals and Complexity*, Springer Proceedings in Complexity, https://doi.org/10.1007/978-3-031-37404-3_19

1 Introduction

Given a partial differential equation (PDE), supplemented with appropriate data, three basic questions arise:

(1) Existence. Does a solution exist in a "good" function space?
(2) Uniqueness. Is the solution unique in this space?
(3) Stability. Is the solution "stable"? i.e. do small errors (disturbances) in the data result in small errors in the solution?

If the answer to the above three properties is yes, then the given PDE problem is called **well-posed**. Otherwise it is called **ill-posed**. Well-posedness is very important for PDE problems modeling physical or socioeconomic situations.

Well-posedness study—a story of analogy. Here we will see that, thanks to Fokas unified transform method, the well-posedness study of initial-boundary value problems (ibvp) is analogous to the well-posedness study of initial value problems (ivp) based on the Fourier transform method.

Studying nonlinear ivp—The three steps:

Step 1: We derive the **solution formula** to the linear forced ivp thanks! to Fourier transform.

Step 2: We derive **linear estimates** for data and forcing in "good" spaces, using the solution formula, and classical analysis.

Step 3: We prove that the iteration map defined by the Fourier solution formula when the forcing is replaced with the nonlinearity is a **contraction** in a "good" solution space, by deriving appropriate **multilinear estimates.**

In each of the three sections that follow, we number the equations starting from (1.1) to help the reader appreciate them as separate entities. All equation referencing within each section, concerns only equations of the section itself.

The Korteweg-de Vries equation ivp story. Next, we focus on the celebrated Korteweg-de Vries (KdV) equation [22, 39, 40, 46], and apply the three steps above to study the well-posedness of its initial value problem (ivp)

$$\partial_t u + \partial_x^3 u + uu_x = 0, \tag{1.1}$$

$$u(x, 0) = u_0(x), \quad t \in \mathbb{R}, \quad x \in \mathbb{R}. \tag{1.2}$$

Step 1. Solving the forced linear ivp thanks to Fourier. To solve the linear KdV ivp with forcing f

$$\partial_t u + \partial_x^3 u = f(x, t), \tag{1.3}$$

$$u(x, 0) = u_0(x), \quad t \in \mathbb{R}, \quad x \in \mathbb{R}, \tag{1.4}$$

we take x-Fourier transform, which for a function $\varphi(x)$ on $\mathbb{R}$ is defined by

$$\widehat{\varphi}(\xi) \doteq \int_{\mathbb{R}} e^{-ix\xi}\varphi(x)dx, \quad \xi \in \mathbb{R}, \tag{1.5}$$

and we get the time ivp: $\partial_t \widehat{u} + (i\xi)^3\widehat{u} = \widehat{f}(\xi, t),\ \widehat{u}(\xi, 0) = \widehat{\varphi}(\xi)$. Then, solving it gives the Duhamel **solution formula**:

$$\begin{aligned} u(x,t) &= S[u_0, f](x,t) \\ &= \frac{1}{2\pi}\int_{-\infty}^{\infty} e^{i(\xi x+\xi^3 t)}\widehat{u_0}(\xi)d\xi + \frac{1}{2\pi}\int_0^t\int_{-\infty}^{\infty} e^{i[\xi x+\xi^3(t-t')]}\widehat{f}(\xi, t')d\xi dt'. \end{aligned} \tag{1.6}$$

Step 2. Deriving Linear Estimates. For data in Sobolev spaces $H^s(\mathbb{R})$, defined by the norm

$$\|u_0\|^2_{H^s} \doteq \int_{\mathbb{R}} (1+|\xi|)^{2s}|\widehat{u_0}(\xi)|^2 d\xi, \tag{1.7}$$

and forcing in Bourgain (solution) space $X^{s,b}(\mathbb{R}^2)$ defined by the norm [5, 6]

$$\|f\|^2_{X^{s,b}} = \|f\|^2_{s,b} \doteq \int_{\mathbb{R}}\int_{\mathbb{R}} (1+|\xi|)^{2s}(1+|\tau-\xi^3|)^{2b}|\widehat{f}(\xi,\tau)|^2 d\xi d\tau, \tag{1.8}$$

we derive the following estimates for the solution $S[u_0, f]$ localized.

Lemma 1 (KdV Linear Estimates) *For any $s \in \mathbb{R}$ and $\frac{1}{2} < b < 1$, there is $c = c(\psi, s, b)$ such that:*

$$\|\psi(t)S[u_0, f]\|_{s,b} \le c\|u_0\|_{H^s} + c\|f\|_{s,b-1}. \tag{1.9}$$

Here and in the sequel ψ is a **time localizer**, which is defined as follows:

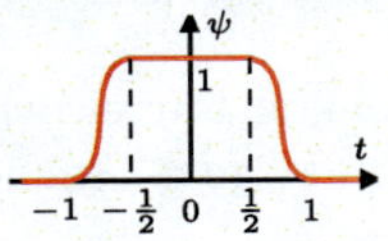

$$\psi \in C_0^\infty(-1, 1),\ 0 \le \psi \le 1 \text{ and } \psi(t) = 1 \text{ for } |t| \le 1/2. \tag{1.10}$$

Step 3. Proving that the iteration map is contraction. Replacing in solution formula (1.6) the forcing f by the nonlinearity $-uu_x$ we obtain the iteration map of the KdV ivp

$$\begin{aligned} \Phi(u)(x,t) &\doteq \psi(t)S[u_0, -uu_x](x,t) \\ &= \frac{\psi(t)}{2\pi}\int_{-\infty}^{\infty} e^{i(\xi x+\xi^3 t)}\widehat{u_0}(\xi)d\xi - \frac{\psi(t)}{2\pi}\int_0^t\int_{-\infty}^{\infty} e^{i[\xi x+\xi^3(t-t')]}\widehat{uu_x}(\xi, t')d\xi dt'. \end{aligned} \tag{1.11}$$

Then, from the linear estimates (1.9) we see that we need the following crucial bilinear estimates for the KdV nonlinearity, proved for $s \geq 0$ by Bourgain [5] and for $s > -\frac{3}{4}$ by Kenig, Ponce and Vega [36].

Theorem 1 (KdV bilinear estimates) *Given $s > -3/4$, there exists $b \in (1/2, 1)$ such that*

$$\|\partial_x(f \cdot g)\|_{X^{s,b-1}} \leq c\|f\|_{X^{s,b}}\|g\|_{X^{s,b}}, \quad f, g \in X^{s,b}. \tag{1.12}$$

Combining bilinear estimate (1.12) with linear estimate (1.9) gives that the iteration map Φ is a contraction with fixed point (local solution) in Bourgain spaces $X^{s,b}$, thus obtaining the next result, proved for $s \geq 0$ in [5], and for $s > -\frac{3}{4}$ in [36].

Theorem 2 (KdV Well-posedness) *If $s > -3/4$, then for any data $u_0 \in H^s(\mathbb{R})$ the KdV ivp* **has a unique solution** *in the space $X^{s,b}$ for appropriate $b > 1/2$. Moreover, the data to solution map is* **Lipschitz continuous**.

For additional results on the well-posedness of the KdV initial value problem with rough data we refer the reader to [7, 32–35, 37, 38, 44] and the references therein.

2 Initial-Boundary Value Problems via the Fokas Method

Now, in **analogy** to ivp, we describe the three steps for studying nonlinear initial-boundary value problems (ibvp) based on the Fokas method [13].

Step 1: We derive the **solution formula** to the forced linear ibvp thanks! to **Fokas** unified transform **method** (UTM).

Step 2: We derive **linear estimates** for data and forcing in "good" spaces, using the Fokas solution formula and classical analysis.

Step 3: We prove that the iteration map defined by the Fokas solution formula when the forcing is replaced by the nonlinearity is a **contraction** in a "good" solution space, where appropriate **multilinear estimates** hold!

The KdV ibvp story. Next we demonstrate the analogues to ivp three steps for studying the well-posedness of the following KdV equation ibvp

$$u_t + u_{xxx} + uu_x = 0, \qquad x \in (0, \infty),\ t \in (0, T), \tag{2.1a}$$
$$u(x, 0) = u_0(x), \qquad x \in (0, \infty), \tag{2.1b}$$
$$u(0, t) = g_0(t), \qquad t \in (0, T). \tag{2.1c}$$

Step 1. Solving the forced linear KdV ibvp thanks! to the Fokas method,

$$u_t + u_{xxx} = f(x, t), \qquad x \in (0, \infty),\ t \in (0, T), \tag{2.2a}$$
$$u(x, 0) = u_0(x), \qquad x \in (0, \infty), \tag{2.2b}$$
$$u(0, t) = g_0(t), \qquad t \in (0, T), \tag{2.2c}$$

we get the UTM **solution formula** via a deformation in the complex plane utilizing the analyticity of the half-line Fourier transform defined below

$$
\begin{aligned}
u(x,t) =& S[u_0, g_0; f] \qquad \text{(to be used with} \quad f = -uu_x) \qquad (2.3)\\
\doteq& \frac{1}{2\pi}\int_{-\infty}^{\infty} e^{i\xi x + i\xi^3 t}[\widehat{u}_0(\xi) + F(\xi,t)]d\xi \qquad \boxed{\sigma = e^{2i\pi/3}}\\
&+\frac{1}{2\pi}\int_{\partial D^+} e^{i\xi x + i\xi^3 t}\{\sigma[\widehat{u}_0(\sigma\xi) + F(\sigma\xi,t)] + \sigma^2[\widehat{u}_0(\sigma^2\xi) + F(\sigma^2\xi,t)]\}d\xi\\
&-\frac{3}{2\pi}\int_{\partial D^+} e^{i\xi x + i\xi^3 t}\xi^2\tilde{g}_0(\xi^3, T)d\xi.
\end{aligned}
$$

$$\widehat{u}_0(\xi) \doteq \int_0^\infty e^{-i\xi x}u_0(x)dx, \quad \text{Im}(\xi) \leq 0, \quad \text{(Half-line Transform)},$$

$$F(\xi,t) \doteq \int_0^t e^{-i\xi^3\tau}\int_0^\infty e^{-i\xi x}f(x,\tau)dxd\tau, \quad \text{Im}(\xi) \leq 0,$$

$$\tilde{g}_0(\xi,t) \doteq \int_0^t e^{-i\xi\tau}g_0(\tau)d\tau, \quad \text{(Temporal Transform)}.$$

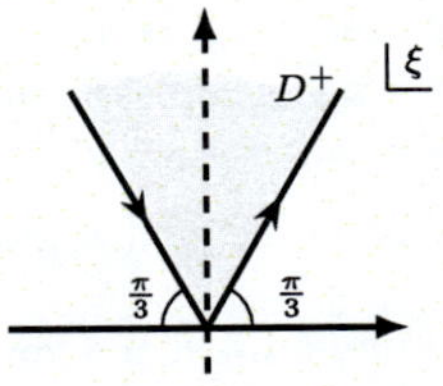

Step 2. Decomposing the linear ibvp (2.2) into simpler problems and using Fokas solution formula (2.3) we derive the **linear estimates** for initial data in Sobolev spaces $H^s(0,\infty)$ and boundary data $H^{(s+1)/3}(0,T)$, where $(s+1)/3$ is the KdV time regularity. More precisely, we obtain the following result.

Theorem 3 (Linear Estimates in modified Bourgain spaces) *For rough data, i.e.* $-\frac{3}{2} < s < \frac{3}{2}$, $s \neq \frac{1}{2}$, *the Fokas formula* (2.3) *defines a solution* u *to the ibvp* (2.2) *with the compatibility condition*

$$u_0(0) = g_0(0), \qquad 1/2 < s < 3/2, \tag{2.4}$$

which is in the space $X^{s,b,\alpha}_{\mathbb{R}^+\times(0,T)}$ *and satisfies the estimates*

$$
\begin{aligned}
&\|S[u_0, g_0; f]\|_{X^{s,b,\alpha}_{\mathbb{R}^+\times(0,T)}}\\
\leq& c\Big[\|u_0\|_{H_x^s(\mathbb{R}^+)} + \|g_0\|_{H_t^{\frac{s+1}{3}}(0,T)} + \|f\|_{X^{s,-b,\alpha-1}_{\mathbb{R}^+\times(0,T)}} + \|f\|_{Y^{s,-b}_{\mathbb{R}^+\times(0,T)}}\Big],
\end{aligned}
\tag{2.5}
$$

where $1/2 < \alpha < \min\{\frac{s}{3}+1, 1\}$, *and* $0 < b < 1/2$.

For smooth data, i.e. $1/2 < s < 3/2$, we have the linear estimates in the Hadamard space $C([0,T]; H_x^s(0,\infty))$ (see [16])

$$\sup_{t\in[0,T]} \|S\big[u_0, g_0; f\big](t)\|_{H_x^s(0,\infty)}$$

$$\leq c_s\Big[\|u_0\|_{H_x^s(0,\infty)} + \|g_0\|_{H_t^{\frac{s+1}{3}}(0,T)} + \max\{T^{\frac{2-s}{3}}, T^{\frac{3-2s}{3}}\}\Big(\int_0^T \|f(t)\|^2_{H_x^s(0,\infty)}dt\Big)^{\frac{1}{2}}\Big].$$

Modified and Temporal Bourgain spaces. The modified Bourgain space $X^{s,b,\alpha}(\mathbb{R}^2)$ is defined by the norm [5]

$$\|u\|^2_{X^{s,b,\alpha}} \doteq \iint_{\mathbb{R}^2}\Big[(1+|\xi|)^{2s}(1+|\tau-\xi^3|)^{2b} + \chi_{|\xi|<1}(1+|\tau|)^{2\alpha}\Big]|\hat{u}(\xi,\tau)|^2 d\xi d\tau,$$

where $\alpha > \frac{1}{2}$ and $\chi_{|\xi|<1}$ is the characteristic function of the interval $(-1,1)$. The "temporal" Bourgain space $Y^{s,b}(\mathbb{R}^2)$ is defined by the norm [11]

$$\|u\|^2_{Y^{s,b}} \doteq \iint_{\mathbb{R}^2}(1+|\tau|)^{\frac{2s}{3}}(1+|\tau-\xi^3|)^{2b}|\widehat{u}(\xi,\tau)|^2 d\xi d\tau. \tag{2.6}$$

Step 3. Proving that the Iteration map is Contraction. From the above linear estimates (2.5) with the forcing f being replaced by the nonlinearity $\partial_x(u^2)$, we see that to show that iteration map is a contraction, we need to estimate $\|\partial_x(fg)\|_{X^{s,-b,\alpha-1}_{\mathbb{R}^+\times(0,T)}}$ and $\|\partial_x(fg)\|_{Y^{s,-b}_{\mathbb{R}^+\times(0,T)}}$. This is done in the next result.

Theorem 4 (Bilinear estimates in modified and temporal Bourgain spaces) *For $s > -\frac{3}{4}$, we have the bilinear estimates in the modified Bourgain spaces*

$$\|\partial_x(f\,g)\|_{X^{s,-b,\alpha-1}} \leq c_1\|f\|_{X^{s,b',\alpha'}}\|g\|_{X^{s,b',\alpha'}}. \tag{2.7}$$

For $-\frac{3}{4} < s < 3$, we have the bilinear estimates in the "temporal" Bourgain spaces

$$\|\partial_x(f\,g)\|_{Y^{s,-b}} \leq c_1\|\partial_x(f\,g)\|_{X^{s,-b}} + c_1\|f\|_{X^{s,b'}}\|g\|_{X^{s,b'}}. \tag{2.8}$$

In estimate (2.7), it suffices to have $\frac{1}{2}-\beta_1 \leq b' \leq b < \frac{1}{2} < \alpha' \leq \alpha \leq \frac{1}{2}+\beta_1$, where

$$\beta_1 = \begin{cases} \frac{1}{36}, & s \geq 0, \\ \frac{1}{96}\big[s+\frac{3}{4}\big], & -\frac{3}{4} < s < 0. \end{cases} \tag{2.9}$$

In estimate (2.8), it suffices to have $\frac{1}{2}-\beta \leq b' \leq b < \frac{1}{2} < \alpha' \leq \alpha \leq \frac{1}{2}+\beta$, where

$$\beta \doteq \min\Big\{\beta_1, \frac{3-s}{36}\Big\}. \tag{2.10}$$

Finally, using iteration map defined by the Fokas solution formula, the linear estimates (2.5), and the bilinear estimates above, we show the three properties of well-posedness in Bourgain spaces (existence, uniqueness and stability) for the same (optimal) critical exponent $s = -3/4$ as in the case of the line.

Theorem 5 (Well-posedness of KdV ibvp via Fokas Method) *If* $-\frac{3}{4} < s < \frac{3}{2}$, $s \neq \frac{1}{2}$, *then for any initial data* $u_0 \in H^s(0,\infty)$, *boundary data* $g_0 \in H_t^{\frac{s+1}{3}}(0,T)$ *and some lifespan* $0 < T_0 \leq T < \frac{1}{2}$, **there is a solution** *for the KdV ibvp* (2.1)*, which is in* $X^{s,b,\alpha}_{\mathbb{R}^+\times(0,T_0)}$ *and which satisfies the size estimate*

$$\|u\|_{X^{s,b,\alpha}_{\mathbb{R}^+\times(0,T_0)}} \leq C(s,b,\alpha)\Big[\|u_0\|_{H^s(\mathbb{R}^+)} + \|g_0\|_{H_t^{\frac{s+1}{3}}(0,T)}\Big], \tag{2.11}$$

for some $b \in (0, \frac{1}{2})$ *and* $\alpha \in (\frac{1}{2}, 1)$*. Also, an estimate for the lifespan is given by* $T_0 = c_0\Big[1 + \|u_0\|_{H^s(\mathbb{R}^+)} + \|g_0\|_{H_t^{\frac{s+1}{3}}(0,T)}\Big]^{-4/\beta}$*, where* β *is defined in* (2.10)*. Furthermore, the* **solution is unique** *in the space* $X^{s,b,\alpha}_{\mathbb{R}^+\times(0,T_0)}$*. Finally, the* **data to solution map** $\{u_0, g_0\} \mapsto u$ *is locally* **Lipschitz continuous.**

For smooth data, i.e. $\frac{1}{2} < s < \frac{3}{2}$, we have well-posedness in the Hadamard space $C([0,T]; H_x^s(0,\infty))$ (see [16]).

Reduced pure ibvp. Next we outline the proof of the linear estimates for the most fundamental linear KdV ibvp – the reduced pure ibvp:

$$\partial_t v + \partial_x^3 v = 0, \quad 0 < x < \infty, \ 0 < t < 2, \tag{2.12a}$$

$$v(x,0) = 0, \tag{2.12b}$$

$$v(0,t) = h(t), \quad \text{supp}\, h \subset (0,2). \tag{2.12c}$$

The time transform of data $h(t)$ over $(0,2)$ is its Fourier transform over $\mathbb{R}$ $\tilde{h}(\xi,2) \doteq \int_0^2 e^{-i\xi\tau}h(\tau)d\tau = \int_{\mathbb{R}} e^{-i\xi\tau}h(\tau)d\tau = \widehat{h}(\xi)$. So, the Fokas solution formula for ibvp (2.12) on $\mathbb{R}^+ \times (0,2)$ reads as follows

$$v(x,t) \doteq S[0,h;0] = -\frac{3}{2\pi}\int_{\partial D^+} e^{i\xi x + i\xi^3 t}\xi^2\widehat{h}(\xi^3)d\xi, \tag{2.13}$$

and satisfies the following **key estimate** in the modified Bourgain spaces.

Theorem 6 (Linear Estimates for reduced pure ibvp) *For rough data, i.e.* $s > -\frac{3}{2}$, *we have:*

$$\|S[0,h;0]\|_{X^{s,b,\alpha}_{\mathbb{R}^+\times(0,2)}} \leq c_{s,b,\alpha}\|h\|_{H_t^{\frac{s+1}{3}}(\mathbb{R})}, \quad 0 \leq b < \frac{1}{2} < \alpha \leq \frac{s}{3} + 1. \tag{2.14}$$

For smooth data, i.e. $s > \frac{1}{2}$, we have (see [16])

$$\sup_{t\in[0,2]} \|S[0,h;0](t)\|_{H^s_x(\mathbb{R}^+)} \le c_s \|h\|_{H_t^{\frac{s+1}{3}}(\mathbb{R})}. \tag{2.15}$$

Proof of estimate (2.14). Using the parametrization $[0,\infty) \ni \xi \mapsto a\xi$ (or $a^2\xi$) $\in \partial D^+$, with $a = e^{i\frac{\pi}{3}}$, we write the solution as $v(x,t) \simeq v_r(x,t) + v_\ell(x,t)$, where v_r corresponds to integration on the right contour and v_ℓ on the left contour

$$v_r(x,t) = \int_0^\infty e^{ia\xi x - i\xi^3 t}\xi^2 \hat{h}(-\xi^3)d\xi, \tag{2.16}$$

$$v_\ell(x,t) = \int_0^\infty e^{ia^2\xi x + i\xi^3 t}\xi^2 \hat{h}(\xi^3)d\xi. \tag{2.17}$$

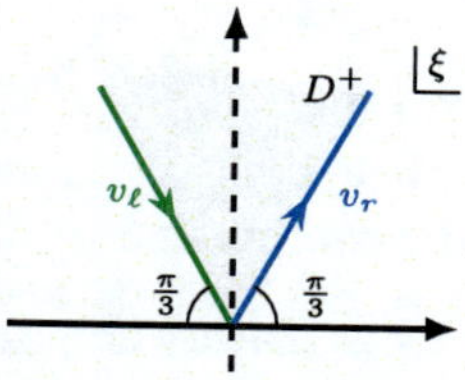

Since the estimation of v_ℓ is similar to v_r, here we only estimate v_r. For v_r we split the ξ-integral in formula (2.16) for ξ near 0 and for ξ near ∞. That is, we write $v_r = v_0 + v_1$, where $a = a_R + ia_I = \frac{1}{2} + i\frac{\sqrt{3}}{2}$,

$$v_0(x,t) \doteq \int_0^1 e^{-i\xi^3 t} e^{ia_R\xi x} e^{-a_I\xi x} \xi^2 \widehat{h}(-\xi^3)d\xi, \quad x \in \mathbb{R}^+, \quad t \in (0,2), \tag{2.18}$$

$$v_1(x,t) \doteq \int_1^\infty e^{-i\xi^3 t} e^{ia_R\xi x} e^{-a_I\xi x} \xi^2 \widehat{h}(-\xi^3)d\xi, \quad x \in \mathbb{R}^+ \quad t \in (0,2). \tag{2.19}$$

Estimate for v_0. Using the smooth version $\varphi_1(x)$ of $|x|$ defined below

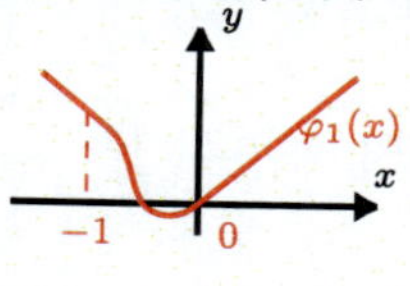

$$\varphi_1(x) = \begin{cases} x, & x \ge 0 \\ -x, & x \le -1 \\ \text{smooth on } \mathbb{R}, \end{cases} \tag{2.20}$$

we extend v_0 as an "almost" even function of x from $\mathbb{R}^+ \times (0,2)$ to $\mathbb{R} \times \mathbb{R}$ via

$$V_0(x,t) \doteq \int_0^1 e^{ia\xi\varphi_1(x)} e^{-i\xi^3 t} \xi^2 \widehat{h}(-\xi^3)d\xi, \quad x \in \mathbb{R}, \ t \in \mathbb{R}. \tag{2.21}$$

Multiplying the extension V_0 by the cutoff function $\psi_4(t) \doteq \psi(t/4)$, we obtain

$$\|v_0\|_{X^{s,b,\alpha}_{\mathbb{R}^+\times(0,2)}} \leq \|\psi_4 V_0\|_{X^{s,b,\alpha}} \lesssim \|h\|_{H_t^{\frac{s+1}{3}}}, \quad \forall s, b, \alpha \in \mathbb{R},$$

which completes the proof of estimate (2.14) for v_0.

Estimate for v_1. Using the identity $\xi[e^{ia_R\xi x}e^{-a_I\xi x}] = \frac{1}{ia}\partial_x[e^{ia_R\xi x}e^{-a_I\xi x}]$ and the fact that $e^{-a_I\xi x}$ is exponentially decaying in ξ (since $x > 0$), we can take the ∂_x-derivative outside the integral sign in (2.19) to rewrite $v_1(x, t)$ as follows

$$v_1(x,t) = \frac{1}{ia}\partial_x \int_1^\infty e^{-i\xi^3 t} e^{ia_R\xi x} e^{-a_I\xi x} \xi \widehat{h}(-\xi^3) d\xi, \quad x \in \mathbb{R}^+ \quad t \in (0,2). \quad (2.22)$$

Next, we extend v_1 from $\mathbb{R}^+ \times (0, 2)$ to $\mathbb{R} \times \mathbb{R}$ via the formula below

$$V_1(x,t) \doteq \frac{\partial_x}{ia} \int_1^\infty e^{-i\xi^3 t} e^{ia_R\xi x} e^{-a_I\xi x} \rho(a_I\xi x) \xi \widehat{h}(-\xi^3) d\xi, \quad x \in \mathbb{R}, \quad t \in \mathbb{R}, \tag{2.23}$$

where $\rho(x)$, $0 \leq \rho(x) \leq 1$, $x \in \mathbb{R}$, is an one-sided C^∞ cutoff function defined as follows outside the interval $(-1, 0)$, and is increasing inside $(-1, 0)$

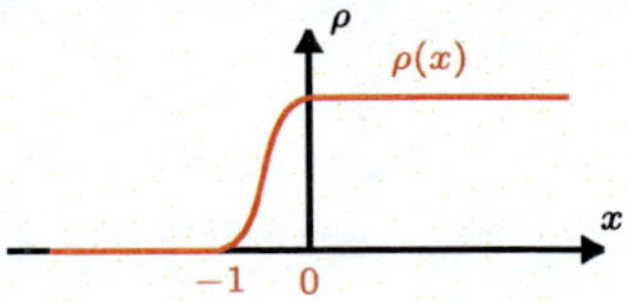

$$\rho(x) \doteq \begin{cases} 1, & x \geq 0, \\ 0, & x \leq -1. \end{cases} \tag{2.24}$$

Using the extension V_1, for $s > -\frac{3}{2}$ we get the desired estimate (2.14) for v_1

$$\|v_1\|_{X^{s,b,\alpha}_{\mathbb{R}^+\times(0,2)}} \leq \|V_1\|_{X^{s,b,\alpha}} \leq c_{s,b}\|h\|_{H_t^{\frac{s+1}{3}}(\mathbb{R})}, \quad 0 \leq b < \frac{1}{2} < \alpha \leq \frac{s}{3} + 1. \qquad \square$$

Other approaches for studying ibvp. There are two other approaches for studying ibvp. The first method, developed by Bona, Sun and Zhang [2–4], uses the Laplace transform in the time variable for solving the forced linear ibvp (see also [10]). The second method, developed by Colliander and Kenig [8], and by Holmer [30, 31], expresses an ibvp as a superposition of ivp.

3 A Higher Dispersion KdV on the Half-Line

Next, we demonstrate the Fokas method for proving well-posedness for the following mth order KdV (KdVm) initial-boundary value problem

$$\partial_t u + (-1)^{j+1}\partial_x^m u + uu_x = 0, \quad x > 0,\ 0 < t < T, \tag{3.1a}$$
$$u(x, 0) = u_0(x), \quad x > 0, \tag{3.1b}$$
$$u(0, t) = g_0(t), \quad \ldots\ldots, \quad \partial_x^{j-1}u(0, t) = g_{j-1}(t), \quad 0 < t < T, \tag{3.1c}$$

where $m = 2j + 1$, $j = 1, 2, 3, \ldots$, and $T < 1$.

Step 1. Solving the forced linear ibvp. For KdVm this ibvp is

$$\partial_t u + \partial_x^{2j+1}u = f(x, t), \quad x > 0,\ 0 < t < T, \tag{3.2a}$$
$$u(x, 0) = u_0(x), \quad x > 0, \tag{3.2b}$$
$$u(0, t) = g_0(t), \quad \ldots\ldots, \quad \partial_x^{j-1}u(0, t) = g_{j-1}(t), \quad 0 < t < T. \tag{3.2c}$$

Applying the Fokas unified transform method we get the solution formula

$$\begin{aligned} u(x, t) =& S[u_0, g_0, \ldots, g_{j-1}; f] \doteq \frac{1}{2\pi}\int_{-\infty}^{\infty} e^{i\xi x + i\xi^m t}[\widehat{u}_0(\xi) + F(\xi, t)]d\xi \\ &+ \sum_{p=1}^{j}\sum_{n=1}^{j+1} c_{p,n}\int_{\partial D_{2p}^+} e^{i\xi x + i\xi^m t}[\widehat{u}_0(\alpha_{p,n}\xi) + F(\alpha_{p,n}\xi, t)]d\xi \\ &+ \sum_{p=1}^{j}\sum_{\ell=0}^{j-1} c'_{p,\ell}\int_{\partial D_{2p}^+} e^{i\xi x + i\xi^m t}(i\xi)^{2j-\ell}\tilde{g}_\ell(\xi^m, T)d\xi. \end{aligned} \tag{3.3}$$

$$\boxed{\alpha_{p,n} \doteq e^{i[m-(2p+1)+2n]\frac{\pi}{m}}}$$

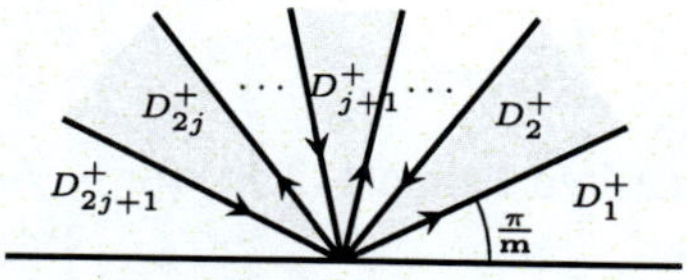

D_{2p}^+ (j is odd)

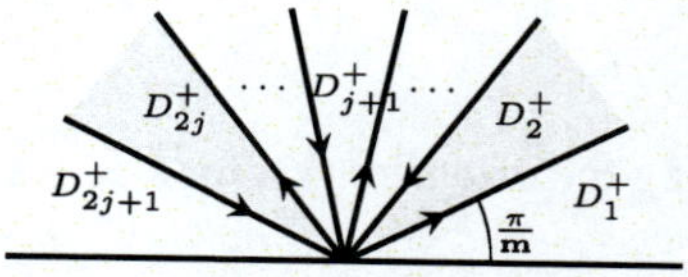

D_{2p}^+ (j is even)

We recall that the half-line Fourier transform $\widehat{u}_0(\xi)$ and the temporal Fourier transform $\tilde{g}_\ell$ are defined as for the KdV (see (2.3)), while the time integral of the half-line Fourier transform of the forcing $f(\cdot, t)$ is defined by

$$F(\xi, t) \doteq \int_0^t e^{-i\xi^m \tau}\hat{f}(\xi, \tau)d\tau = \int_0^t e^{-i\xi^m \tau}\int_0^\infty e^{-i\xi x} f(x, \tau)dxd\tau, \quad \mathrm{Im}(\xi) \le 0.$$

The modified Bourgain space $X^{s,b,\alpha}(\mathbb{R}^2)$ for KdVm is defined by the norm

$$\|u\|^2_{X^{s,b,\alpha}} \doteq \iint_{\mathbb{R}^2} \Big[(1+|\xi|)^{2s}(1+|\tau-\xi^m|)^{2b} + \chi_{|\xi|<1}(1+|\tau|)^{2\alpha}\Big] |\hat{u}(\xi,\tau)|^2 d\xi d\tau,$$

and the "temporal" Bourgain space $Y^{s,b}$ is defined by the norm

$$\|u\|^2_{Y^{s,b}} \doteq \iint_{\mathbb{R}^2} (1+|\tau|)^{\frac{2s}{m}}(1+|\tau-\xi^m|)^{2b}|\widehat{u}(\xi,\tau)|^2 d\xi d\tau. \tag{3.4}$$

Step 2. Deriving Linear Estimates. For the Fokas solution formula (3.3), we have the following result [29].

Theorem 7 (KdVm Linear estimates) *For $-j-\frac{1}{2} < s \le j+1$ the Fokas formula (3.3) defines a solution u to the KdVm ibvp (3.2) (under appropriate compatibility conditions) which is in the space $X^{s,b,\alpha}_{\mathbb{R}^+\times(0,T)}$ and satisfies the estimates*

$$\begin{aligned} &\|S[u_0, g_0, \ldots, g_{j-1}; f]\|_{X^{s,b,\alpha}(\mathbb{R}^+\times(0,T))} \\ &\lesssim \Big[\|u_0\|_{H^s_x(0,\infty)} + \sum_{\ell=0}^{j-1}\|g_\ell\|_{H_t^{\frac{s+j-\ell}{m}}(0,T)} + \|f\|_{X^{s,-b,\alpha-1}_{\mathbb{R}^+\times(0,T)}} + \|f\|_{Y^{s,-b}_{\mathbb{R}^+\times(0,T)}}\Big], \end{aligned} \tag{3.5}$$

for some $0 < b < \frac{1}{2}$ and $\frac{1}{2} < \alpha < 1$ (that can be described more precisely [29]).

Step 3. Proving that the iteration map is contraction. Like in the proof of KdV ibvp, to show that the iteration map defined by the Fokas solution formula is contraction, we need to derive the bilinear estimates indicated by the KdVm nonlinearity and the linear estimates (3.5) above. These are contained in the following result, which is proved in [29].

Theorem 8 (KdVm (optimal) Bilinear estimates) *For $s > -j + \frac{1}{4}$, we have the bilinear estimates in the modified Bourgain spaces*

$$\|\partial_x(f\cdot g)\|_{X^{s,-b,\alpha-1}} \le c_{s,b,\alpha}\|f\|_{X^{s,b',\alpha'}}\|g\|_{X^{s,b',\alpha'}}. \tag{3.6}$$

For $-j+\frac{1}{4} < s < m$, we have the bilinear estimates in the "temporal" Bourgain spaces

$$\|\partial_x(fg)\|_{Y^{s,-b}} \le \|\partial_x(fg)\|_{X^{s,-b}} + c_{s,b}\|f\|_{X^{s,b'}}\|g\|_{X^{s,b'}}. \tag{3.7}$$

In the above estimates, $b' \le b$ and $\alpha' \le \alpha$ are some numbers in $(0, 1)$.

Finally, using the linear and bilinear Estimates we show that the Fokas iteration map has a fixed point in modified Bourgain spaces, thus establishing the well-posedness of our KdVm ibvp (3.1) for the same critical Sobolev exponent as in case of the ivp on the line proved in [12]. More precisely, we obtain the following optimal result [29].

Theorem 9 (KdVm ibvp (optimal) well-posedness) *If* $-j+\frac{1}{4}<s\le j+1$, $s\neq \frac{1}{2},\frac{3}{2},\dots,j-\frac{1}{2}$, *then for any initial data* $u_0\in H^s(0,\infty)$, *boundary data* $g_\ell\in H^{\frac{1}{m}(s+j-\ell)}(0,T)$, $\ell=0,1,\dots,j-1$, *and some lifespan* $0<T_0\le T<\frac{1}{2}$, **there is a solution** *for the KdVm ibvp* (3.1), *which is in* $X^{s,b,\alpha}_{\mathbb{R}^+\times(0,T_0)}$ *and which satisfies the size estimate*

$$\|u\|_{X^{s,b,\alpha}_{\mathbb{R}^+\times(0,T_0)}}\le C\left(\|u_0\|_{H^s(\mathbb{R}^+)}+\sum_{\ell=0}^{j-1}\|g_\ell\|_{H^{\frac{s+j-\ell}{m}}(0,T)}\right), \tag{3.8}$$

for some $b\in(0,\frac{1}{2})$ *and* $\alpha\in(\frac{1}{2},1)$. *Also, an estimate for the lifespan is given by* $T_0=c_0\Big(1+\|u_0\|_{H^s(\mathbb{R}^+)}+\sum_{\ell=0}^{j-1}\|g_\ell\|_{H^{\frac{s+j-\ell}{m}}(0,T)}\Big)^{-4/\beta}$, *where* $\beta>0$ *is depending on* s, b *and* m. *Furthermore, the* **solution is unique** *in* $X^{s,b,\alpha}_{\mathbb{R}^+\times(0,T_0)}$. *Finally, the* **data to solution map** $\{u_0,g_0,\dots,g_{j-1}\}\mapsto u$ *is locally* **Lipschitz continuous.**

Progress in ibvp and Conclusion. Via the Fokas method, for the KdV ibvp we have obtained analogues to ivp optimal well-posedness results. The Fokas method works equally well for higher order nonlinear equations, like the KdVm, providing again optimal ibvp results. Also, via the Fokas method we have studied the NLS ibvp on the half-line [17, 25], and the NLS ibvp on the half-plane [24, 26]. Concluding, we recall that the Fokas method [13, 15], is motivated by the inverse scattering transform and provides a novel approach for solving initial-boundary value problems for linear and integrable nonlinear partial differential equations. It is the first and crucial step of our work here. It provides the solution formula for the forced linear ibvp and opens the way for deriving good linear estimates analogues to those for ivp. For further results on ibvp for KdV, NLS, Boussinesq, heat, and related equations we refer to [1, 9, 13, 14, 14, 18–21, 23–25, 27, 28, 41–43, 45] and the references therein.

Acknowledgements The author was partially supported by a grant from the Simons Foundation (#524469 to Alex Himonas).

References

1. Batal, A., Fokas, A.S., Özsari, T.: Fokas method for linear boundary value problems involving mixed spatial derivatives. Proc. A. **476**(2239), 20200076, 15 pp (2020)
2. Bona, J.L., Sun, S.M., Zhang, B.-Y.: A non-homogeneous boundary-value problem for the Korteweg-de Vries equation in a quarter plane. Trans. Amer. Math. Soc. **354**(2), 427–490 (2002)
3. Bona, J.L., Sun, S.M., Zhang, B.-Y.: A nonhomogeneous boundary-value problem for the Korteweg-de Vries equation posed on a finite domain. Commun. Partial Diff. Equ. **28**(7–8), 1391–1436 (2003)
4. Bona, J.L., Sun, S.M., Zhang, B.-Y.: Nonhomogeneous boundary value problems of one-dimensional nonlinear Schrödinger equations. J. Math. Pures Appl. **109**, 1–66 (2018)

5. Bourgain, J.: Fourier transform restriction phenomena for certain lattice subsets and applications to nonlinear evolution equations. Part II: The KdV equation. Geom. Funct. Anal. **3**(3), 209–262 (1993)
6. Bourgain, J.: Fourier transform restriction phenomena for certain lattice subsets and applications to nonlinear evolution equations. Geom. Funct. Anal. **3**(2), 107–156 (1993)
7. Colliander, J., Keel, M., Staffilani, G., Takaoka, H., Tao, T.: Sharp global well-posedness for KdV and modified KdV on $\mathbb{R}$ and $\mathbb{T}$. J. Amer. Math. Soc. **16**(3), 705–749 (2003)
8. Colliander, J.E., Kenig, C.E.: The generalized Korteweg-de Vries equation on the half-line. Commun. Partial Diff. Equ. **27**(11–12), 2187–2266 (2002)
9. Deconinck, B., Trogdon, T., Vasan, V.: The method of Fokas for solving linear partial differential equations. SIAM Rev. **56**(1), 159–186 (2014)
10. Erdoğan, M.B., Tzirakis, N.: Regularity properties of the cubic nonlinear Schrödinger equation on the half line. J. Funct. Anal. **271**, 2539–2568 (2016)
11. Faminskii, A.V.: An initial boundary-value problem in a half-strip for the Korteweg-de Vries equation in fractional-order Sobolev spaces (English summary). Commun. Partial Diff. Equ. **29**(11–12), 1653–1695 (2004)
12. Figueira, R., Himonas, A., Yan, F.: A higher dispersion KdV equation on the line. Nonlinear Anal. **199**, 112055, 38 pp (2020)
13. Fokas, A.S.: A unified transform method for solving linear and certain nonlinear PDEs. Proc. Roy. Soc. Lond. Ser. A **453**(1962), 1411–1443 (1997)
14. Fokas, A.S.: A new transform method for evolution partial differential equations. IMA J. Appl. Math. **67**(6), 559–590 (2002)
15. Fokas, A.S.: A unified approach to boundary value problems. In: CBMS-NSF Regional Conference Series in Applied Mathematics, 78, xvi+336 pp. Society for Industrial and Applied Mathematics (SIAM), Philadelphia, PA (2008)
16. Fokas, A.S., Himonas, A.A., Mantzavinos, D.: The Korteweg-de Vries equation on the half-line. Nonlinearity **29**(2), 489–527 (2016)
17. Fokas, A.S., Himonas, A.A., Mantzavinos, D.: The nonlinear Schrödinger equation on the half-line. Trans. Amer. Math. Soc. **369**(1), 681–709 (2017)
18. Fokas, A.S., Its, A.R., Sung, L.-Y.: The nonlinear Schrödinger equation on the half-line. Nonlinearity **18**(4), 1771–1822 (2005)
19. Fokas, A.S., Lenells, J.: The unified method: I. Nonlinearizable problems on the half-line. J. Phys. A **45**(19), 195201, 38 pp (2012)
20. Fokas, A.S., Pelloni, B.: Introduction. Unified Transform for Boundary Value Problems, pp. 1–9. SIAM, Philadelphia, PA (2015)
21. Fokas, A.S., Spence, E.: Synthesis, as opposed to separation, of variables. SIAM Rev. **54**(2), 291–324 (2012)
22. Gardner, C.S., Greene, J.M., Kruskal, M.D., Miura, R.M.: Method for solving the Korteweg-de Vries equation. Phys. Rev. Lett. **19**(19), 1095–1097 (1967)
23. Himonas, A.A., Mantzavinos, D.: The "good" Boussinesq equation on the half-line. J. Differ. Equ. **258**(9), 3107–3160 (2015)
24. Himonas, A.A., Mantzavinos, D.: Well-posedness of the nonlinear Schrödinger equation on the half-plane. Nonlinearity **33**, 5567–5609 (2020)
25. Himonas, A.A., Mantzavinos, D.: The nonlinear Schrödinger equation on the half-line with a Robin boundary condition. Anal. Math. Phys. **11**(4), Paper No. 157 (2021)
26. Himonas, A.A., Mantzavinos, D.: The Robin and Neumann Problems for the Nonlinear Schrödinger Equation on the Half-Plane. Proc. A. **478**(2265), Paper No. 279, 20 pp (2022)
27. Himonas, A.A., Mantzavinos, D., Yan, F.: The Korteweg-de Vries equation on an interval. J. Math. Phys. **60**(5), 051507, 26 pp (2019)
28. Himonas, A.A., Yan, F.: The Korteweg-de Vries equation on the half-line with Robin and Neumann data in low regularity spaces. Nonlinear Anal. **222**, 113008, 31 pp (2022)
29. Himonas, A.A., Yan, F.: A higher dispersion KdV equation on the half-line. J. Diff. Equ. **333**(5), 55–102 (2022)

30. Holmer, J.: The initial-boundary-value problem for the 1D nonlinear Schrödinger equation on the half-line. Diff. Integral Equ. **18**(6), 647–668 (2005)
31. Holmer, J.: The initial-boundary-value problem for the Korteweg-de Vries equation. Commun. Partial Diff. Equ. **31**(8), 1151–1190 (2006)
32. Kenig, C.E., Ponce, G., Vega, L.: On the (generalized) Korteweg-de Vries equation. Duke Math. J. **59**(3), 585–610 (1989)
33. Kenig, C.E., Ponce, G., Vega, L.: Oscillatory integrals and regularity of dispersive equations. Indiana Univ. Math. J. **40**(1), 33–69 (1991)
34. Kenig, C.E., Ponce, G., Vega, L.: Well-posedness of the initial value problem for the Korteweg-de Vries equation. J. Amer. Math. Soc. **4**(2), 323–347 (1991)
35. Kenig, C.E., Ponce, G., Vega, L.: Well-posedness and scattering results for the generalized Korteweg-de Vries equation via the contraction principle. Commun. Pure Appl. Math. **46**(4), 527–620 (1993)
36. Kenig, C.E., Ponce, G., Vega, L.: A bilinear estimate with applications to the KdV equation. J. Amer. Math. Soc. **9**(2), 573–603 (1996)
37. Kenig, C.E., Ponce, G., Vega, L.: On the ill-posedness of some canonical dispersive equations. Duke Math. J. **106**(3), 617–633 (2001)
38. Killip, R., Visan, M.: KdV is well-posed in H^{-1}. Ann. of Math. (2) **190**(1), 249–305 (2019)
39. Korteweg, D.J., de Vries, G.: On the change of form of long waves advancing in a rectangular canal, and on a new type of long stationary waves. Phil. Mag. Ser. 5 **39**(240), 422–443 (1895)
40. Lax, P.D.: Integrals of nonlinear equations of evolution and solitary waves. Commun. Pure Appl. Math. **21**, 467–490 (1968)
41. Lenells, J.: The KdV equation on the half-line: the Dirichlet to Neumann map. J. Phys. A **46**(34), 345203, 20 pp (2013)
42. Lenells, J., Fokas, A.S.: The unified method: II. NLS on the half-line with t-periodic boundary conditions. J. Phys. A **45**(19), 195202, 36 pp (2012)
43. Özsari, T., Yolcu, N.: The initial-boundary value problem for the biharmonic Schrödinger equation on the half-line. Commun. Pure Appl. Anal. **18**(6), 3285–3316 (2019)
44. Saut, J.-C., Temam, R.: Remarks on the Korteweg-de Vries equation. Israel J. Math. **24**(1), 78–87 (1976)
45. Yan, F.: Well-posedness of a higher dispersion KdV equation on the half-line. J. Math. Phys. **61**(8), 081506, 29 pp (2020)
46. Zabusky, N.J., Kruskal, M.D.: Interaction of solitons in a collisionless plasma and the recurrence of initial states. Phys. Rev. Lett. **15**(6), 240–243 (1965)

Instabilities of Linear Evolution PDEs via the Fokas Method

A. Chatziafratis, L. Grafakos, S. Kamvissis, and I. G. Stratis

Dedicated to Professor A. S. Fokas on the occasion of his 70th birthday

Abstract In this short article, we use the formula provided by the Fokas method for initial-boundary-value problems (ibvp) for the linearised KdV equation on the half-line for positive time. Depending on the sign of the dispersive term, the long range asymptotics can depend in a very sensitive way on the behavior of the data at the point $(0, 0)$. Such instabilities have apparently not been noticed before and they are expected to appear for a large set of equations. As to which equations are unstable and which are not, this is an open question worthy of further investigation.

Keywords Fokas formula · IBVPs for evolution PDE · Half-line linearized KdV equation · Long-space asymptotics · Conditional decay · Instability effects

A. Chatziafratis · I. G. Stratis
Department of Mathematics, National and Kapodistrian University of Athens, 157 84 Athens, Greece
e-mail: chatziafrati@math.uoa.gr

I. G. Stratis
e-mail: istratis@math.uoa.gr

A. Chatziafratis · S. Kamvissis (✉)
Institute of Applied and Computational Mathematics, FORTH, 700 13 Heraklion, Greece
e-mail: kamvissis@uoc.gr

L. Grafakos
Department of Mathematics, University of Missouri, Columbia, MO 65211, USA
e-mail: grafakosl@missouri.edu

S. Kamvissis
Department of Pure and Applied Mathematics, University of Crete, Rethymno, Greece

T. Bountis et al. (eds.), *Chaos, Fractals and Complexity*, Springer Proceedings in Complexity, https://doi.org/10.1007/978-3-031-37404-3_20

1 Introduction

When the Fokas method was introduced by Fokas about 25 years ago [5] (see also [6–8]), it was initially conceived as a method for solving initial-boundary value problems for completely integrable nonlinear equations like KdV, NLS, or more generally equations that can be formulated as evolutions in time of a linear differential operator $L(t)$ governed by the famous Lax pair equation $dL/dt = BL - LB$, where $B(L)$ is usually some auxiliary anti-symmetric linear differential operator.

While the study of an initial-value problem for KdV involves the study of the scattering transform for the associated linear Schrödinger operator L, the role of B being somewhat trivialised, the study of the initial—boundary value problem involves the joint study of scattering data for both operators L, B; thus the term "Unified Transform". The interdependence of the two operators renders this new method a highly nontrivial extension of the standard scattering method.

Even though this method was initially proposed for nonlinear problems, it soon became evident that it was also applicable to linear problems. While, before the new method, the existing tools for boundary value problems of linear PDEs (like the Laplace or the sine transform) were explicitly applicable to very specific equations, the new method has been spectacularly successful in a much wider class of problems, of any order, even elliptic [1], even with non-constant coefficients and in all sorts of domains in the (x, t)-plane; see, for instance, [2–4] and references cited therein. In fact, the linear method even offered some insights to the nonlinear integrability theory by helping to realize that Lax pairs provide the generalization of the divergence formulation from a separable linear to an integrable nonlinear PDE [9]. Not only very explicit formulae for the solutions are provided, but such formulae are very efficient numerically. It is fair to say that the Fokas method has thus rejuvenated the study of linear equations.

In this paper, we focus on one simple consequence of the Fokas theory. We report on the discovery of an instability phenomenon, apparently not noticed before. Let us, for example, consider the very specific initial-boundary value problem for the two linear KdV equations:

$$\begin{cases} \partial_t u + \partial_{xxx} u = 0, \ (x,t) \in \mathbb{R}^+ \times \mathbb{R}^+ \\ u(x,0) = u_0(x), \quad x \in \mathbb{R}^+ \\ u(0,t) = g_0(t), \quad t \in \mathbb{R}^+, \end{cases} \tag{1}$$

and

$$\begin{cases} \partial_t u - \partial_{xxx} u = 0, \ (x,t) \in \mathbb{R}^+ \times \mathbb{R}^+ \\ u(x,0) = u_0(x), \quad x \in \mathbb{R}^+ \\ u(0,t) = g_0(t), \quad t \in \mathbb{R}^+ \\ u_x(0,t) = g_1(t), \quad t \in \mathbb{R}^+, \end{cases} \tag{2}$$

where the initial and boundary data u_0, g_0 and g_1 are functions defined in $\mathbb{R}^+$ and satisfy appropriate conditions (see Theorem 1 and Sect. 3).

2 The Equation $\partial_t u + \partial_{xxx} u = 0$

The Fokas formula for the solution of (1) is

$$u(x,t) = \frac{1}{2\pi} \int_{\lambda=-\infty}^{\infty} e^{i\lambda x - \omega(\lambda)t} \hat{u}_0(\lambda) d\lambda$$
$$+ \frac{1}{2\pi} \int_{\lambda \in \Gamma} e^{i\lambda x - \omega(\lambda)t} [\alpha \hat{u}_0(\alpha\lambda) + \alpha^2 \hat{u}_0(\alpha^2\lambda)] d\lambda$$
$$- \frac{1}{2\pi} \int_{\lambda \in \Gamma} e^{i\lambda x - \omega(\lambda)t} 3\lambda^2 \tilde{g}_0(\omega(\lambda), t) d\lambda, \quad (3)$$

where $\hat{u}_0(\lambda) = \int_{y=0}^{\infty} e^{-i\lambda y} u_0(y) dy$ (defined for $\lambda \in \mathbb{C}$ with $\Im \lambda \le 0$), $\tilde{g}_0(\omega(\lambda), t) = \int_{\tau=0}^{t} e^{\omega(\lambda)\tau} g_0(\tau) d\tau$ with $\omega(\lambda) = -i\lambda^3$, $\alpha = e^{2\pi i/3}$, and $\Gamma = \partial\Omega^-$ with $\Omega^- = \{\lambda \in \mathbb{C} : \operatorname{Im}\lambda \ge 0 \text{ and } \operatorname{Re}\omega(\lambda) \le 0\}$ (Fig. 1).

Theorem 1 (see [3]) *If $u_0(x) \in \mathcal{S}([0,\infty))$ and $g_0(t) \in C^\infty([0,\infty))$ then the function $u(x,t)$, defined by (3), satisfies the following relation*

$$\lim_{x \to +\infty} \frac{\partial^k u(x,t)}{\partial x^k} = 0 \quad (4)$$

for every nonnegative integer k, uniformly for t in compact subsets of $(0,\infty)$.

Proof Firstly, let us fix a $t > 0$. By appropriate deformation of the contours, we have that

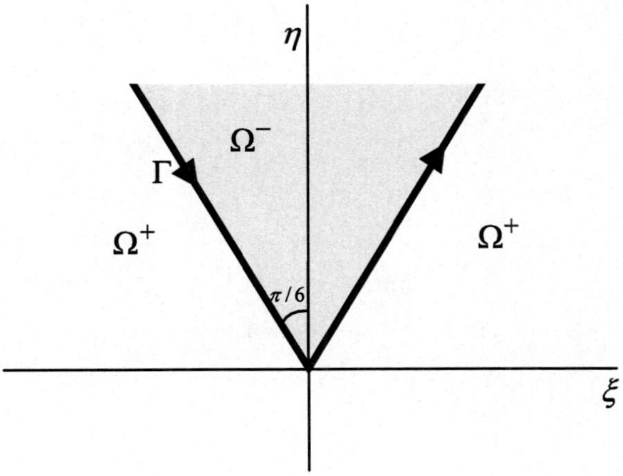

Fig. 1 The contour Γ is the boundary of Ω^-

$$\frac{1}{i^k}\frac{\partial^k}{\partial x^k}\left[\int_{\lambda=-\infty}^{\infty} e^{i\lambda x-\omega(\lambda)t}\hat{u}_0(\lambda)d\lambda\right] =$$

$$\int_{\lambda=-1}^{1} \lambda^k e^{i\lambda x-\omega(\lambda)t}\hat{u}_0(\lambda)d\lambda$$

$$+\left(\int_{\lambda=-\infty}^{-1} + \int_{\lambda=1}^{\infty}\right)\lambda^k e^{i\lambda x-\omega(\lambda)t}[\hat{u}_0(\lambda)-\sigma_N(\lambda)]d\lambda$$

$$+\int_{\substack{\operatorname{Im}\lambda=1\\ -\infty<\operatorname{Re}\lambda\le -1 \text{ or } 1\le \operatorname{Re}\lambda<\infty}} \lambda^k e^{i\lambda x-\omega(\lambda)t}\sigma_N(\lambda)d\lambda$$

$$+\int_{\lambda\in[-1+i,-1]\cup[1,1+i]} \lambda^k e^{i\lambda x-\omega(\lambda)t}\sigma_N(\lambda)d\lambda \tag{5}$$

provided that $N > k$.

We claim that

$$\lim_{x\to\infty}\frac{\partial^k}{\partial x^k}\left[\int_{\lambda=-\infty}^{\infty} e^{i\lambda x-\omega(\lambda)t}\hat{u}_0(\lambda)d\lambda\right] = 0. \tag{6}$$

For its proof, it suffices to show the following:

$$\lim_{x\to\infty}\int_{\lambda=-1}^{1} \lambda^k e^{i\lambda x-\omega(\lambda)t}\hat{u}_0(\lambda)d\lambda = 0 \tag{7}$$

$$\lim_{x\to\infty}\left(\int_{\lambda=-\infty}^{-1} + \int_{\lambda=1}^{\infty}\right)\left(\lambda^k e^{i\lambda x-\omega(\lambda)t}[\hat{u}_0(\lambda)-\sigma_N(\lambda)]d\lambda\right) = 0 \tag{8}$$

$$\lim_{x\to\infty}\int_{\substack{\operatorname{Im}\lambda=1\\ -\infty<\operatorname{Re}\lambda\le -1 \text{ or } 1\le \operatorname{Re}\lambda<\infty}} \lambda^k e^{i\lambda x-\omega(\lambda)t}\sigma_N(\lambda)d\lambda = 0 \tag{9}$$

$$\lim_{x\to\infty}\int_{\lambda\in[-1+i,-1]\cup[1,1+i]} \lambda^k e^{i\lambda x-\omega(\lambda)t}\sigma_N(\lambda)d\lambda = 0. \tag{10}$$

Applying the Riemann–Lebesgue lemma to the function

$$\varphi(\lambda) = \begin{cases} \lambda^k e^{-\omega(\lambda)t}\hat{u}_0(\lambda) & \text{for} \quad -1 \le \lambda \le 1 \\ 0 & \text{for } \lambda \in \mathbb{R} - [-1, 1] \end{cases}$$

which is clearly L^1 in $\mathbb{R}$, we obtain (7).

Similarly, (8) follows from the Riemann–Lebesgue lemma applied to the function

$$\Phi(\lambda) = \begin{cases} \lambda^k e^{-\omega(\lambda)t}[\hat{u}_0(\lambda) - \sigma_N(\lambda)] & \text{for } \lambda \in \mathbb{R} - [-1, 1] \\ 0 & \text{for} \quad -1 \le \lambda \le 1 \end{cases}$$

which is also L^1 in $\mathbb{R}$, since $N > k$.

Now, for $\lambda = \xi + i\eta$ with $\eta = 1$, $\left|e^{i\lambda x}\right| = e^{-x}$. Therefore, the absolute value of the integral in (9) is

$$\le e^{-x} \int\limits_{\substack{\operatorname{Im}\lambda=1 \\ -\infty<\operatorname{Re}\lambda\le-1 \text{ or } 1\le\operatorname{Re}\lambda<\infty}} \left|\lambda^k e^{-\omega(\lambda)t}\sigma_N(\lambda)\right| d|\lambda|,$$

and (9) follows.

Finally, for $\lambda = \xi + i\eta$, $\left|e^{i\lambda x}\right| = e^{-\eta x}$. It follows that if $\lambda = \xi + i\eta \in [-1 + i, -1] \cup [1, 1 + i]$ and $\lambda \ne \pm 1$ then $\eta > 0$, whence $\lim\limits_{x \to \infty} [\lambda^k e^{i\lambda x - \omega(\lambda)t}\sigma_N(\lambda)] = 0$. Therefore, (10) follows from Lebesgue's dominated convergence theorem.

Also,

$$\lim_{x \to \infty} \int\limits_{\Gamma} \lambda^k e^{i\lambda x - \omega(\lambda)t}[\alpha\hat{u}_0(\alpha\lambda) + \alpha^2\hat{u}_0(\alpha^2\lambda)]d\lambda = 0, \tag{11}$$

since the factor $e^{i\lambda x}$ in the above integral has absolute value $e^{-\sqrt{3}x|\lambda|/2}$ when $\lambda \in \Gamma$, and, for $x \ge 1$, the integrand is dominated by $|\lambda|^{k-1}e^{-\sqrt{3}|\lambda|/2}$—up to a constant—and $\int\limits_{\Gamma} |\lambda|^{k-1}e^{-\frac{\sqrt{3}}{2}|\lambda|}d|\lambda| < +\infty$.

Similarly,

$$\lim_{x \to \infty} \int\limits_{\Gamma} \lambda^k e^{i\lambda x - \omega(\lambda)t} 3\lambda^2 \widetilde{g}_0(\omega(\lambda), t)d\lambda = 0. \tag{12}$$

Now, (4) follows from (3), (6), (11) and (12).

Finally, it is easy to see that all the above limits are uniform for t in compact subsets of $(0, \infty)$. □

Theorem 2 ([3]) *With the assumptions as in Theorem 1, the function $u(x, t)$, defined by (3), satisfies the following equation*

$$\lim_{x\to+\infty}[xu(x,t)]=0 \tag{13}$$

uniformly for t in compact subsets of $(0,\infty)$.

Proof With N sufficiently large, integration by parts gives

$$\begin{aligned}
&ix\int_{\lambda=-\infty}^{\infty} e^{i\lambda x-\omega(\lambda)t}\hat{u}_0(\lambda)d\lambda\\
&=\int_{\lambda=-1}^{1}\frac{d}{d\lambda}(e^{i\lambda x})e^{-\omega(\lambda)t}\hat{u}_0(\lambda)d\lambda\\
&\quad+\left(\int_{\lambda=-\infty}^{-1}+\int_{\lambda=1}^{\infty}\right)\left(\frac{d}{d\lambda}(e^{i\lambda x})e^{-\omega(\lambda)t}[\hat{u}_0(\lambda)-\sigma_N(\lambda)]d\lambda\right)\\
&\quad+\int_{\substack{\mathrm{Im}\,\lambda=1\\ -\infty<\mathrm{Re}\lambda\le-1 \text{ or } 1\le\mathrm{Re}\,\lambda<\infty}}\frac{d}{d\lambda}(e^{i\lambda x})e^{-\omega(\lambda)t}\sigma_N(\lambda)d\lambda\\
&\quad+\int_{\lambda\in[-1+i,-1]\cup[1,1+i]}\frac{d}{d\lambda}(e^{i\lambda x})e^{-\omega(\lambda)t}\sigma_N(\lambda)d\lambda\\
&=-\int_{\lambda=-1}^{1}e^{i\lambda x}\frac{d}{d\lambda}[e^{-\omega(\lambda)t}\hat{u}_0(\lambda)]d\lambda\\
&\quad-\left(\int_{\lambda=-\infty}^{-1}+\int_{\lambda=1}^{\infty}\right)\left(e^{i\lambda x}\frac{d}{d\lambda}\{e^{-\omega(\lambda)t}[\hat{u}_0(\lambda)-\sigma_N(\lambda)]\}d\lambda\right)\\
&\quad+\int_{\substack{\mathrm{Im}\lambda=1\\ -\infty<\mathrm{Re}\,\lambda\le-1 \text{ or } 1\le\mathrm{Re}\,\lambda<\infty}}e^{i\lambda x}\frac{d}{d\lambda}[e^{-\omega(\lambda)t}\sigma_N(\lambda)]d\lambda\\
&\quad-\int_{\lambda\in[-1+i,-1]\cup[1,1+i]}e^{i\lambda x}\frac{d}{d\lambda}[e^{-\omega(\lambda)t}\sigma_N(\lambda)d\lambda],
\end{aligned}\tag{14}$$

since the "intermediate evaluations" cancel each other.

Now, the integrals in RHS of (14) tend to zero, as $x\to+\infty$, by the Riemann–Lebesgue lemma, as in the proof of Theorem 1. Therefore, (14) implies that

$$\lim_{x\to\infty}\left(x\int_{\lambda=-\infty}^{\infty}e^{i\lambda x-\omega(\lambda)t}\hat{u}_0(\lambda)d\lambda\right)=0. \tag{15}$$

Next, by the presence of the factor $e^{i\lambda x}$ and the fact that integration is taken on $\Gamma \cap \{|\lambda| \geq 1\}$,

$$\lim_{x \to \infty} \left[x \int\limits_{\Gamma\cap\{|\lambda|\geq 1\}} e^{i\lambda x - \omega(\lambda)t} [\alpha \hat{u}_0(\alpha\lambda) + \alpha^2 \hat{u}_0(\alpha^2 \lambda)] d\lambda \right] = 0,$$

$$\lim_{x \to \infty} \left[x \int\limits_{\Gamma\cap\{|\lambda|\geq 1\}} e^{i\lambda x - \omega(\lambda)t} 3\lambda^2 \widetilde{g}_0(\omega(\lambda), t) d\lambda \right] = 0. \quad (16)$$

On the other hand, writing $xe^{i\lambda x} = d(e^{i\lambda x})/id\lambda$ and integrating by parts, we obtain

$$\lim_{x \to \infty} \left[x \int\limits_{\Gamma\cap\{|\lambda|\leq 1\}} e^{i\lambda x - \omega(\lambda)t} [\alpha \hat{u}_0(\alpha\lambda) + \alpha^2 \hat{u}_0(\alpha^2 \lambda)] d\lambda \right] = 0,$$

$$\lim_{x \to \infty} \left[x \int\limits_{\Gamma\cap\{|\lambda|\leq 1\}} e^{i\lambda x - \omega(\lambda)t} 3\lambda^2 \widetilde{g}_0(\omega(\lambda), t) d\lambda \right] = 0, \quad (17)$$

Now, (13) follows from (15), (16) and (17). □

Examining the proofs of the previous two theorems, we easily see that we can prove the following more general theorem.

Theorem 3 ([3]) *With the assumptions as in Theorem 1, the function $u(x,t)$, defined by (3), satisfies the following equation:*

$$\lim_{x \to +\infty} \left(x^\ell \frac{\partial^k u(x,t)}{\partial x^k} \right) = 0$$

for nonnegative integers k and ℓ, uniformly for t in compact subsets of $(0, \infty)$.

3 The Equation $\partial_t u - \partial_{xxx} u = 0$

Assuming that $u_0(x) \in \mathcal{S}\big([0,\infty)\big)$ and $g_0(t), g_1(t) \in C^\infty\big([0,\infty)\big)$, the Fokas solution for (2) is

$$u(x,t) = \frac{1}{2\pi}\left[\int\limits_{\lambda=-\infty}^{\infty} e^{i\lambda x - \omega(\lambda)t} \hat{u}_0(\lambda) d\lambda - \int\limits_{\partial\Omega_1^-} e^{i\lambda x - \omega(\lambda)t} \hat{u}_0(\alpha\lambda) d\lambda - \int\limits_{\partial\Omega_2^-} e^{i\lambda x - \omega(\lambda)t} \hat{u}_0(\alpha^2\lambda) d\lambda \right]$$
$$+ \frac{1}{2\pi}\left[\int\limits_{\partial\Omega_1^-} e^{i\lambda x - \omega(\lambda)t} (1-\alpha^2)\lambda^2 \widetilde{g}_0(\omega(\lambda), t) d\lambda + \int\limits_{\partial\Omega_2^-} e^{i\lambda x - \omega(\lambda)t} (1-\alpha)\lambda^2 \widetilde{g}_0(\omega(\lambda), t) d\lambda \right]$$

$$-\frac{i}{2\pi}\left[\int_{\partial\Omega_1^-} e^{i\lambda x-\omega(\lambda)t}(1-\alpha)\lambda\widetilde{g}_1(\omega(\lambda),t)]d\lambda + \int_{\partial\Omega_2^-} e^{i\lambda x-\omega(\lambda)t}(1-\alpha^2)\lambda\widetilde{g}_1(\omega(\lambda),t)]d\lambda\right], \tag{18}$$

for $x > 0$ and $t > 0$, where $\omega(\lambda) = i\lambda^3$, $\alpha = e^{2\pi i/3}$,

$$\begin{aligned}\Omega_1^- &= \{\lambda \in \mathbb{C} : \operatorname{Im}\lambda \geq 0,\ \operatorname{Re}\lambda \leq 0, \text{ and } \operatorname{Re}\omega(\lambda) \leq 0\}\\ &= \{\lambda \in \mathbb{C} : (2\pi/3) \leq \arg\lambda \leq \pi\},\end{aligned}$$

and

$$\begin{aligned}\Omega_2^- &= \{\lambda \in \mathbb{C} : \operatorname{Im}\lambda \geq 0,\ \operatorname{Re}\lambda \geq 0, \text{ and } \operatorname{Re}\omega(\lambda) \leq 0\}\\ &= \{\lambda \in \mathbb{C} : 0 \leq \arg\lambda \leq \pi/3\}.\end{aligned}$$

See also Fig. 2.

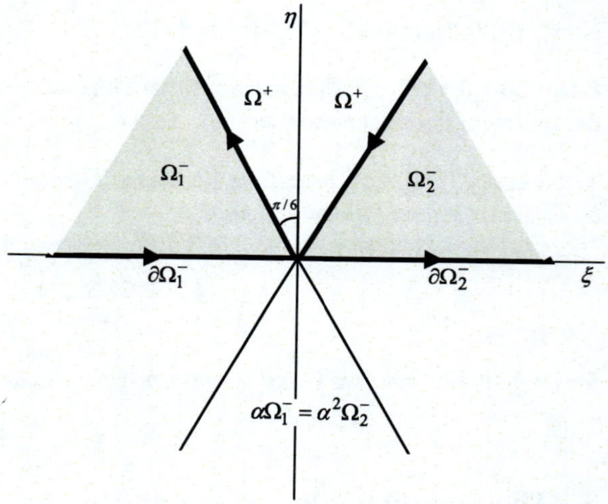

Fig. 2 The sets Ω_1^- and Ω_2^-, and their boundaries

For the function $u(x, t)$, defined by (18), the following theorems hold. (Detailed proofs will appear in [2].)

Theorem 4 ([2]) *With fixed* $t_1 > t_0 > 0$, *the solution* $u(x, t)$, *given by* (18), *satisfies the following:*
As $x \to +\infty$ *and uniformly for* $t_0 \leq t \leq t_1$,
1st $u(x,t) = [u_0(0) - g_0(0)]\dfrac{\sqrt{3}\sqrt[4]{3}}{\sqrt{\pi}}\dfrac{t^{1/4}}{x^{3/4}}\sin\left(\dfrac{2}{3\sqrt{3}}\dfrac{x^{3/2}}{\sqrt{t}} - \dfrac{5\pi}{12}\right) + O(1/x)$,

2nd $$\frac{\partial u(x,t)}{\partial x} = [u_0(0) - g_0(0)]\frac{2\sqrt{3}\sqrt[4]{3}}{\sqrt{\pi}}\frac{1}{t^{1/4}x^{1/4}}\cos\left(\frac{2}{3\sqrt{3}}\frac{x^{3/2}}{\sqrt{t}} - \frac{5\pi}{12}\right)$$
$$+[u_0'(0) - g_1(0)]\frac{\sqrt{3}\sqrt[4]{3}}{\sqrt{\pi}}\frac{t^{1/4}}{x^{3/4}}\sin\left(\frac{2}{3\sqrt{3}}\frac{x^{3/2}}{\sqrt{t}} - \frac{\pi}{12}\right) + O(1/x),$$

3rd $$\frac{\partial^2 u(x,t)}{\partial x^2} = -[u_0(0) - g_0(0)]\frac{2\sqrt[4]{3}}{\sqrt{\pi}}\frac{x^{1/4}}{t^{3/4}}\sin\left(\frac{2}{3\sqrt{3}}\frac{x^{3/2}}{\sqrt{t}} - \frac{5\pi}{12}\right)$$
$$+[u_0'(0) - g_1(0)]\frac{2\sqrt{3}\sqrt[4]{3}}{\sqrt{\pi}}\frac{1}{t^{1/4}x^{1/4}}\cos\left(\frac{2}{3\sqrt{3}}\frac{x^{3/2}}{\sqrt{t}} - \frac{\pi}{12}\right) + O(1/x),$$

4th $$\frac{\partial^3 u(x,t)}{\partial x^3} = -[u_0(0) - g_0(0)]\frac{2}{\sqrt[4]{3}\sqrt{\pi}}\frac{x^{3/4}}{t^{5/4}}\cos\left(\frac{2}{3\sqrt{3}}\frac{x^{3/2}}{\sqrt{t}} - \frac{5\pi}{12}\right)$$
$$+[u_0'(0) - g_1(0)]\frac{2\sqrt{3}\sqrt[4]{3}}{\sqrt{\pi}}\frac{x^{1/4}}{t^{3/4}}\sin\left(\frac{2}{3\sqrt{3}}\frac{x^{3/2}}{\sqrt{t}} - \frac{\pi}{12}\right)$$
$$+[u_0'''(0) - g_0'(0)]\frac{\sqrt{3}\sqrt[4]{3}}{\sqrt{\pi}}\frac{t^{1/4}}{x^{3/4}}\sin\left(\frac{2}{3\sqrt{3}}\frac{x^{3/2}}{\sqrt{t}} - \frac{5\pi}{12}\right) + O(1/x).$$

Theorem 5 ([2]) *With the notation and in the sense of Theorem 4, we have:*
1st If $u_0(0) = g_0(0)$ and $u_0'(0) = g_1(0)$, then

$$\frac{\partial^3 u(x,t)}{\partial x^3} = [u_0'''(0) - g_0'(0)]\frac{\sqrt{3}\sqrt[4]{3}}{\sqrt{\pi}}\frac{t^{1/4}}{x^{3/4}}\sin\left(\frac{2}{3\sqrt{3}}\frac{x^{3/2}}{\sqrt{t}} - \frac{5\pi}{12}\right) + O(1/x).$$

2nd If $\lim_{x\to\infty}\frac{\partial^2 u(x,t)}{\partial x^2}$ exists for some $t > 0$, then $u_0(0) = g_0(0)$. Conversely, if $u_0(0) = g_0(0)$ then

$$\frac{\partial^2 u(x,t)}{\partial x^2} = [u_0'(0) - g_1(0)]\frac{2\sqrt{3}\sqrt[4]{3}}{\sqrt{\pi}}\frac{1}{t^{1/4}x^{1/4}}\cos\left(\frac{2}{3\sqrt{3}}\frac{x^{3/2}}{\sqrt{t}} - \frac{\pi}{12}\right) + O(1/x),$$

uniformly for t in compact subsets of $(0, +\infty)$.
3rd For $n \geq 4$,

$$\frac{\partial^n u(x,t)}{\partial x^n} = [u_0(0) - g_0(0)]\frac{2}{3^{(2n-3)/4}\sqrt{\pi}}\frac{x^{(2n-3)/4}}{t^{(2n-1)/4}}\cdot$$
$$\cdot \operatorname{Re}\left\{i^{n-1}\exp\left[i\left(\frac{2}{3\sqrt{3}}\frac{x^{3/2}}{\sqrt{t}} - \frac{5\pi}{12}\right)\right]\right\} + O(x^{(2n-5)/4}).$$

4th Let $k \in \mathbb{N}$. If the limit $\lim_{x\to\infty}\frac{\partial^{4k-1}u(x,t)}{\partial x^{4k-1}}$ exists for some $t > 0$, then

$$u_0^{(3\ell-3)}(0) = g_0^{(\ell-1)}(0) \text{ and } u_0^{(3\ell-2)}(0) = g_1^{(\ell-1)}(0), \text{ for } \ell = 1, 2, \ldots, k. \quad (19)$$

Conversely, (19) implies that

$$\lim_{x \to \infty} \frac{\partial^n u(x,t)}{\partial x^n} = 0,$$

uniformly for t in compact subsets of $(0, +\infty)$, *for* $n = 0, 1, 2, \ldots, 4k$.

Comment: The above theorems show that the behavior of the solution, for large x, depends in a very sensitive way on the given data at the point $(x, t) = (0, 0)$.

4 Conclusion

Given the enormous power of today's computers, the role of PDE theory is partly relegated to the qualitative study of solutions, with particular attention to instabilities. An interesting consequence of the spectacularly successful Fokas theory for the solution of initial-boundary value problems for linear PDEs is the observation of instabilities. For some (not all) equations, the behavior of the solution, for large x, depends in a very sensitive way on the compatibility conditions at the point $(x, t) = (0, 0)$. Apparently this is a phenomenon not observed before. In the nonlinear case, the stable/unstable dichotomy is apparent mostly in the zero dispersion (semiclassical) limit and is related to the self-adjoint/non-self-adjoint dichotomy for the associated (spatial) Lax operator ([10]). It would be interesting to study what kind of linear evolution equations exhibit long-range instabilities and which equations do not. Also, it will be interesting to study whether there is a similar effect on long-time asymptotics. Such investigations for large values of spatial and temporal variables, for a variety of dispersive equations, are in progress and results will appear in subsequent publications. For the particular case of the linearized NLS with t-periodic boundary data, we refer to [11] where the large-t behavior of the solution is considered.

Acknowledgements A. Chatziafratis gratefully acknowledges funding through a research grant awarded by the State Scholarships Foundation. L. Grafakos would like to acknowledge the support of the Simons Fellowship 819503 and of the Simons Grant 624733.

References

1. Ashton, A.C.L.: On the rigorous foundations of the Fokas method for linear elliptic partial differential equations. Proc. R. Soc. A **468**, 1325–1331 (2012)
2. Chatziafratis, A., Grafakos, L., Kamvissis, S.: Explicit solution to the Airy equation on the half-line and its boundary and asymptotic behavior (preprint) (2022)
3. Chatziafratis, A., Kamvissis, S., Stratis, I.G.: Boundary behavior of the solution to the linear Korteweg-de Vries equation on the half-line. Stud. Appl. Math. **150**, 339–379 (2023). https://doi.org/10.1111/sapm.12542
4. Chatziafratis, A., Mantzavinos, D.: Boundary behavior for the heat equation on the half-line. Math. Meth. Appl. Sci. **45**, 7364–7393 (2022). https://doi.org/10.1002/mma.8245
5. Fokas, A.S.: A unified transform method for solving linear and certain nonlinear PDEs. Proc. R. Soc. Lond. A **453**, 1411–1443 (1997)

6. Fokas, A.S.: On the integrability of linear and nonlinear partial differential equations. J. Math. Phys. **41**, 4188–4237 (2000)
7. Fokas, A.S.: A new transform method for evolution partial differential equations. IMA J. Appl. Math. **67** (2002)
8. Fokas, A.S.: A unified approach to boundary value problems. In: CBMS-NSF Regional Conference Series in Applied Mathematics 78. SIAM, Philadelphia, PA (2008)
9. Fokas, A.S., Spence, E.A.: Synthesis, as opposed to separation, of variables. SIAM Rev. **54**(2), 291–324 (2012)
10. Kamvissis, S.: From stationary phase to steepest descent. Contem. Math. **458**, 145–162 (2008)
11. Lenells, J., Fokas, A.S.: The unified method: II. NLS on the half-line with t-periodic boundary conditions. J. Phys. A: Math. Theor. **45** (2012)

Fokas Diagonalization

D. A. Smith

Abstract A method for solving linear initial boundary value problems was recently reimplemented as a true spectral transform method. As part of this reformulation, the precise sense in which the spectral transforms diagonalize the underlying spatial differential operator was elucidated. That work concentrated on two point initial boundary value problems and interface problems on networks of finite intervals. In the present work, we extend these results, by means of three examples, to new classes of problems: problems on semiinfinite domains, problems with nonlocal boundary conditions, and problems in which the partial differential equation features mixed derivatives. We show that the transform pair derived via the Fokas transform method features the same Fokas diagonalization property in each of these new settings, and we argue that this weak diagonalization property is precisely that needed to ensure success of a spectral transform method.

Keywords Spectral method for PDE · Fourier transform · Unified transform method · Initial boundary value problem

1 Introduction

Fourier transform methods for solving initial boundary value problems for partial differential equations were instrumental to the advances of mathematical physics in the 19th century. They essentially all work by reexpressing a spatial differential operator as a diagonal multiplication operator acting in the spectral domain. This diagonalization reduces the spatiotemporal partial differential equation to an ordinary differential equation in the time variable only. After determining the solution of the ordinary differential equation, an inverse transform is used to map back from the

D. A. Smith (✉)
Division of Science, Yale-NUS College, Singapore, Singapore
e-mail: dave.smith@yale-nus.edu.sg

Department of Mathematics, National University of Singapore, Singapore, Singapore

T. Bountis et al. (eds.), *Chaos, Fractals and Complexity*, Springer Proceedings in Complexity, https://doi.org/10.1007/978-3-031-37404-3_21

spectral domain to coordinate space, yielding the solution of the original initial boundary value problem.

In the present work, we describe an advance on this method. Specifically, we explain how a family of solution methods for initial boundary value problems, known collectively as the Fokas transform method, or unified transform method, and developed over the past quarter century, can be interpreted as an advance on the classical Fourier transform method. We identify, within the Fokas transform method, pairs of transforms and inverse transforms that can be used to solve each problem. We show that, although these transforms do not diagonalize the relevant spatial differential operators in the usual sense, they each posses precisely the diagonalization property that is required for their use in a spectral transform method. This weak diagonalization property is motivated in Sects. 1.1–1.4, and characterized informally in Criterion 2. In Sect. 2, it is precisely stated and proved for three examples which are all beyond the class covered in the recent article [1].

1.1 The Classical Spectral Transform Method

Suppose we wish to solve a problem like

Problem (*Half line Dirichlet problem for the heat equation*)

$$[\partial_t - \partial_x^2]q(x,t) = 0 \qquad (x,t) \in (0,\infty) \times (0,T), \tag{1.1.PDE}$$

$$q(x,0) = Q(x) \qquad x \in [0,\infty), \tag{1.1.IC}$$

$$q(0,t) = 0 \qquad t \in [0,T], \tag{1.1.BC}$$

in which $Q \in \mathcal{S}[0,\infty)$, the space of smooth functions on the half line, rapidly decaying along with all of their derivatives. The differential operator L defined by

$$L\phi = -\phi', \qquad \operatorname{Dom} L = \{\phi \in \mathcal{S}[0,\infty) : \phi(0) = 0\} \tag{1.2}$$

is such that, interpreting L as always acting in the spatial variable, we may express (1.1.PDE) as $[\partial_t + L]q(x,t) = 0$.

Consider some hypothetical transform F which accepts a function of the spatial variable x and returns a function of a new "spectral" variable λ. We will apply this transform in the spatial variable to functions of space and timeto yield functions of λ and t. We suppose that this can be done in such a way that the transform commutes with the temporal derivative operator. Suppose also that the transform is linear.

We can apply such a transform, to (1.1.PDE) to get

$$\frac{\mathrm{d}}{\mathrm{d}t}F[q](\lambda;t) + F[Lq](\lambda;t) = 0.$$

If F has further the *diagonalization* property that $F[L\phi](\lambda) = \lambda^2 F[\phi](\lambda)$, then this simplifies to the temporal ordinary differential equation

$$\frac{\mathrm{d}}{\mathrm{d}t} F[q](\lambda; t) + \lambda^2 F[q](\lambda; t) = 0$$

for $F[q](\lambda; t)$. Its solution is

$$F[q](\lambda; t) = \mathrm{e}^{-\lambda^2 t} C(\lambda),$$

and evaluation at $t = 0$ combined with appeal to (1.1.IC) establishes that $C(\lambda) = F[Q](\lambda)$. Finally, we suppose that there exists another transform F^{-1} which allows us to map back from functions of the spectral variable to functions of x, and that this second transform is an inverse of the original transform in the sense that, for all x, $F^{-1}[F[\phi]](x) = \phi(x)$. Then

$$q(x, t) = F^{-1}\big[F[q](\,\cdot\,; t)\big](x) = F^{-1}\left[\mathrm{e}^{-\cdot^2 t} F[Q]\right](x).$$

This is a representation of the solution of problem (1.1) which relies only on the initial datum Q and this speculative spectral transform. So it remains only to find such a transform and the method is complete. It is worth noting that, because we have not explicitly used (1.1.BC) in the solution method, they must be connected to the transform itself. Consequently, if problem (1.1) and another with different boundary conditions are to have different solutions, then they must have different associated transforms. However, this makes the method very general; by simply selecting a different transform, the method is immediately applicable to the Neumann or Robin problems for the heat equation on the half line. The method also requires little modification before its application to partial differential equations with different L, or even different domains.

For this problem, of course, such F is well known: the Fourier sine transform. For the Neumann problem, the F should be the Fourier cosine transform. For the finite interval Dirichlet heat problem, the discrete Fourier sine transform, whose inverse is more commonly called the Fourier sine series, is the right transform. But with a new transform required for each problem, we quickly exhaust the classical spectral transforms.

1.2 Derivation of Transforms

For finite interval problems, the appropriate transform is typically derived by separation of variables and solution of an appropriate Sturm–Liouville problem. The classical Sturm–Liouville theory does not apply to problems of higher order, nonselfadjoint problems, or problems with more general types of boundary conditions, but much is known for many such problems. Two point boundary value problems have seen the most attention, where Birkhoff's work [4] was instrumental in showing that the crucial completeness and orthogonality results of Sturm–Liouville theory may be extended to a broad class. This work was greatly expanded on over the next few decades, but Jackson and Hopkins soon identified examples where completeness of eigenfunctions fails [22, 23], dooming the usual approach to definition of transform pairs using eigenfunctions of the spatial differential operator and those of its adjoint. Operators have been classified as "regular", "irregular", and "degenerate", with various definitions of each class, but the general theme of regular operators having all the properties we need to construct transform pairs, irregular operators having some impediments that make the construction more delicate, and degenerate operators, such as those identified by Jackson and Hopkins, usually considered beyond scope. Full surveys are given by Locker [25, 26], with updates for irregular operators and beyond two point operators provided by Freiling [20].

An alternative approach is required to derive the appropriate transform pair, at least for degenerate irregular operators. As the Fokas transform method is a spectral method, it is reasonable to hope that it may be a source for the appropriate transforms. As we shall argue, the situation is slightly more complex: the transform method itself must be modified to admit the transforms derived via the Fokas transform method, but the modification is natural. A brief discussion of the problems solved via the Fokas transform method is appropriate. Rather than attempting a full survey, we shall refer to a few examples that emphasize the classes of problems solved, because such variations may impact on the spectral transform method. The Fokas transform method has been used to solve problems on the finite interval [15, 31], the half line [19], and interface domains [8–10]. It has also been used to solve problems with multipoint [30] and nonlocal [27] conditions replacing the usual boundary conditions. Extension to partial differential equations involving mixed derivatives have been addressed [12, 24], and the method is also well understood for linear systems of partial differential equations [7]. The Fokas transform method is much broader than the examples listed so far, having been applied to elliptic and hyperbolic equations without a time variable [5, 6, 13] and integrable semilinear equations. These, along with semidiscrete problems [3], and problems where the boundaries [16, 17] or boundary conditions [21, 33] move in time, are not discussed in this paper, because spectral methods in which the spectrum is time dependent are necessarily more complex. See [14, 28] and their references and citations for a picture of the broader Fokas transform method and [11] for an introduction.

1.3 Example Problems

We list here three problems which cannot be treated using the classical spectral transforms. All of these problems have been solved using the Fokas transform method, and references are provided for the derivation of their corresponding Fokas transforms. We select these particular problems for attention because their solution representations are relatively simple, yet the problems are sufficiently varied both to highlight the broad applicability of the Fokas transform method itself and to demonstrate that our reinterpretation of this method is equally universal. Among these examples appear no two point boundary value problems, multipoint problems on finite intervals, nor interface problems on networks of finite intervals, because there already exists a complete charecterization of Fokas diagonalization in these settings [1].

Problem (*Half line Neumann problem for the Stokes equation*)

$$[\partial_t + \partial_x^3]q(x,t) = 0 \qquad (x,t) \in (0,\infty) \times (0,T), \tag{1.3.PDE}$$
$$q(x,0) = Q(x) \qquad x \in [0,\infty), \tag{1.3.IC}$$
$$q_x(0,t) = 0 \qquad t \in [0,T], \tag{1.3.BC}$$

in which $Q \in \mathcal{S}[0,\infty)$. We also define the differential operator L by

$$L\phi = \phi''', \qquad \text{Dom } L = \left\{\phi \in \mathcal{S}[0,\infty) : \phi'(0) = 0\right\} \tag{1.4}$$

to represent the spatial part of the problem.

The Dirichlet version of problem (1.3) was solved using the Fokas transform method in [11, Sect. 3.3]. It is straightforward to adapt their argument to the Neumann case. The finite interval two point analogue of L, with two supplementary boundary conditions provided at the other end of the spatial interval is precisely the operator Jackson and Hopkins identified as having incomplete eigenfunctions [22, 23].

Problem (*Finite interval problem for the heat equation with a nonlocal condition*)

$$[\partial_t - \partial_x^2]q(x,t) = 0 \qquad (x,t) \in (0,\infty) \times (0,T), \tag{1.5.PDE}$$
$$q(x,0) = Q(x) \qquad x \in [0,1], \tag{1.5.IC}$$
$$\int_0^1 K(y)q(y,t)\mathrm{d}y = 0 = q_x(1,t) \qquad t \in [0,T], \tag{1.5.BC}$$

in which $Q \in \mathrm{C}[0,1]$ and $K : [0,1] \to \mathbb{R}$ is both supported in a neighbourhood of 0 and globally sufficiently smooth. The corresponding differential operator L is given by

$$L\phi = -\phi'', \qquad \text{Dom } L = \left\{\phi \in C[0,1] : \int_0^1 K(y)\phi(y)\mathrm{d}y = 0 = \phi'(1)\right\}. \tag{1.6}$$

Problem (1.5) was solved using the Fokas transform method in [27]. It has a physical application in the problem of determining the concentration of a translucent mixture using a light sensor of finite width.

Problem (*Half line Dirichlet problem for the linearized BBM equation*)

$$[\partial_t(1-\partial_x^2)+\partial_x]q(x,t)=0 \qquad (x,t)\in(0,\infty)\times(0,T), \tag{1.7.PDE}$$
$$q(x,0)=Q(x) \qquad x\in[0,\infty), \tag{1.7.IC}$$
$$q(0,t)=0 \qquad t\in[0,T], \tag{1.7.BC}$$

where $Q\in\mathcal{S}[0,\infty)$ and $Q(0)=0$. We also define the differential operators L and M by

$$L\phi=\phi', \qquad \operatorname{Dom} L=\{\phi\in\mathcal{S}[0,\infty):\phi(0)=0\}, \tag{1.8a}$$
$$M\phi=1-\phi'', \qquad \operatorname{Dom} M=\operatorname{Dom} L, \tag{1.8b}$$

so that the partial differential and boundary conditions can be represented as $[\partial_t M+L]q(\cdot,t)=0$.

Problem (1.7) was solved in [12] using the Fokas transform method, as part of the full class of Robin problems. We specialise here to the Dirichlet problem so that the exposition may be aided by simpler formulae. Problem (1.7) represents the small amplitude linearization of the bidirectional water wave model derived by Benjamin, Bona, and Mahoney [2].

1.4 A More General Spectral Transform Method

Suppose we aim to solve one of the above problems. The classical Fourier sine and cosine transforms (or their discrete analogues for the finite interval problem) will not work. Indeed, for the above problems, there are no known transforms that have both properties of diagonalizing the spatial differential operator and being invertible. Therefore, our simple transform method will not succeed. We provide here the archetype of a more general transform method which, as we argue in Sect. 2, is applicable to these problems. We describe the method for problem (1.3), but the method is identical for problem (1.5) and requires only natural generalization to problem (1.7).

Suppose we have a transform F and apply it in the spatial variable to (1.3.PDE), obtaining

$$\frac{\mathrm{d}}{\mathrm{d}t}F[q](\lambda;t)+F[Lq](\lambda;t)=0,$$

for a certain set of complex λ. Because our transform may not diagonalize the differential operator L, we must admit the possibility that L is diagonalized with a remainder:

$$F[Lq](\lambda;t)=\omega(\lambda)F[q](\lambda;t)+R[q](\lambda;t). \tag{1.9}$$

Then

$$\frac{\mathrm{d}}{\mathrm{d}t}F[q](\lambda;t)+\omega(\lambda)F[q](\lambda;t)+R[q](\lambda;t)=0.$$

Integration by parts tells us that the "eigenvalue" $\omega(\lambda)$ must be $-\mathrm{i}\lambda^3$ and the boundary terms are collected into the remainder transform $R[q](\lambda;t)$.

We solve the ordinary differential equation for $F[q](\lambda;t)$, treating the remainder transform as if it were an inhomogeneity. Indeed, multiplying by $\mathrm{e}^{\omega(\lambda)t}$ yields

$$\frac{\mathrm{d}}{\mathrm{d}t}\left(\mathrm{e}^{\omega(\lambda)t}F[q](\lambda;t)\right)+\mathrm{e}^{\omega(\lambda)t}R[q](\lambda;t),$$

and integration in time from 0 to t, followed by application of (1.3.IC) and rearrangement yields

$$F[q](\lambda;t)=\mathrm{e}^{-\omega(\lambda)t}F[Q](\lambda)-\int_0^t \mathrm{e}^{\omega(\lambda)(s-t)}R[q](\lambda;s). \tag{1.10}$$

Arriving at (1.10) has required no special properties of the transform other than linearity and commutativity with the temporal derivative, both in the first step. We have not yet assumed that the transform diagonalizes L. At this point, so that we may obtain an expression for q itself, we must assume that the transform is invertible. We denote the inverse by F^{-1}, and assume it too is linear. Then equation (1.10) simplifies to

$$q(x,t)=F^{-1}\left[\mathrm{e}^{-\omega t}F[Q]\right](x)-F^{-1}\left[\int_0^t \mathrm{e}^{(s-t)\omega}R[q](\,\cdot\,;s)\mathrm{d}s\right](x). \tag{1.11}$$

Unfortunately, our earlier pretence notwithstanding, $R[q]$ is not data, so (1.11) does not provide an effective representation of the solution to problem (1.3). To proceed, let us suppose further that the latter term evaluates to zero whenever q satisfies the boundary conditions. Then the solution of the initial boundary value problem has been obtained:

$$q(x,t)=F^{-1}\left[\mathrm{e}^{-\omega t}F[Q]\right](x). \tag{1.12}$$

The extra assumption is a restriction on R, so may be seen as part of the replacement for the diagonalization property of the transform. Therefore, the requirements on the transform F to make the spectral transform method work are:

Criterion 1 F is invertible, and both F and its inverse F^{-1} are linear.

Criterion 2 F diagonalises the differential operator L (which describes the spatial part of the initial boundary value problem) in the sense of (1.9), with remainder R having the property that, provided q is sufficiently smooth and satisfies the boundary conditions, the latter term of (1.11) evaluates to 0.

Clearly, each initial boundary value problem will require its own transform F. But transforms obeying analogues of both criteria 1 and 2 have been constructed for problems posed on the finite interval, with arbitrary constant coefficient differential operator and any linear boundary conditions [1]. Because these transform pairs were discovered via the Fokas transform method, we call the weak diagonalization criterion 2 *Fokas diagonalization*. As we shall argue below, there exist transforms that are tailored to each of the problems introduced above, which also exhibit Fokas diagonalization, so the above generalized transform method may be applied without hindrance. Specifically, for each problem, we shall define the transform pair and prove Criteria 1 and 2 as theorems. Beyond [1], the papers [18, 29, 32] provide an earlier view of Fokas diagonalization and the spectral transform method for two point problems and half line problems in which the spatial operator has monomial character.

2 Results

2.1 Half Line Neumann Problem for the Stokes Equation

Adapting [11, Sect. 3.3], the Fokas transform pair

$$F[\phi](\lambda) = \begin{cases} \int_0^\infty \mathrm{e}^{-\mathrm{i}\lambda y}\phi(y)\mathrm{d}y & \lambda \in \mathbb{R}, \\ \int_0^\infty \phi(y)\left[\alpha^2\mathrm{e}^{-\mathrm{i}\alpha\lambda y} + \alpha\mathrm{e}^{-\mathrm{i}\alpha^2\lambda y}\right]\mathrm{d}y & \lambda \in \partial D^+, \end{cases} \tag{2.1a}$$

$$F^{-1}[f](x) = \frac{1}{2\pi}\int_{\mathbb{R}\cup\partial D^+} \mathrm{e}^{\mathrm{i}\lambda x} f(\lambda)\mathrm{d}\lambda \qquad x \in [0,\infty), \tag{2.1b}$$

may be derived. Here, D^+ is the sector $\arg(\lambda) \in (\frac{\pi}{3}, \frac{2\pi}{3})$, ∂D^+ is the positively oriented contour traversing its boundary, the contour denoted $\mathbb{R}$ is oriented in the increasing sense, and the primitive cube root of unity $\alpha = \mathrm{e}^{2\pi\mathrm{i}/3}$.

Proposition 3 *If F, F^{-1} are defined by (2.1), then they are both linear and, for all ϕ sufficiently smooth and all $x \in (0, 1)$, $F^{-1}[F[\phi]](x) = \phi(x)$.*

Proof These are integral transforms, so they inherit linearity from the improper definite integrals and contour integrals of which they are composed.

By definition,

$$\begin{aligned} F^{-1}[F[\phi]](x) &= \frac{1}{2\pi}\int_{-\infty}^{\infty} \mathrm{e}^{\mathrm{i}\lambda x} F[\phi](\lambda)\mathrm{d}\lambda + \frac{1}{2\pi}\int_{\partial D^+} \mathrm{e}^{\mathrm{i}\lambda x} F[\phi](\lambda)\mathrm{d}\lambda \\ &= \phi(x) + \frac{1}{2\pi}\int_{\partial D^+} \mathrm{e}^{\mathrm{i}\lambda x} F[\phi](\lambda)\mathrm{d}\lambda, \end{aligned} \tag{2.2}$$

where the second equality holds for all $x \in (0, \infty)$ at which ϕ is continuous, and is justified by the usual Fourier inversion theorem for piecewise smooth ϕ. It remains only to show that the remaining contour integral term evaluates to 0.

The definition of $F[\phi](\lambda)$ is analytically extensible from ∂D^+ to a neighbourhood of the sector D^+. Doing so will necessarily overwrite the definition of $F[\phi](\lambda)$ on part of $\mathbb{R}$, but that is inconsequential as we have already removed that part of the domain from consideration; we, for the purposes of the rest of this proof, consider $F[\phi](\lambda)$ as being defined by its ∂D^+ formula everywhere on $\mathbb{C}$. Integrating by parts, we see that, as $\lambda \to \infty$ from within clos D^+,

$$F[\phi](\lambda) = \frac{\mathrm{i}}{\lambda}\left\{\left[\phi(y)\left(\alpha\mathrm{e}^{-\mathrm{i}\alpha\lambda y} + \alpha^2\mathrm{e}^{-\mathrm{i}\alpha^2\lambda y}\right)\right]_0^\infty - \int_0^\infty \phi'(y)\left(\alpha\mathrm{e}^{-\mathrm{i}\alpha\lambda y} + \alpha^2\mathrm{e}^{-\mathrm{i}\alpha^2\lambda y}\right)\mathrm{d}y\right\} = \mathcal{O}\left(\lambda^{-1}\right),$$

and this decay is uniform in $\arg(\lambda)$ within the given sector. Hence, by Jordan's lemma and Cauchy's theorem, the remaining integral on the right of (2.2) evaluates to 0. □

Theorem 4 *Suppose F, F^{-1} are defined by (2.1). There exists a remainder transform R for which, for all $\lambda \in \mathbb{R} \cup \partial D^+$ and all $\phi \in$ Dom L,*

$$F[L\phi](\lambda) = -\mathrm{i}\lambda^3 F[\phi](\lambda) + R[\phi](\lambda). \tag{2.3}$$

Moreover, if, for all $t \in [0, T]$ $q(\cdot, t) \in$ Dom L then, for all $t \in [0, T]$ and all $x \in (0, \infty)$,

$$F^{-1}\left[\int_0^t \mathrm{e}^{-\mathrm{i}\cdot^3(s-t)} R[q](\cdot\,; s)\mathrm{d}s\right](x) = 0. \tag{2.4}$$

Proof Integrating by parts thrice in the definition of $F[\phi](\lambda)$ and applying the boundary condition $\phi'(0) = 0$, we find that (2.3) holds with

$$R[\phi](\lambda) = \begin{cases} -r(\lambda; \phi) & \text{if } \lambda \in \mathbb{R}, \\ r(\lambda; \phi) & \text{if } \lambda \in \partial D^+, \end{cases} \qquad r(\lambda; \phi) = \phi''(0) - \lambda^2\phi(0),$$

in which we think of the polynomial $r(\cdot\,; \phi)$ as having domain all of $\mathbb{C}$. Note that it is entire, and is $o(\lambda^3)$ as $\lambda \to \infty$.

Let E^+ be the union of sectors $\arg(\lambda) \in (0, \frac{\pi}{3}) \cup (\frac{2\pi}{3}, \pi)$. Then, integrating by parts, as $\lambda \to \infty$ from within clos E^+,

$$\begin{aligned}&\int_0^t \mathrm{e}^{-\mathrm{i}\lambda^3(s-t)} r(\lambda; q(\cdot\,, s))\mathrm{d}s \\ &= \frac{\mathrm{i}}{\lambda^3}\left\{\left[\mathrm{e}^{-\mathrm{i}\lambda^3(s-t)} r(\lambda; q(\cdot\,, s))\right]_0^t + \int_0^t \mathrm{e}^{-\mathrm{i}\lambda^3(s-t)} r(\lambda; q_t(\cdot\,, s))\mathrm{d}s\right\} = \mathcal{O}\left(\lambda^{-1}\right),\end{aligned}$$

uniformly in $\arg(\lambda)$ within those closed sectors. Therefore, by Jordan's lemma and Cauchy's theorem,

$$\int_{\partial E^+} e^{i\lambda x} \int_0^t e^{-i\lambda^3(s-t)} r(\lambda; q(\,\cdot\,, s)) ds d\lambda = 0.$$

Hence, by comparing the sector boundaries ∂E^+ and ∂D^+,

$$\begin{aligned}\int_{\partial D^+} e^{i\lambda x} \int_0^t e^{-i\lambda^3(s-t)} r ds d\lambda &= \int_{\partial D^+ \cup \partial E^+} e^{i\lambda x} \int_0^t e^{-i\lambda^3(s-t)} r ds d\lambda \\ &= \int_{-\infty}^{\infty} e^{i\lambda x} \int_0^t e^{-i\lambda^3(s-t)} r ds d\lambda,\end{aligned}$$

in which we have suppressed the dependence of r on $\lambda, q(\,\cdot\,, s)$. Therefore,

$$\begin{aligned}&F^{-1}\left[\int_0^t e^{-i\cdot^3(s-t)} R[q](\,\cdot\,; s) ds\right](x) \\ &\qquad = \frac{1}{2\pi} \int_0^{\infty} e^{i\lambda x} \int_0^t e^{-i\lambda^3(s-t)} \big(r - r\big) ds d\lambda = 0.\end{aligned}$$

□

In Proposition 3 we have fulfilled Criterion 1, and Theorem 4 establishes Criterion 2. Therefore, the general transform method of Sect. 1.4 can be implemented for this problem to derive solution (1.12).

The argument is presented above in such a way that it requires minimal modification for the Dirichlet problem; the transform pair may be found in [11, Sect. 3.3] and $r(\lambda; \phi)$ is replaced with $\phi''(0) + i\lambda\phi'(0)$, which is still entire and $o(\lambda^3)$.

2.2 *Finite Interval Problem for the Heat Equation with a Nonlocal Condition*

As derived in [27], the Fokas transform pair is

$$F[\phi](\lambda) = \begin{cases} \int_0^1 e^{-i\lambda y} \phi(y) dy & \lambda \in \mathbb{R}, \\ -\zeta^+(\lambda; \phi)/\Delta(\lambda) & \lambda \in \partial D_\rho^+, \\ -e^{-i\lambda}\zeta^-(\lambda; \phi)/\Delta(\lambda) & \lambda \in \partial D_\rho^-, \end{cases} \tag{2.5a}$$

$$F^{-1}[f](x) = \frac{1}{2\pi} \int_{\mathbb{R} \cup D_\rho^+ \cup D_\rho^-} e^{i\lambda x} f(\lambda) d\lambda \qquad x \in [0, 1], \tag{2.5b}$$

where

$$\Delta(\lambda) = \int_0^1 K(y)\cos([1-y]\lambda)\mathrm{d}y, \tag{2.6}$$

$$\zeta^+(\lambda;\phi) = \int_0^1 K(y)\cos([1-y]\lambda)\int_0^y \mathrm{e}^{-\mathrm{i}\lambda z}\phi(z)\mathrm{d}z\mathrm{d}y + \int_0^1 K(y)\mathrm{e}^{-\mathrm{i}\lambda y}\int_y^1 \cos([1-z]\lambda)\phi(z)\mathrm{d}z\mathrm{d}y, \tag{2.7}$$

$$\zeta^-(\lambda;\phi) = \int_0^1 K(y)\int_y^1 \sin([z-y]\lambda)\phi(z)\mathrm{d}z\mathrm{d}y, \tag{2.8}$$

and $D_\rho^\pm = \{\lambda \in \mathbb{C}^\pm$ such that $\Re(\lambda^2) < 0$ and $|\lambda| > \rho\}$ for ρ sufficiently large to ensure all zeros of Δ have imaginary part bounded between $\pm\rho/\sqrt{2}$, and $\partial D_\rho^\pm$ indicates the positively oriented contour traversing the boundary of the region $D_\rho^\pm$. The existence of such ρ was proved in [27, Lemma 2.1].

Proposition 5 *If F, F^{-1} are defined by (2.5), then they are both linear and, for all φ sufficiently smooth and all $x \in (0,1)$, $F^{-1}[F[\phi]](x) = \phi(x)$.*

Proof Linearity follows from linearity of the definite real integral and the contour integral. With $\zeta^\pm$ defined by (2.7) and (2.8), it was shown in [27, Lemma 2.2] that, provided K is of bounded variation, K is continuous and supported in a neighbourhood of 0, and $\|\phi'\|_\infty$ is bounded, then, as $\lambda \to \infty$ from within clos $D_\rho^\pm$, $\zeta^\pm(\lambda;\phi)/\Delta(\lambda) = \mathcal{O}\left(\lambda^{-1}\right)$, uniformly in $\arg(\lambda)$. The ratios $\zeta^\pm/\Delta$ are, because of the choice of ρ sufficiently large, analytic in neighbourhoods of clos $D_\rho^\pm$. It follows by Jordan's lemma and Cauchy's theorem that the integrals along the boundaries of $D_\rho^\pm$ appearing in $F^{-1}[F[\phi]](x)$ both evaluate to 0. Therefore,

$$F^{-1}[F[\phi]](x) = \frac{1}{2\pi}\int_{-\infty}^{\infty} \mathrm{e}^{\mathrm{i}\lambda x} F[\phi](\lambda)\mathrm{d}\lambda.$$

The proposition follows from the usual Fourier inversion theorem. □

Theorem 6 *Suppose F, F^{-1} are defined by (2.5). There exists a remainder transform R for which, for all $\lambda \in \mathbb{R} \cup D_\rho^+ \cup D_\rho^-$ and all $\phi \in$ Dom L,*

$$F[L\phi](\lambda) = \lambda^2 F[\phi](\lambda) + R[\phi](\lambda). \tag{2.9}$$

Moreover, if $q : [0,1]\times[0,T] \to \mathbb{C}$ is such that, for all $t \in [0,T]$, $q(\cdot,t) \in$ Dom L, and q is sufficiently smooth in t, then, for all $t \in [0,T]$ and all $x \in (0,1)$,

$$F^{-1}\left[\int_0^t \mathrm{e}^{\cdot^2(s-t)} R[q](\cdot\,;s)\mathrm{d}s\right](x) = 0. \tag{2.10}$$

Proof Integration by parts and application of the boundary and nonlocal conditions yield that

$$R[\phi](\lambda) = \begin{cases} r_-(\lambda;\phi)\mathrm{e}^{-\mathrm{i}\lambda} - r_+(\lambda;\phi) & \text{if } \lambda \in \mathbb{R}, \\ r_+(\lambda;\phi) & \text{if } \lambda \in \partial D_\rho^+, \\ r_-(\lambda;\phi) & \text{if } \lambda \in \partial D_\rho^-, \end{cases}$$

where

$$r_+(\lambda;\phi) := -\phi'(0) - \mathrm{i}\lambda\phi(0), \qquad r_-(\lambda;\phi) := -\mathrm{i}\lambda\phi(1).$$

Note that, for each of the three contours on which $R[\phi]$ is defined, it may be analytically extended to all of $\mathbb{C}$, yielding a triply defined function on $\mathbb{C}$, with each definition entire.

We denote by $E_\rho^\pm$ the sets $\{\lambda \in \mathbb{C}^\pm$ such that $\Re(\lambda^2) > 0$ and $|\lambda| > \rho\}$. Then, integrating by parts, and suppressing the dependence of $r_\pm$ on λ, $q(\cdot\,, s)$,

$$\int_0^t \mathrm{e}^{\lambda^2(s-t)} r_+ \mathrm{d}s = \mathcal{O}\left(\lambda^{-2}\right)$$

as $\lambda \to \infty$ from within clos $E_\rho^\pm$, uniformly in $\arg(\lambda)$. Hence, by Jordan's lemma,

$$\int_{\partial E_\rho^+} \mathrm{e}^{\mathrm{i}\lambda x} \int_0^t \mathrm{e}^{\lambda^2(s-t)} r_+ \mathrm{d}s\mathrm{d}\lambda = 0.$$

It follows, by comparing the paths of the contours, that

$$\int_{\partial D_\rho^+} \mathrm{e}^{\mathrm{i}\lambda x} \int_0^t \mathrm{e}^{\lambda^2(s-t)} r_+ \mathrm{d}s\mathrm{d}\lambda = \int_{\gamma_\rho^+} \mathrm{e}^{\mathrm{i}\lambda x} \int_0^t \mathrm{e}^{\lambda^2(s-t)} r_+ \mathrm{d}s\mathrm{d}\lambda,$$

in which γ_ρ^+ is the contour that extents from $-\infty$ to $-\rho$, then follows the semicircular path in $\mathbb{C}^+$ from $-\rho$ to ρ, then proceeds from ρ to ∞. Because the integrand is entire, the contour γ_ρ^+ may be deformed to the real line. Similarly, but noting that the real part of λ decreases as λ traverses ∂D_ρ^-, whereas it increased when λ followed ∂D_ρ^+,

$$\int_{\partial D_\rho^-} \mathrm{e}^{\mathrm{i}\lambda(x-1)} \int_0^t \mathrm{e}^{\lambda^2(s-t)} r_- \mathrm{d}s\mathrm{d}\lambda = -\int_{-\infty}^{\infty} \mathrm{e}^{\mathrm{i}\lambda(x-1)} \int_0^t \mathrm{e}^{\lambda^2(s-t)} r_- \mathrm{d}s\mathrm{d}\lambda.$$

Using the previous two displayed equations to justify the second equality, it follows that

$$
\begin{aligned}
F^{-1}\left[\int_0^t e^{\lambda^2(s-t)} R[q](\,\cdot\,; s)\mathrm{d}s\right](x) &= \int_{-\infty}^{\infty} e^{\mathrm{i}\lambda x} \int_0^t e^{\lambda^2(s-t)}[r_- e^{-\mathrm{i}\lambda} - r_+]\mathrm{d}s\mathrm{d}\lambda \\
&\quad + \int_{\partial D_\rho^+} e^{\mathrm{i}\lambda x} \int_0^t e^{\lambda^2(s-t)} r_+ \mathrm{d}s\mathrm{d}\lambda + \int_{\partial D_\rho^-} e^{\mathrm{i}\lambda(x-1)} \int_0^t e^{\lambda^2(s-t)} r_- \mathrm{d}s\mathrm{d}\lambda \\
&= \int_{-\infty}^{\infty} e^{\mathrm{i}\lambda x} \int_0^t e^{\lambda^2(s-t)}[r_- e^{-\mathrm{i}\lambda} - r_+]\mathrm{d}s\mathrm{d}\lambda + \int_{-\infty}^{\infty} e^{\mathrm{i}\lambda x} \int_0^t e^{\lambda^2(s-t)} r_+ \mathrm{d}s\mathrm{d}\lambda \\
&\quad - \int_{-\infty}^{\infty} e^{\mathrm{i}\lambda(x-1)} \int_0^t e^{\lambda^2(s-t)} r_- \mathrm{d}s\mathrm{d}\lambda = 0.
\end{aligned}
$$

□

Because Proposition 5 and Theorem 6 follow exactly the archetypes of Criteria 1 and 2, the transform method is effective at finding the solution of problem (1.5), with $\omega(\lambda) = \lambda^2$.

2.3 *Half Line Dirichlet Problem for the Linearized BBM Equation*

Deconinck and Vasan show in [12, Sect. 3] that the Fokas transform pair is

$$
F[\phi](\lambda) = \begin{cases} \displaystyle\int_0^\infty e^{-\mathrm{i}\lambda y}\phi(y)\mathrm{d}y & \lambda \in \mathbb{R}, \\ \displaystyle\frac{-1}{\lambda^2}\int_0^\infty e^{\frac{-\mathrm{i}}{\lambda}y}\phi(y)\mathrm{d}y & \lambda \in \partial C, \end{cases} \tag{2.11a}
$$

$$
F^{-1}[f](x) = \frac{1}{2\pi}\int_{\mathbb{R}\cup C} e^{\mathrm{i}\lambda x} f(\lambda)\mathrm{d}\lambda \qquad x \in [0, \infty), \tag{2.11b}
$$

where C is a small positively oriented simple closed contour enclosing $\lambda = \mathrm{i}$.

Proposition 7 *If F, F^{-1} are defined by (2.11), then they are both linear and, for all ϕ sufficiently smooth and all $x \in (0, \infty)$, $F^{-1}[F[\phi]](x) = \phi(x)$.*

Proof These are integral transforms, so they are linear. By definition,

$$
F^{-1}[F[\phi]](x) = \frac{1}{2\pi}\int_{-\infty}^{\infty} e^{\mathrm{i}\lambda x}\int_0^\infty e^{-\mathrm{i}\lambda y}\phi(y)\mathrm{d}y\mathrm{d}\lambda - \frac{1}{2\pi}\int_C \frac{1}{\lambda^2}\int_0^\infty e^{\frac{-\mathrm{i}}{\lambda}y}\phi(y)\mathrm{d}y\mathrm{d}\lambda.
$$

The integral of the second contour integral is analytic in $\mathbb{C}^+$, in which C lies. Therefore, by Cauchy's theorem, the second contour integral evaluates to 0. The proposition follows by the usual Fourier inversion theorem. □

Theorem 8 *Suppose F, F^{-1} are defined by (2.11). There exists a remainder transform R for which, for all $\lambda \in \mathbb{R} \cup C$ and all $\phi \in$ Dom $L =$ Dom M,*

$$F[L\phi](\lambda) = \omega_L(\lambda) F[\phi](\lambda) + R_L[\phi](\lambda), \tag{2.12a}$$
$$F[M\phi](\lambda) = \omega_M(\lambda) F[\phi](\lambda) + R_M[\phi](\lambda), \tag{2.12b}$$

where

$$\omega_L(\lambda) = \begin{cases} \mathrm{i}\lambda & \text{if } \lambda \in \mathbb{R}, \\ \frac{\mathrm{i}}{\lambda} & \text{if } \lambda \in C, \end{cases} \qquad \omega_M(\lambda) = \begin{cases} 1+\lambda^2 & \text{if } \lambda \in \mathbb{R}, \\ 1+\frac{1}{\lambda^2} & \text{if } \lambda \in C. \end{cases}$$

Moreover, if q is sufficiently smooth, $q(\cdot, t) \in$ Dom $L =$ Dom M then, for all $t \in [0, T]$ and all $x \in (0, \infty)$,

$$F^{-1}\left[\int_0^t \mathrm{e}^{(s-t)\omega} \frac{R_L[q](\cdot\,; s) + R_M[q_t](\cdot\,; s)}{\omega_M} \mathrm{d}s\right](x) = 0, \tag{2.13}$$

where $\omega(\lambda) = \omega_L(\lambda)/\omega_M(\lambda) = \mathrm{i}\lambda/(1+\lambda^2)$.

Proof Integration by parts and application of the boundary condition demonstrates (2.12) with $R_L = 0$ and

$$R_M[\phi](\lambda) = \begin{cases} \phi'(0) & \text{if } \lambda \in \mathbb{R}, \\ -\phi'(0)\frac{1}{\lambda^2} & \text{if } \lambda \in C. \end{cases}$$

Therefore, the left side of (2.13) is

$$\frac{1}{2\pi}\int_{-\infty}^{\infty} \mathrm{e}^{\mathrm{i}\lambda x} \int_0^t \mathrm{e}^{\frac{\mathrm{i}\lambda}{1+\lambda^2}(s-t)} q_{xt}(0; s)\mathrm{d}s \frac{1}{1+\lambda^2} \mathrm{d}\lambda$$
$$- \frac{1}{2\pi}\int_{C} \mathrm{e}^{\mathrm{i}\lambda x} \int_0^t \mathrm{e}^{\frac{\mathrm{i}\lambda}{1+\lambda^2}(s-t)} q_{xt}(0; s)\mathrm{d}s \frac{1}{1+\frac{1}{\lambda^2}} \left(\frac{1}{\lambda^2}\right) \mathrm{d}\lambda. \tag{2.14}$$

The integrand of the first contour integral in expression (2.14) is analytic on $\mathbb{C} \setminus \{\pm\mathrm{i}\}$, It is straightforward to show that $\Re(\omega(\lambda)) \geqslant 0$ on clos $\mathbb{C}^+$ except on the closed disc with radius 1 and center at 0. Hence, as $\lambda \to \infty$ from within clos $\mathbb{C}^+$,

$$\int_0^t \mathrm{e}^{\frac{\mathrm{i}\lambda}{1+\lambda^2}(s-t)} q_{xt}(0; s)\mathrm{d}s \frac{1}{1+\lambda^2} = \mathcal{O}\left(\lambda^{-2}\right),$$

uniformly in $\arg(\lambda)$, and the same function is analytic everywhere but $\pm\mathrm{i}$. Hence, by Jordan's lemma and Cauchy's theorem, the first contour integral in expression (2.14) may be deformed from $\mathbb{R}$ to C, whereupon it cancels with the other contour integral in that expression. Thereby, (2.13) is established. □

Transform Method Theorem 8 does not follow the archetype of Criterion 2. But (1.7.PDE) has form different from (1.3.PDE), for which the general transform method of Sect. 1.4 was developed, so it is unsurprising that the form of Fokas diagonalization is different. Below, we implement the transform method applicable to problem (1.7), noting how Theorem 8 is used.

Applying the transform F to (1.7.PDE), we obtain

$$0 = \frac{\mathrm{d}}{\mathrm{d}t} F[Mq](\lambda; t) + F[Lq](\lambda; t),$$

for all $\lambda \in \mathbb{R} \cup C$. It follows from (2.12) that

$$0 = \frac{\mathrm{d}}{\mathrm{d}t}\Big(\omega_M(\lambda) F[q](\lambda; t) + R_M[q](\lambda; t)\Big) + \omega_L(\lambda) F[q](\lambda; t) + R_L[q](\lambda; t).$$

Rearranging and multiplying by $\mathrm{e}^{\omega(\lambda)t}/\omega_M(\lambda)$, we find

$$0 = \frac{\mathrm{d}}{\mathrm{d}t}\left(\mathrm{e}^{\omega(\lambda)t} F[q](\lambda; t)\right) + \mathrm{e}^{\omega(\lambda)t} \frac{R_L[q](\lambda; t) + R_M[q_t](\lambda; t)}{\omega_M(\lambda)}.$$

Integrating in time from 0 to t, applying (1.7.IC) and rearranging, we find

$$F[q](\lambda; t) = \mathrm{e}^{-\omega(\lambda)t} F[Q](\lambda) - \int_0^t \mathrm{e}^{\omega(\lambda)(s-t)} \frac{R_L[q](\lambda; s) + R_M[q_t](\lambda; s)}{\omega_M(\lambda)} \mathrm{d}s.$$

When we apply the inverse transform, because q obeys (1.7.BC), Theorem 8 guarantees that the latter term evaluates to 0. Hence the solution of problem (1.7) is

$$q(x, t) = F^{-1}\left[\mathrm{e}^{-\omega t} F[Q](\,\cdot\,; t)\right](x).$$

3 Conclusion

We aimed both to show that the Fokas transform method can be reformulated to appear as a generalized spectral transform method for initial boundary value problems and to provide a unified characterization of Fokas diagonalization.

With the detailed work [1] already demonstrating the former claim for finite interval problems with two point, multipoint and interface type boundary conditions, the ambit of the present work is to demonstrate the generalization to other settings by means of a few examples. There exist important classes of initial boundary value problems for which the Fokas transform method has been implemented that are not represented among the examples studied above. But significant progress has been made towards this aim.

Regarding the second aim, Fokas diagonalization in the first two examples exactly matches Criterion 2, which also matches Fokas diagonalization for finite interval problems [1]. The final example demonstrates that the statement of Fokas diagonalization is properly considered as a property of the boundary value problem under study, rather than any particular (local) differential operator. But it also reinforces the point that the statement of Fokas diagonalization may be read directly from the spectral transform method; it is precisely the necessary and sufficient condition for the spectral transform method to succeed. Although, for the sake of brevity, not demonstrated in the current work, this extends to Fokas diagonalization for system problems such as those studied in [7, 24].

3.1 On Pedagogy

Having taught a few cohorts of undergraduates a typical course on boundary value problems, the author has observed that those learning spectral transform methods for the first time find it helpful to have an alternative viewpoint, or even introductory experience, in which the following two concepts are presented as separate:

C1. How a transform pair is used in a spectral method to solve an initial boundary value problem.
C2. How the transform pair suitable for any particular problem may be derived.

It is to the detriment of students' learning, when the particular discrete Fourier transform appropriate for a given initial boundary value problem is derived by separation of variables only as part of a long solution method, and the definitions of the forward and inverse transforms are not explicitly identified as such. Students may see solving a particular Sturm Liouville problem as part of solving a boundary value problem, but less commonly understand that they are deriving the spectral transform that will diagonalize the spatial differential operator for that problem.

Because C2 is quite difficult and technical, the relative simplicity and astonishingly broad applicability of C1 remain mysterious to many students. The author has had some success in teaching first C1, with a hypothetical transform, and gently guiding students to discover for themselves C2: separation of variables and enough Sturm–Liouville theory to explicitly construct the desired transform.

In a similar way, the Fokas transform method is usually presented in the literature as a monolithic method, at the end of which a solution has been derived, but the transform itself is rarely emphasized. It is the author's intention that concept C1 for the Fokas transform method be essentially captured in Sect. 1.4. It is the author's hope that, by drawing attention to the crucial properties of the Fokas transform, Criteria 1 and 2, this work may lighten the burden of learning the Fokas transform method, and inspire more to study it.

References

1. Aitzhan, S.A., Bhandari, S., Smith, D.A.: Fokas diagonalization of piecewise constant coefficient linear differential operators on finite intervals and networks. Acta Appl. Math. **177**(2), 1–66 (2022)
2. Benjamin, T.B., Bona, J.L., Mahony, J.J.: Model equations for long waves in nonlinear dispersive systems. Philos. Trans. R. Soc. Lond. Ser. A Math. Phys. Eng. Sci. **272**, 47–78 (1972)
3. Biondini, G., Hwang, G.: Initial-boundary-value problems for discrete evolution equations: discrete linear Schrödinger and integrable discrete nonlinear Schrödinger equations. Inverse Probl. **24**(6), 65011 (2008)
4. Birkhoff, G.D.: Boundary value and expansion problems of ordinary linear differential equations. Trans. Amer. Math. Soc. **9**, 373–395 (1908)
5. Crowdy, D.: Fourier-Mellin transforms for circular domains. Comput. Methods Funct. Theory **15**, 655–687 (2015)
6. Crowdy, D.G., Luca, E.: A transform method for the biharmonic equation in multiply connected circular domains. IMA J. Appl. Math. **83**, 942–976 (2018)
7. Deconinck, B., Guo, Q., Shlizerman, E., Vasan, V.: Fokas's unified transform method for linear systems. Quart. Appl. Math. **76**(3), 463–488 (2018)
8. Deconinck, B., Pelloni, B., Sheils, N.E.: Non-steady-state heat conduction in composite walls. Proc. R. Soc. Lond. Ser. A Math. Phys. Eng. Sci. **470**(2165), 20130605 (2014)
9. Deconinck, B., Sheils, N.: The time-dependent Schrödinger equation with piecewise constant potentials. Europ. J. Appl. Math. **31**, 57–83 (2020)
10. Deconinck, B., Sheils, N.E., Smith, D.A.: The linear KdV equation with an interface. Commun. Math. Phys. **347**, 489–509 (2016)
11. Deconinck, B., Trogdon, T., Vasan, V.: The method of Fokas for solving linear partial differential equations. SIAM Rev. **56**(1), 159–186 (2014)
12. Deconinck, B., Vasan, V.: Well-posedness of boundary-value problems for the linear Benjamin-Bona-Mahony equation. Discrete & Contin. Dyn. Sys. A **33**(7), 3171–3188 (2013)
13. Fokas, A.S.: Two dimensional linear partial differential equations in a convex polygon. Proc. R. Soc. Lond. Ser. A Math. Phys. Eng. Sci. **457**, 371–393 (2001)
14. Fokas, A.S.: A unified approach to boundary value problems, CBMS-SIAM (2008)
15. Fokas, A.S., Pelloni, B.: Two-point boundary value problems for linear evolution equations. Math. Proc. Cambridge Philos. Soc. **131**, 521–543 (2001)
16. Fokas, A.S., Pelloni, B.: Generalized Dirichlet to Neumann map for moving initial-boundary value problems. J. Math. Phys. **48** (2007)
17. Fokas, A.S., Pelloni, B., Xia, B.: Evolution equations on time-dependent intervals. IMA J. Appl. Math. **84**, 1044–1060 (2019)
18. Fokas, A.S., Smith, D.A.: Evolution PDEs and augmented eigenfunctions. Finite interval. Adv. Diff. Equ. **21**(7/8), 735–766 (2016)
19. Fokas, A.S., Wang, Z.: Generalised Dirichlet to Neumann maps for linear dispersive equations on half-line. Math. Proc. Camb. Philos. Soc. **164**(2), 297–324 (2018)
20. Freiling, G.: Irregular boundary value problems. Results Math. **62**, 265–294 (2012)
21. Govindarajan, R., Prasath, S.G., Vasan, V.: Accurate solution method for the Maxey-Riley equation, and the effects of Basset history. J. Fluid Mech. **868**, 428–460 (2019)
22. Hopkins, J.W.: Some convergent developments associated with irregular boundary conditions. Trans. Amer. Math. Soc. **20**, 245–259 (1919)
23. Jackson, D.: Expansion problems with irregular boundary conditions. Proc. Amer. Acad. Arts Sci. **51**(7), 383–417 (1915)
24. Johnston, C.M., Gartman, C.T., Mantzavinos, D.: The linearized classical Boussinesq system on the half-line. Stud. Appl. Math. **145**(3), 635–657 (2021)
25. Locker, J.: Spectral theory of non-self-adjoint two-point differential operators. Mathematical Surveys and Monographs, vol. 73. American Mathematical Society, Providence, Rhode Island (2000)

26. Locker, J.: Eigenvalues and completeness for regular and simply irregular two-point differential operators. Memoirs of the American Mathematical Society, no. 911, vol. 195. American Mathematical Society, Providence, Rhode Island (2008)
27. Miller, P.D., Smith, D.A.: The diffusion equation with nonlocal data. J. Math. Anal. Appl. **466**(2), 1119–1143 (2018)
28. Pelloni, B.: Advances in the study of boundary value problems for nonlinear integrable PDEs. Nonlinearity **28**, R1–R38 (2015)
29. Pelloni, B., Smith, D.A.: Evolution PDEs and augmented eigenfunctions. Half line. J. Spectr. Theory **6**, 185–213 (2016)
30. Pelloni, B., Smith, D.A.: Nonlocal and multipoint boundary value problems for linear evolution equations. Stud. Appl. Math. **141**(1), 46–88 (2018)
31. Smith, D.A.: Well-posed two-point initial-boundary value problems with arbitrary boundary conditions. Math. Proc. Cambridge Philos. Soc. **152**, 473–496 (2012)
32. Smith, D.A.: The unified transform method for linear initial-boundary value problems: a spectral interpretation. Unified Transform Method for Boundary Value Problems: Applications and Advances. SIAM, Philadelphia, PA (2015)
33. Smith, D.A., Toh, W.-Y.: Linear evolution equations on the half line with dynamic boundary conditions. Eur. J. Appl. Math. **33**(3), 505–537 (2021)

A Novel Difference-Integral Equation Satisfied Asymptotically by the Riemann Zeta Function

Athanassios S. Fokas, Konstantinos Kalimeris, and J. Lenells

Abstract In 2009 Fokas began a program of study of the investigation of the large t-asymptotics of the Riemann zeta function, $\zeta(\sigma + it)$. In the current work we present a novel difference-integral equation which is satisfied asymptotically by $\zeta(1/2 + it)$. This equation is obtained starting with a singular integral equation presented for the first time in 2019 and using a finite Fourier transform representation of the Riemann zeta function. The relevant analysis involves a plethora of tools and techniques developed by Fokas and collaborators during the last decade.

1 Introduction

In 2009, one of the authors, motivated by the understanding of the importance of complex analysis in the investigation of asymptotics, begun a program of study of the investigation of the asymptotics of the Riemann zeta function $\zeta(s)$, $s \in \mathbb{C}$.

It is well known that the leading asymptotics of $\zeta(s)$ as $t = \text{Im}s \to \infty$, is expressed in terms of two transcendental sums whose ranges of summation are from 0 to x and from 0 to y, where x and y satisfy the constraint $xy = t/2\pi$. Siegel, in his classical paper [8] presented the asymptotics of $\zeta(s)$ to all orders in the important particular case of $x = y = \sqrt{t/2\pi}$. In a recent publication in the Memoirs of the

A. S. Fokas (✉)
Department of Applied Mathematics and Theoretical Physics, University of Cambridge, Cambridge, UK
e-mail: tf227@cam.ac.uk

Viterbi School of Engineering, University of Southern California, Los Angeles, Los Angeles, CA, USA

A. S. Fokas · K. Kalimeris
Mathematics Research Center, Academy of Athens, Athens, Greece

J. Lenells
Department of Mathematics, KTH Royal Institute of Technology, Stockholm, Sweden

T. Bountis et al. (eds.), *Chaos, Fractals and Complexity*, Springer Proceedings in Complexity, https://doi.org/10.1007/978-3-031-37404-3_22

American Mathematical Society [5], two of the authors presented analogous results for $\zeta(s)$, as well as for a novel two parameter generalization of $\zeta(s)$, for any x and y to all orders.

Fokas pioneered a new approach to the asymptotics of $\zeta(s)$ based on the derivation of a novel integral equation satisfied by $|\zeta(s)|^2$, see Eq. (2.1). The large t analysis of this equation led to an interesting asymptotic result, namely, it provided the analogue of the famous Lindelöf hypothesis for a certain variation of $\zeta(s)$ [3]. Additional results were derived in [6, 7].

The analysis of the novel integral equation mentioned above is based on the following: the interval of integration of the associated integral is decomposed into four subintervals. For the first two of the resulting integrals it is possible to obtain explicit estimates, whereas for the remaining two integrals one needs to use an appropriate representation for $\zeta(s)$. In all our earlier works, we replaced $\zeta(s)$ by its leading asymptotics. This has two limitations. First, it makes it very difficult to control the relevant errors, and second, it introduces sums, for which it is difficult to obtain rigorous estimates.

Here we introduce a new idea: we express $\zeta(s)$ in terms of its the Fourier transform representation. It is worth noting that this development was motivated by the so-called unified transform, also known as the Fokas method [2, 4]. Indeed, if a function is defined on the full line it is well known that it can be represented in terms of the Fourier transform, whereas if it is defined on the half line it is often represented in terms of the Laplace transform, which is equivalent to the Fourier transform defined on the half-line. If a function is defined on a finite interval, traditionally, it is expressed in terms of a Fourier series. However, the unified transform suggests a paradigm shift: such a function should be expressed in terms of the Fourier transform defined on a finite domain. Using this idea and employing some earlier results of [3] we are led to the following difference-integral equation satisfied by the Riemann zeta function of $\sigma = 1/2$:

$$\begin{aligned}\left|\zeta\left(\frac{1}{2}+it\right)\right|^2 \sim &\operatorname{Re}\left\{\sqrt{\frac{2}{\pi}}\left(c-e^{-i\frac{\pi}{4}}e^{-i}\right)t^{-i}\right\}\left|\zeta\left(\frac{1}{2}+i(t-1)\right)\right|^2\left(1+O\left(\frac{1}{\ln t}\right)\right)\\ &+\operatorname{Re}\left\{\sqrt{\frac{2}{\pi}}e^{-i\frac{\pi}{4}}t^{-it}\int_{t^{\delta_2}}^{t-1}\frac{e^{i(t-\rho)\ln(t-\rho)+i\rho\ln\rho}}{\sqrt{t-\rho}}\left|\zeta\left(\frac{1}{2}+i\rho\right)\right|^2 d\rho\right\}\\ &+\ln t+2\gamma-\ln 2\pi, \qquad t\to\infty, \end{aligned} \tag{1.1}$$

where c is a complex constant given by

$$c=\int_1^{+\infty}\frac{e^{-ix}}{x^{1-i}}dx \approx -0.0713-1.0417i. \tag{1.2}$$

This paper is organised as follows. In Sect. 2 we review some of the basic results of [3], which includes decomposing the integral appearing in (2.1) into 4 integrals, I_j, $j = 1, 2, 3, 4$. In Sect. 3 we express the leading asymptotic behaviour of I_3 and

I_4 in terms of the finite Fourier transform of $\zeta(s)$. In Sect. 4 we sketch the derivation of (1.1). In Sect. 5 we present numerical evidence of the validity of (1.1).

2 Review of Some of the Results of [3]

In this section we review the singular integral equation for the Riemann zeta function, as well as associated results which were derived in [3].

We start with the singular integral equation for all $t > 0$:

$$\frac{t}{\pi}\oint_{-\infty}^{\infty}\text{Re}\left\{\frac{\Gamma(it-i\tau t)}{\Gamma\left(\frac{1}{2}+it\right)}\Gamma\left(\frac{1}{2}+i\tau t\right)\right\}\left|\zeta\left(\frac{1}{2}+i\tau t\right)\right|^2 \mathrm{d}\tau+\mathcal{G}(t)=0, \quad (2.1)$$

where the principal value integral is defined with respect to $\tau = 1$, and the function $\mathcal{G}(t)$ is defined by the formula

$$\mathcal{G}(t)=\text{Re}\left\{\Psi\left(\frac{1}{2}+it\right)\right\}+2\gamma-\ln 2\pi+\frac{2}{1+4t^2}, \quad (2.2)$$

with $\Psi(z)$ denoting the digamma function, i.e.,

$$\Psi(z)=\frac{\frac{\mathrm{d}}{\mathrm{d}z}\Gamma(z)}{\Gamma(z)}, \quad z\in\mathbb{C},$$

and γ denoting the Euler constant.

It is shown in [3] that for $\delta_1 > 0,\ \delta_4 > 0,\ \delta_{14} = \min(\delta_1, \delta_4)$, Eq. (2.1) simplifies to the equation

$$\frac{t}{\pi}\oint_{-t^{\delta_1-1}}^{1+t^{\delta_4-1}}\text{Re}\left\{\frac{\Gamma(it-i\tau t)}{\Gamma\left(\frac{1}{2}+it\right)}\Gamma\left(\frac{1}{2}+i\tau t\right)\right\}\left|\zeta\left(\frac{1}{2}+i\tau t\right)\right|^2 \mathrm{d}\tau+\mathcal{G}(t)$$
$$+O\left(e^{-\pi t^{\delta_{14}}}\right)=0, \quad t\to\infty, \quad (2.3)$$

where the principal value integral is defined with respect to $\tau = 1$.

We split the above interval of integration into the following subintervals:

$$L_1=[-t^{\delta_1-1},t^{-1}],\ L_2=[t^{-1},t^{\delta_2-1}],\ L_3=[t^{\delta_2-1},1-t^{\delta_3-1}],$$
$$L_4=[1-t^{\delta_3-1},1+t^{\delta_4-1}],\quad \delta_2>0,\ \delta_3>0. \quad (2.4)$$

Denote by I_j the integrals along the intervals L_j. It is shown in [3] that

$$I_1(t,\delta_1)=O\left(t^{-\frac{1}{2}+\frac{4}{3}\delta_1}\right),\qquad I_2(t,\delta_2)=O\left(t^{-\frac{1}{2}+\delta_2}\ln t\right),$$
$$\mathcal{G}(t)=\ln t+2\gamma-\ln 2\pi=O(\ln t),\qquad t\to\infty. \tag{2.5}$$

Thus, if $\delta_1\leq\frac{3}{8}$ and $\delta_2<\frac{1}{2}$, Eq. (2.3) becomes

$$I_3+I_4=-\ln t-2\gamma+\ln 2\pi+o(1)\qquad t\to\infty. \tag{2.6}$$

Let $\check{I}_3$ and $\check{I}_4$ denote the leading order terms as $t\to\infty$ of I_3 and I_4, respectively. Then, Eq. (2.6) is

$$\check{I}_3+\check{I}_4=O(\ln t)-\left(I_3-\check{I}_3\right)-\left(I_4-\check{I}_4\right),\qquad t\to\infty. \tag{2.7}$$

A rigorous treatment of the RHS of (2.7), which will be presented in forthcoming publication, yields

$$\left(I_3-\check{I}_3\right)+\left(I_4-\check{I}_4\right)=o(1),\qquad t\to\infty. \tag{2.8}$$

Employing (2.8) into (2.6) yields

$$\check{I}_3+\check{I}_4=-\ln t-2\gamma+\ln 2\pi+o(1)\qquad t\to\infty. \tag{2.9}$$

In what follows we will present arguments suggesting that (2.9) leads to the difference-integral Eq. (1.1) for $\left|\zeta\left(\frac{1}{2}+it\right)\right|^2$. The rigorous derivation of (1.1) will be presented in forthcoming publication.

3 Computation of $\check{I}_3$ and $\check{I}_4$

Preliminaries for I_3

Letting $\tau=\frac{\rho}{t}$ in the definition of I_3, we find

$$I_3=\frac{1}{\pi}\int_{t^{\delta_2}}^{t-t^{\delta_3}}\mathrm{Re}\left\{\frac{\Gamma(it-i\rho)}{\Gamma\left(\frac{1}{2}+it\right)}\Gamma\left(\frac{1}{2}+i\rho\right)\right\}\left|\zeta\left(\frac{1}{2}+i\rho\right)\right|^2d\rho. \tag{3.1}$$

Using

$$\frac{\Gamma(it-i\rho)}{\Gamma\left(\frac{1}{2}+it\right)}\Gamma\left(\frac{1}{2}+i\rho\right)\sim\frac{\sqrt{2\pi}e^{-i\frac{\pi}{4}}e^{itF\left(\frac{\rho}{t}\right)}}{\sqrt{t-\rho}},\qquad t\to\infty, \tag{3.2}$$

where $F(x)$ is defined by

$$F(x) = (1 - x)\ln(1 - x) + x \ln x, \tag{3.3}$$

we find that the leading contribution of I_3 is given by

$$\check{I}_3 := \sqrt{\frac{2}{\pi}} \int_{t^{\delta_2}}^{t-t^{\delta_3}} \frac{\mathrm{Re}\left\{e^{-i\frac{\pi}{4}} e^{itF\left(\frac{\rho}{t}\right)}\right\}}{\sqrt{t-\rho}} \left|\zeta\left(\frac{1}{2} + i\rho\right)\right|^2 d\rho. \tag{3.4}$$

Preliminaries for I_4

Letting $\tau = \frac{\rho}{t}$ in the definition of I_4, we find

$$I_4 = \frac{1}{\pi} \oint_{t-t^{\delta_3}}^{t+t^{\delta_4}} \mathrm{Re}\left\{\frac{\Gamma(it - i\rho)}{\Gamma\left(\frac{1}{2} + it\right)} \Gamma\left(\frac{1}{2} + i\rho\right)\right\} \left|\zeta\left(\frac{1}{2} + i\rho\right)\right|^2 d\rho, \tag{3.5}$$

where the principal value integral is defined with respect to $\rho = t$. Letting $x = t - \rho$, we obtain

$$I_4 = \frac{1}{\pi} \oint_{-t^{\delta_4}}^{t^{\delta_3}} \mathrm{Re}\left\{\Gamma(ix) \frac{\Gamma\left(\frac{1}{2} + it - ix\right)}{\Gamma\left(\frac{1}{2} + it\right)}\right\} \left|\zeta\left(\frac{1}{2} + it - ix\right)\right|^2 dx, \tag{3.6}$$

where the principal value integral is defined with respect to $x = 0$. It is well-known that the Gamma function admits the integral representation

$$\Gamma(ix) = \frac{1}{e^{-\pi x} - e^{\pi x}} \int_{H_1} \frac{e^z}{z} z^{ix} \mathrm{d}z, \tag{3.7}$$

with H_1 denoting the Hankel contour with a branch cut along the negative real axis, see Fig. 1, defined by

$$H_1 = \left\{re^{-i\pi} | 1 < r < \infty\right\} \cup \left\{e^{i\theta} | -\pi < \theta < \pi\right\} \cup \left\{re^{i\pi} | 1 < r < \infty\right\}. \tag{3.8}$$

Using the asymptotic formula

$$\frac{\Gamma\left(\frac{1}{2} + it - ix\right)}{\Gamma\left(\frac{1}{2} + it\right)} = e^{\frac{\pi x}{2}} t^{-ix} e^{ix} \left(1 - \frac{x}{t}\right)^{i(t-x)} \left[1 + O\left(\frac{1}{t}\right)\right], \quad t \to \infty, \tag{3.9}$$

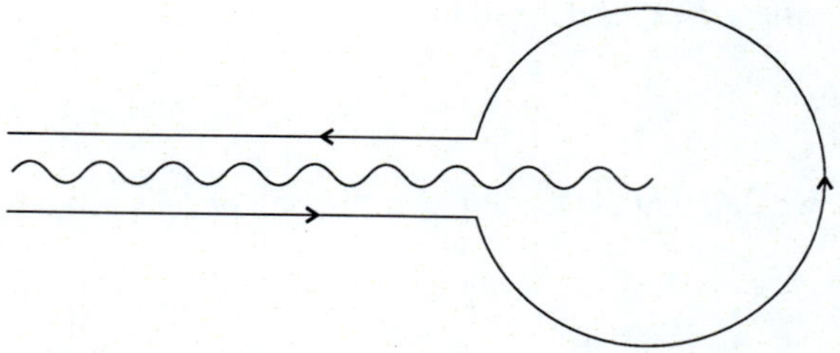

Fig. 1 The Hankel contour H_1

as well as the estimate

$$e^{ix}\left(1-\frac{x}{t}\right)^{i(t-x)}=1+O\left(\frac{x^2}{t}\right),\quad \frac{x}{t}\to 0,$$

we find that the leading contribution of I_4 is given by

$$\check{I}_4=\mathrm{Re}\left\{\frac{1}{\pi}\int_{H_1}\frac{e^z}{z}\oint_{-t^{\delta_4}}^{t^{\delta_3}}\frac{e^{\frac{\pi x}{2}}}{e^{-\pi x}-e^{\pi x}}\left(\frac{z}{t}\right)^{ix}\left|\zeta\left(\frac{1}{2}+it-ix\right)\right|^2 dxdz\right\},\quad (3.10)$$

with the principal value integral defined with respect to $x=0$.

The Finite Fourier Transform

In order to compute the large t asymptotics of the RHS of (3.4) and (3.10) we will employ the finite Fourier transform, where it turns out that it will be more convenient to integrate from $t=1$:

$$\Phi(\nu):=\int_1^T\left|\zeta\left(\frac{1}{2}+it\right)\right|^2 e^{i\nu t}dt,\qquad \nu\in\mathbb{C}.\quad (3.11)$$

Then,

$$\left|\zeta\left(\frac{1}{2}+it\right)\right|^2=\frac{1}{2\pi}\int_{-\infty}^{\infty}\Phi(\nu)e^{-i\nu t}d\nu,\qquad t\in[1,T].\quad (3.12)$$

Remark 1 Equations (3.11) and (3.12) imply

$$\int_{-\infty}^{\infty}\Phi(\nu)\left(\frac{1}{2\pi}\int_1^T e^{i\tau(k-\nu)}d\tau\right)d\nu=\Phi(k).\quad (3.13)$$

Note that

$$\frac{1}{2\pi}\int_1^T e^{i\tau(k-\nu)}d\tau=\frac{1}{2i\pi}\frac{e^{iT(k-\nu)}-e^{i(k-\nu)}}{k-\nu}.\quad (3.14)$$

In the case of the Fourier transform on the full line the analogue of the LHS in (3.14) equals $\delta(k-\nu)$. In the case of the finite Fourier transform, Eq. (3.13) is a direct consequence of analyticity: $\Phi(\nu)$ is an entire function for which

$$\Phi(\nu) \sim \frac{1}{i\nu}\left[e^{i\nu T}\left|\zeta\left(\frac{1}{2}+iT\right)\right|^2 - e^{i\nu}\left|\zeta\left(\frac{1}{2}+i\right)\right|^2\right], \qquad \nu \to \infty. \tag{3.15}$$

Thus, $e^{-i\nu T}\Phi(\nu)$ is an entire function for which

$$e^{-i\nu T}\Phi(\nu) \sim \frac{1}{i\nu}\left[\left|\zeta\left(\frac{1}{2}+iT\right)\right|^2 - e^{-i\nu(T-1)}\left|\zeta\left(\frac{1}{2}\right)\right|^2\right].$$

Hence we rewrite (3.13) as

$$\frac{1}{2i\pi}\int_{\tilde{R}}\Phi(\nu)\frac{e^{iT(k-\nu)}}{k-\nu}d\nu - \frac{1}{2i\pi}\int_{\tilde{R}}\Phi(\nu)\frac{e^{i(k-\nu)}}{k-\nu}d\nu,$$

where $\tilde{R}$ is the real line slightly deformed at the point $\nu = k$, with a small semicircle of radius $\epsilon \to 0$ contained in the lower half complex plane. Thus, the first integral vanishes, by closing at $\mathbb{C}^-$, whereas the second integral gives $\Phi(k)$, by closing at $\mathbb{C}^+$.

Remark 2 Using in (3.11) the fact that $\left|\zeta\left(\frac{1}{2}+it\right)\right|^2$ is real yields the condition $\overline{\Phi(\nu)} = \Phi(-\nu)$.

The Derivation of $\check{I}_4$

Equation (3.12) yields

$$\left|\zeta\left(\frac{1}{2}+i(t-x)\right)\right|^2 = \frac{1}{2\pi}\int_{-\infty}^{\infty}\Phi(\nu)e^{-i\nu t}e^{i\nu x}d\nu, \qquad t \in [1, T]. \tag{3.16}$$

Using the above equation into (3.10), we obtain

$$\check{I}_4 = \mathrm{Re}\left\{\frac{1}{2\pi}\int_{-\infty}^{\infty}\Phi(\nu)e^{-i\nu t}\tilde{I}_4 d\nu\right\}, \tag{3.17}$$

with

$$\tilde{I}_4 = \frac{1}{\pi}\int_{H_1}\frac{e^z}{z}\oint_{-t^{\delta_4}}^{t^{\delta_3}}\frac{e^{\frac{\pi x}{2}}}{e^{-\pi x}-e^{\pi x}}A^{ix}dxdz, \qquad A := \frac{z}{t}e^{\nu}. \tag{3.18}$$

Using the result of Proposition 6.3 of [3], namely the estimate

$$\frac{1}{\pi}\int_{H_1}\frac{e^z}{z}\oint_{-t^{\delta_4}}^{t^{\delta_3}}\frac{e^{\frac{\pi x}{2}}}{e^{-\pi x}-e^{\pi x}}A^{ix}dxdz \sim -1+\frac{1}{\pi}\int_{H_1}\frac{e^z}{z}\frac{e^{it^{\delta_3}\ln A-\frac{\pi t^{\delta_3}}{2}}}{\frac{\pi}{2}-i\ln A}dz, \quad (3.19)$$

we obtain

$$\tilde{I}_4 \sim -1+\frac{1}{\pi}\int_{H_1}\frac{e^z}{z}\frac{e^{it^{\delta_3}\ln(Mz)-\frac{\pi t^{\delta_3}}{2}}}{\frac{\pi}{2}-i\ln(Mz)}dz, \qquad M=\frac{e^{\nu}}{t}, \qquad t\to\infty. \quad (3.20)$$

Furthermore, Proposition 6.4 of [3] yields

$$\frac{1}{\pi}\int_{H_1}\frac{e^z}{z}\frac{e^{it^{\delta_3}\ln(Mz)-\frac{\pi t^{\delta_3}}{2}}}{\frac{\pi}{2}-i\ln(Mz)}dz = 2e^{-\frac{i}{M}}+E_4^{SD}(\nu,t), \quad (3.21)$$

where the first term occurs iff $e^{\nu}\in\left(t^{1-\delta_3},t\right)$ and

$$E_4^{SD}(\nu,t)=\frac{1}{\pi}\int_{H_1}\frac{e^{\lambda[w+i\ln(\frac{iw}{\alpha})]}}{-iw\ln\left(\frac{iw}{\alpha}\right)}dw, \qquad \lambda=t^{\delta_3}, \quad \alpha=\frac{t^{1-\delta_3}}{e^{\nu}}. \quad (3.22)$$

Hence,

$$\tilde{I}_4 \sim -1+2e^{-ite^{-\nu}}+E_4^{SD}, \qquad t\to\infty, \quad (3.23)$$

where the second term occurs iff $e^{\nu}\in\left(t^{1-\delta_3},t\right)$.

Thus, $\check{I}_4$ takes the form

$$\check{I}_4=\mathrm{Re}\Bigg\{-\frac{1}{2\pi}\int_{-\infty}^{\infty}\Phi(\nu)e^{-i\nu t}d\nu+\frac{1}{\pi}\int_{(1-\delta_3)\ln t}^{\ln t}\Phi(\nu)e^{-it\left(\nu+e^{-\nu}\right)}d\nu$$
$$+\frac{1}{2\pi}\int_{-\infty}^{\infty}\Phi(\nu)e^{-i\nu t}E_4^{SD}(\nu,t)d\nu\Bigg\}.$$

Using the fact that $\overline{\Phi(\nu)}=\Phi(-\nu)$, see Remark 2, the term $\check{I}_4$ can also be written in the following form:

$$\check{I}_4(t)=\frac{1}{2\pi}\int_{(1-\delta_3)\ln t}^{\ln t}\Phi(\nu)e^{-it\left(\nu+e^{-\nu}\right)}d\nu+\frac{1}{2\pi}\int_{-\ln t}^{(\delta_3-1)\ln t}\Phi(\nu)e^{-it\left(\nu-e^{\nu}\right)}d\nu$$
$$+\frac{1}{4\pi}\int_{-\infty}^{\infty}\Phi(\nu)e^{-i\nu t}E_4^{SD}(\nu,t)d\nu+\frac{1}{4\pi}\int_{-\infty}^{\infty}\Phi(\nu)e^{-i\nu t}\overline{E_4^{SD}(-\nu,t)}d\nu$$
$$-\frac{1}{2\pi}\int_{-\infty}^{\infty}\Phi(\nu)e^{-i\nu t}d\nu. \quad (3.24)$$

Proposition 1 *Let $\check{I}_4$ be defined by (3.24). Then*

$$\check{I}_4 = -\left|\zeta\left(\frac{1}{2}+it\right)\right|^2 + Re\left\{\frac{1}{\pi}\int_{(1-\delta_3)\ln t}^{\ln t}\Phi(\nu)e^{-it(\nu+e^{-\nu})}d\nu + \frac{1}{2\pi}\int_{-\infty}^{\infty}\Phi(\nu)e^{-i\nu t}E_4^{SD}(\nu,t)d\nu\right\}, \tag{3.25}$$

where

$$E_4^{SD}(\nu,t) = \frac{1}{\pi}\int_{H_1}\frac{e^{\lambda[w+i\ln(\frac{iw}{\alpha})]}}{-iw\ln\left(\frac{iw}{\alpha}\right)}dw, \qquad \lambda = t^{\delta_3},\ \ \alpha = \frac{t^{1-\delta_3}}{e^{\nu}}. \tag{3.26}$$

Proof Employing (3.12) in the last term of the RHS of Eq. (3.24) yields (3.25).

The term $2e^{-i/M}$ appearing in the RHS of (3.21) arises from the evaluation of the contribution of the pole $z_P = -i/M$ and gives rise to the second term in the RHS of (3.25). Thus, we will use the notation $\check{I}_4^P(t)$ for this term. Similarly, we denote the last term (3.25) as $\check{I}_4^{SD}$.

Hence, (3.25) takes the form

$$\check{I}_4 = -\left|\zeta\left(\frac{1}{2}+it\right)\right|^2 + \text{Re}\left\{\check{I}_4^P + \check{I}_4^{SD}\right\}, \tag{3.27}$$

where

$$\check{I}_4^P(t) = \frac{1}{\pi}\int_{(1-\delta_3)\ln t}^{\ln t}\Phi(\nu)e^{-it(\nu+e^{-\nu})}d\nu \tag{3.28}$$

and

$$\check{I}_4^{SD}(t) = \frac{1}{2\pi}\int_{-\infty}^{\infty}\Phi(\nu)e^{-i\nu t}E_4^{SD}(\nu,t)d\nu. \tag{3.29}$$

4 Sketch of the Derivation of (1.1)

It can be shown that $\check{I}_4^{SD}$ is negligible, compared to $\check{I}_4^P$; the rigorous derivation will be presented in forthcoming publication. Thus, in what follows we analyse the contribution of $\check{I}_4^P$.

Using (3.11) in (3.28) yields

$$\begin{aligned}\check{I}_4^P(t) &= \frac{1}{\pi}\int_{(1-\delta_3)\ln t}^{\ln t}\int_1^T\left|\zeta\left(\frac{1}{2}+i\tau\right)\right|^2 e^{-it(\nu+e^{-\nu}-\frac{\tau}{t}\nu)}d\tau d\nu \\ &= \frac{1}{\pi}\int_1^T J(\tau;t,\delta_3)\left|\zeta\left(\frac{1}{2}+i\tau\right)\right|^2 d\tau,\end{aligned} \tag{4.1}$$

where

$$J(\tau; t, \delta_3) = \int_{(1-\delta_3)\ln t}^{\ln t} e^{-itg(\nu, \frac{\tau}{t})} d\nu, \tag{4.2}$$

with

$$g\left(\nu, \frac{\tau}{t}\right) = \nu + e^{-\nu} - \frac{\tau}{t}\nu. \tag{4.3}$$

The stationary phase method yields the estimate

$$\int_{(1-\delta_3)\ln t}^{\ln t} e^{-itg(\nu, \frac{\tau}{t})} d\nu \sim \sqrt{2\pi} e^{-i\frac{\pi}{4}} \frac{e^{-i(t-\tau)} e^{i(t-\tau)\ln(t-\tau)} e^{-i(t-\tau)\ln t}}{\sqrt{t-\tau}},$$
$$\text{if } \tau \in \left(t - t^{\delta_3}, t - 1\right). \tag{4.4}$$

Indeed, the stationary point is given by solving $g_\nu = 0$, which yields $\nu^* = -\ln\left(1 - \frac{\tau}{t}\right)$. Hence,

$$\nu^* \in \left((1-\delta_3)\ln t, \ln t\right) \Longleftrightarrow \tau \in \left(t - t^{\delta_3}, t - 1\right).$$

It is interesting to note that the RHS of (4.4) can be rewritten in the form

$$\frac{1}{\pi} \int_{(1-\delta_3)\ln t}^{\ln t} e^{-itg(\nu, \frac{\tau}{t})} d\nu \sim \sqrt{\frac{2}{\pi}} e^{-i\frac{\pi}{4}} \frac{e^{itF(\frac{\tau}{t})}}{\sqrt{t-\tau}}, \quad \text{if } \tau \in \left(t - t^{\delta_3}, t - 1\right), \tag{4.5}$$

with F defined in (3.3). This can be derived by using

$$e^{-i(t-\tau)} e^{i(t-\tau)\ln(t-\tau)} e^{-i(t-\tau)\ln t} = e^{itF(\frac{\tau}{t})} \left[1 + O\left(t^{2\delta_3 - 1}\right)\right], \quad t \to \infty, \tag{4.6}$$

into (4.4). In order to prove (4.6) we observe that

$$e^{-i(t-\tau)} e^{i(t-\tau)\ln(t-\tau)} e^{-i(t-\tau)\ln t} = e^{itF(\frac{\tau}{t})} \exp\left\{-i\left[t - \tau + \tau \ln\left(\frac{\tau}{t}\right)\right]\right\}.$$

Making the change of variables $\tau = t - x$, $x \in \left(1, t^{\delta_3}\right)$, we find

$$t - \tau + \tau \ln\left(\frac{\tau}{t}\right) = x + (t-x)\ln\left(1 - \frac{x}{t}\right) = x + (t-x)\left(-\frac{x}{t} + O\left(\frac{x}{t}\right)^2\right) \sim \frac{x^2}{t},$$

hence

$$\exp\left\{-i\left[t - \tau + \tau \ln\left(\frac{\tau}{t}\right)\right]\right\} = 1 + O\left(\frac{x^2}{t}\right).$$

The fact that $x \in \left(1, t^{\delta_3}\right)$ yields (4.6).

The stationary point of J coincides with the endpoint $\ln t$ if $\tau = t - 1$. Evaluating the RHS of (4.4) at $\tau = t - 1$ we find $\sqrt{2\pi}e^{-i\frac{\pi}{4}}e^{-i}t^{-i}$. Thus, as $\nu \to \ln t$ and $\tau \to t - 1$, we obtain the following contribution:

$$\frac{1}{\pi}\left[\hat{J}(t;\delta_3) - \sqrt{2\pi}e^{-i\frac{\pi}{4}}e^{-i}t^{-i}\right]\left|\zeta\left(\frac{1}{2} + i(t-1)\right)\right|^2$$

where $\hat{J}(t;\delta_3)$ denotes the contribution of $J(\tau, t; \delta_3)$ in the neighbourhood of $\tau = t - 1$. It turns out that

$$\hat{J}(t;\delta_3) = \sqrt{2\pi}J(t-1, t;\delta_3) = \sqrt{2\pi}\int_{(1-\delta_3)\ln t}^{\ln t} e^{-i(\nu + te^{-\nu})}d\nu.$$

In order to evaluate the above integral we let $\nu = \ln t - \ln x$, and find

$$\int_{(1-\delta_3)\ln t}^{\ln t} e^{-i(\nu+te^{-\nu})}d\nu = t^{-i}\int_1^{t^{\delta_3}} \frac{e^{-ix}}{x^{1-i}}dx = t^{-i}\int_1^{\infty} \frac{e^{-ix}}{x^{1-i}}dx - t^{-i}\int_{t^{\delta_3}}^{\infty} \frac{e^{-ix}}{x^{1-i}}dx. \tag{4.7}$$

Using integration by parts we find that the second integral in the RHS of (4.7) is $O\left(t^{-\delta_3}\right)$.

Similar considerations apply to the case that $\tau = t - t^{\delta_3}$, where the stationary point approaches the other endpoint, $(1 - \delta_3)\ln t$, but now the relevant contribution is $O\left(t^{-\frac{\delta_3}{2}}\right)$. Hence we find,

$$\begin{aligned} \check{I}_4^P \sim\ & \sqrt{\frac{2}{\pi}}e^{-i\frac{\pi}{4}}\int_{t-t^{\delta_3}}^{t-1} \frac{e^{itF(\frac{\tau}{t})}}{\sqrt{t-\tau}}\left|\zeta\left(\frac{1}{2} + i\tau\right)\right|^2 d\tau \qquad (4.8)\\ & + \sqrt{\frac{2}{\pi}}\left(c - e^{-i\frac{\pi}{4}}e^{-i}\right)t^{-i}\left|\zeta\left(\frac{1}{2} + i(t-1)\right)\right|^2\left(1 + O\left(\frac{1}{\ln t}\right)\right), \qquad t \to \infty, \end{aligned}$$

with F and c defined in (3.3) and (1.2), respectively.

Simplifying $tF\left(\frac{\rho}{t}\right)$, we find,

$$\begin{aligned} (t-\rho)\ln\left(1 - \frac{\rho}{t}\right) + \rho\ln\left(\frac{\rho}{t}\right) &= (t-\rho)\ln(t-\rho) - (t-\rho)\ln t + \rho\ln\rho - \rho\ln t \\ &= (t-\rho)\ln(t-\rho) + \rho\ln\rho - t\ln t. \end{aligned}$$

Hence, (1.1) follows by employing (3.4), (3.27) and (4.8) in (2.9).

5 Numerical Evidence

In this section we check numerically the validity of the difference-integral Eq. (1.1), namely

$$\left|\zeta\left(\frac{1}{2}+it\right)\right|^2 \sim \mathrm{Re}\left\{\sqrt{\frac{2}{\pi}}\left(c-e^{-i\frac{\pi}{4}}e^{-i}\right)t^{-i}\right\}\left|\zeta\left(\frac{1}{2}+i(t-1)\right)\right|^2$$
$$+\sqrt{\frac{2}{\pi}}\int_{t^{\delta_2}}^{t-1}\frac{\mathrm{Re}\left\{e^{-i\frac{\pi}{4}}e^{itF\left(\frac{\rho}{t}\right)}\right\}}{\sqrt{t-\rho}}\left|\zeta\left(\frac{1}{2}+i\rho\right)\right|^2 d\rho+\ln t+2\gamma-\ln 2\pi$$
$$+O\left(\frac{\left|\zeta\left(\frac{1}{2}+i(t-1)\right)\right|^2}{\ln t}\right), \qquad t\to\infty, \tag{5.1}$$

with $F(x)$ and c defined in (3.3) and (1.2), respectively.

In Fig. 2 we depict the LHS by the blue curve, and the RHS (ignoring the error term) by the red dashed line, for the range $t \in (357, 440)$. In Fig. 3 we depict the difference of LHS minus the RHS, for the same range of t. In Fig. 4 we observe that this difference is dominated by the error term $O\left(\frac{|\zeta(\frac{1}{2}+i(t-1))|^2}{\ln t}\right)$; we plot the absolute value of the above-mentioned difference in green, and the $\frac{|\zeta(\frac{1}{2}+i(t-1))|^2}{\ln t}$ in black. We

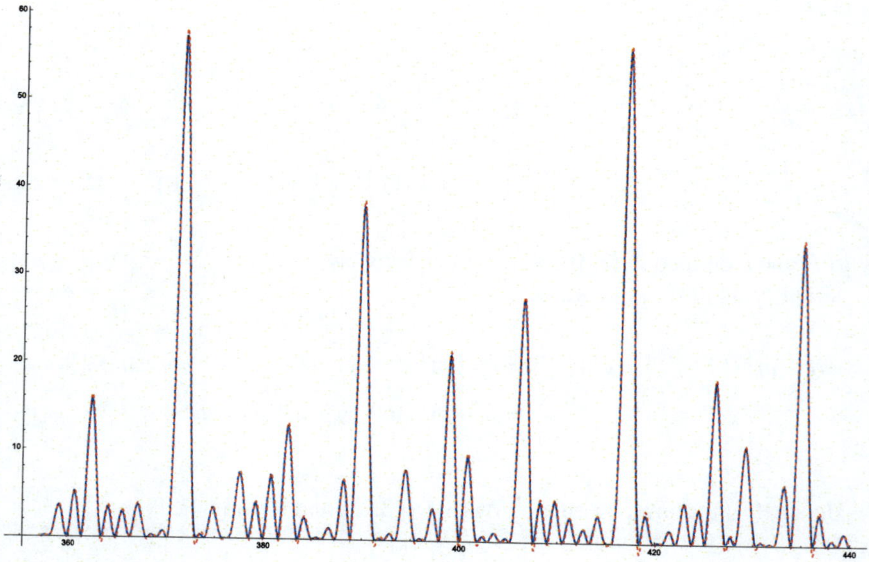

Fig. 2 The LHS (blue) and the RHS (red dashed), for the range $t \in (357, 440)$

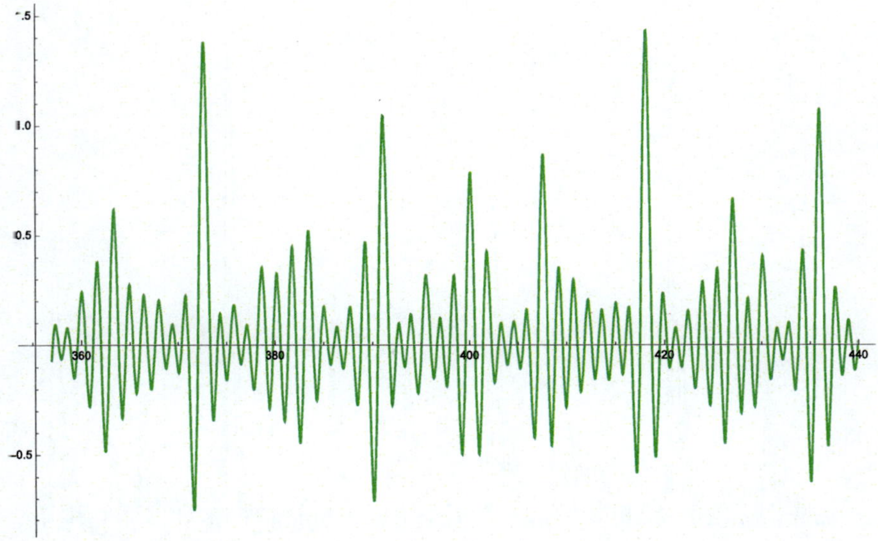

Fig. 3 The difference of LHS minus the RHS

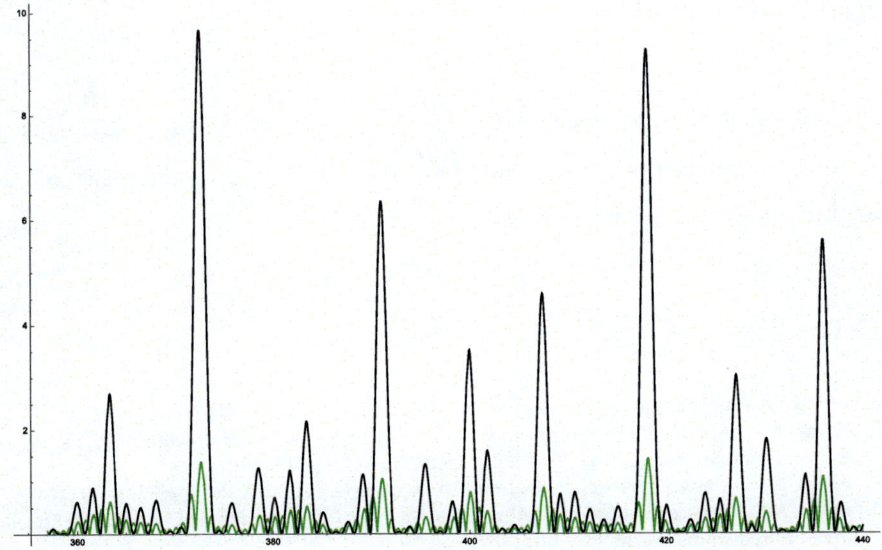

Fig. 4 The absolute difference of LHS minus the RHS (green), versus $\frac{\left|\zeta\left(\frac{1}{2}+i(t-1)\right)\right|^2}{\ln t}$ (black)

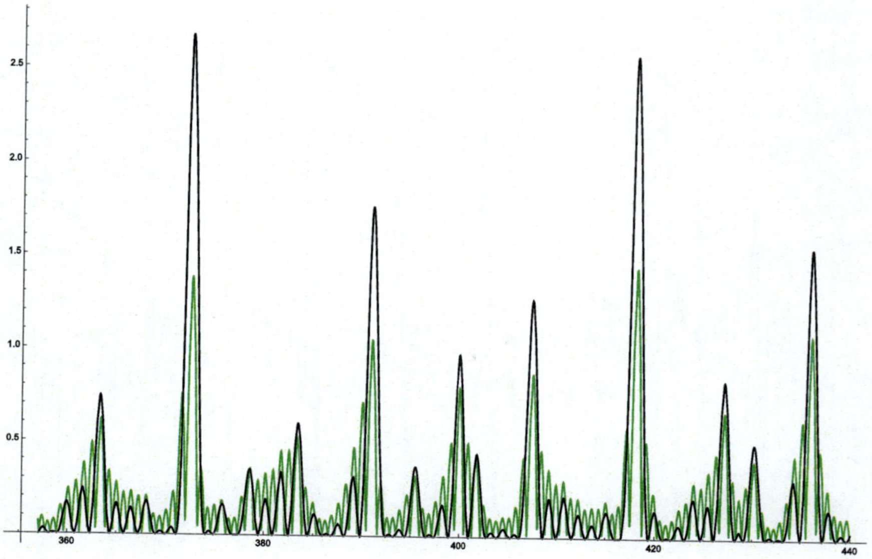

Fig. 5 The absolute difference of LHS minus the RHS (green), versus $\frac{\sqrt{\pi}\left|\left(c-e^{-i\frac{\pi}{4}}e^{-i}\right)\right|}{\ln t}$ $\left|\zeta\left(\frac{1}{2}+i(t-1)\right)\right|^2$ (black)

find interesting that if we scale $\frac{\left|\zeta\left(\frac{1}{2}+i(t-1)\right)\right|^2}{\ln t}$ by $\sqrt{\pi}\left|\left(c-e^{-i\frac{\pi}{4}}e^{-i}\right)\right| \approx 0.276$, Fig. 5 illustrate clearer that the error term $O\left(\frac{\left|\zeta\left(\frac{1}{2}+i(t-1)\right)\right|^2}{\ln t}\right)$ 'captures the peaks' of the explicit difference LHS minus RHS.

References

1. Erdélyi, A.: Asymptotic Expansions (No. 3). Courier Corporation (1956)
2. Fokas, A.S.: A unified transform method for solving linear and certain nonlinear PDEs. Proc. R. Soc. Lond. Ser. Math. Phys. Eng. Sci. *453*(1962), 1411–1443 (1997)
3. Fokas, A.S.: A novel approach to the Lindelöf hypothesis. Trans. Math. Appl. **3**, 1–49 (2019)
4. Fokas, A.S., Kaxiras, E.: Modern Mathematical Methods for Scientists and Engineers: A Street-Smart Introduction. World Scientific (2022)
5. Fokas, A.S., Lenells, J.: On the asymptotics to all orders of the Riemann Zeta function and of a two-parameter generalization of the Riemann Zeta function. Mem. AMS. **275**(1351) (2022)
6. Kalimeris, K., Fokas, A.S.: Explicit asymptotics for certain single and double exponential sums. Proc. R. Soc. Edinb. Math. **150**(2), 607–632 (2020)
7. Kalimeris, K., Fokas, A.S.: A novel integral equation for the Riemann zeta function and large t-asymptotics. Math. **7**(7), 650 (2019)
8. Siegel, C.L., Nachlaßzur, Über Riemanns, analytischen Zahlentheorie, Quellen Studien zur Geschichte der Math. Astron. und Phys. Abt. B: Studien 2: 4580,: reprinted in Gesammelte Abhandlungen, vol. 1, p. 1966. Springer-Verlag, Berlin (1932)

The Role of Periodicity in the Solution of Third Order Boundary Value Problems

B. Pelloni and D. A. Smith

Abstract In this short paper, we elucidate how the solution of certain illustrative boundary value problems for the Airy equation $u_t + u_{xxx} = 0$ on $[0, 1]$ can be expressed as a perturbation of the solution of the purely periodic problem. The motivation is to understand the role boundary conditions play in the properties of the solution. This is particularly important in related work on the solution of linear dispersive problems with discontinuous initial data and the phenomena of revivals and fractalization.

Keywords Initial boundary value problems · Regularity of solutions · Fokas transform method · Revivals and fractalization

1 Introduction

Linear dispersive equations such as the free-space Schrödinger or Airy equations, respectively

$$iu_t - u_{xx} = 0 \quad \text{or} \quad u_t + u_{xxx} = 0, \quad \text{where } u = u(x,t),$$

are important models in the mathematical modelling of reality, as they constitute a powerful way to capture the dominant linear behaviour driving the time evolution of many physical phenomena that propagate in a wave-like manner. We consider them in one space dimension, so $x \in \mathbb{R}$, with $t > 0$ denoting time. These are the simplest equations that, mathematically, capture the main features of the behaviour of even- and odd-order linear dispersive problems.

B. Pelloni (✉)
Heriot-Watt University and Maxwell Institute for the Mathematical Sciences, Edinburgh, Scotland
e-mail: b.pelloni@hw.ac.uk

D. A. Smith
Yale-NUS College and National University of Singapore, Singapore, Island
e-mail: dave.smith@yale-nus.edu.sg

T. Bountis et al. (eds.), *Chaos, Fractals and Complexity*, Springer Proceedings in Complexity, https://doi.org/10.1007/978-3-031-37404-3_23

Our physical reality demands that we pose these equations on a finite interval. We will assume in all that follows that $x \in [0, 1]$, and that we are given initial and boundary conditions that yield a well-posed problem that admits a unique solution.

While the method of solution of these linear boundary value problems, via separation of variables or eigenfunction expansion, has been a standard tool of the mathematical trade for a very long time, these techniques are not universally applicable. In particular, they rely on the full and explicit knowledge of the spectral structure of the spatial linear differential operator. However, when the boundary conditions are such that this operator is not self-adjoint, this spectral structure may not be known or easily determined. This is particularly true for odd-order operators, which is the case we focus on in this note.

We also remark that care must be taken when the given initial or boundary data are not sufficiently regular, as it is then necessary to interpret the solution and its representation in a suitably weak sense. We will not dwell on this aspect in this note, but it is an important consideration for some of the applications, notably the study of revivals and weak revivals, that motivated the considerations in this paper.

More precisely, our motivation is recent work on the so-called *Talbot effect*, or *revival* phenomenon [2, 10]. This phenomenon, first described experimentally in the mid 1850's by scientist and pioneer of photography, William Henry Fox Talbot [14] and rediscovered several times since in other dispersive systems (notably by Olver [9]), occurs when an initial datum with jump discontinuities is propagated periodically. What happens then is that the behaviour of the solution at times that are rational multiples of a certain quantity related to the length of the interval, that we call rational times, is markedly different from the behaviour at generic times. At rational times, the solution is a superposition of translated and dilated copies of the initial conditions, so in particular it is spatially discontinuous, while at generic times the solution is spatially continuous (albeit nowhere differentiable). See [13] for a recent survey of these phenomena.

The revival phenomenon occurs and is fully understood in the case of linear dispersive PDEs with periodic boundary conditions [1, 4, 9]. When studying whether this phenomenon persists in more general situations, for example for more general boundary conditions, or for nonlinear dispersive PDEs, it is natural to consider what echo is left by the solution of the linear periodic problem (for example, if a weaker form of revivals occurs). Whether a specific signature of the periodic problem can be identified within the solution of more general problems is the question that we consider, for selected examples, in this paper.

2 A Motivating Example and the Unified Transform of Fokas

One particularly natural and surprisingly hard problem, illustrating the limitations of the eigenfunction expansion technique, is the following.

Assume $u(x, t)$ solves the following *Dirichlet type* boundary value problem on $[0, 1]$ for the Airy equation:

$$\begin{aligned} &u_t + u_{xxx} = 0, && x \in (0, 1),\ t > 0, \\ &u(x, 0) = f(x), && x \in (0, 1), \\ &u(0, t) = u(1, t) = \partial_x u(1, t) = 0, && t > 0. \end{aligned} \tag{1}$$

Here, and in what follows, we do not make specific assumptions on the regularity of the initial datum $f(x)$. If $f(x)$ is sufficiently smooth, we might expect the solution representation to be valid pointwise, though this depends on the compatibility at the corners $(0, 0)$ and $(1, 0)$ (see also e.g. [15]). The regularity assumptions on $f(x)$ can be relaxed by adopting a weaker notion of solution.

It is remarkable that, while the associated differential operator has an infinite number of discrete eigenvalues, the associated eigenfunctions do not form an unconditional basis, so that there is no generalised Fourier series representation for the solution of this problem. Although this was formally established as far back as 1915 [8], the knowledge fell into obscurity until the problem was rediscovered and solved within a more general setting [5, 6, 11].

However, the problem can be fully and effectively solved using the Fokas transform method [7]. Indeed, using this approach it can be shown that this problem has a unique solution $u(x, t)$, [11]. The solution admits the explicit contour integral representation

$$\begin{aligned} 2\pi u(x, t) = &\int_{\mathbb{R}} e^{ikx+ik^3t} \widehat{u_0}(k) dk \\ &+ \int_{\Gamma^+} e^{ikx+ik^3t} \frac{\zeta^+(k)}{\Delta(k)} dk + \int_{\Gamma^-} e^{ik(x-1)+ik^3t} \frac{\zeta^-(k)}{\Delta(k)} dk, \end{aligned}$$

where $\Gamma^\pm$ are the contours in $\mathbb{C}^\pm$ defined as the locus of $\Re(ik^3) = 0$, $\widehat{u_0}(k)$ denotes the Fourier transform of $u_0(x)$, and $\zeta^\pm(k)$, $\Delta(k)$ are entire functions of k only, fully determined by the function $u_0(x)$. The zeros of Δ are the cube roots of the eigenvalues of the spatial differential operator, and it can be shown that they are not on the integration contours [11, Sect. A]. Hence the integrals are well defined.

This leaves open the question of how this problem relates to the simpler case of periodic boundary conditions, whose solution has a classical representation in terms of a Fourier series. This is the question we consider below, for this example as well as for examples of boundary conditions that couple the two ends of the interval $[0, 1]$.

3 Non-periodic Boundary Value Problems

Let $u(x, t)$ denote the solution of a given boundary value problem of the form

$$u_t + u_{xxx} = 0, \qquad x \in (0, 1),\ t > 0, \tag{2a}$$
$$u(x, 0) = f(x), \qquad x \in (0, 1), \tag{2b}$$
$$\text{three homogeneous boundary conditions on } u(x, t). \tag{2c}$$

We assume that the prescribed boundary conditions are not periodic ones, and that they are such that the solution exists and is unique. For the Airy equation, the boundary conditions for which this well-posedness holds are characterised in [12]. In particular, this is the case for the illustrative examples we examine below. We stress that we rely crucially on the Fokas transform approach to guarantee that such existence results hold for the examples given.

To relate the solution of such a boundary value problem with the solution of the periodic problem, we consider a natural decomposition of the solution $u(x, t)$. Namely, let $v(x, t)$ denote the solution of the purely periodic problem, with the same initial condition $v(x, 0) = f(x)$, so that $v(x, t)$ satisfies

$$v_t + v_{xxx} = 0,\quad x \in (0, 1),\ t > 0; \qquad v(x, 0) = f(x),\quad x \in (0, 1); \tag{3}$$
$$\partial_x^j v(0, t) = \partial_x^j v(1, t), \quad j = 0, 1, 2, \quad t > 0. \tag{4}$$

The function $v(x, t)$ admits the following explicit representation as a Fourier series, pointwise if $v(x, t)$ is sufficiently smooth (e.g. at least Hölder continuous), and in $L^2[0, 1]$ otherwise:

$$v(x, t) = \sum_{n \in \mathbb{Z}} \mathrm{e}^{ik_n x + ik_n^3 t} \hat{f}(k_n), \qquad k_n = 2\pi n. \tag{5}$$

The spectral structure of this problem is entirely understood: the spectrum is fully discrete and given by $\{k_n,\ n \in \mathbb{Z}\}$, while the associated eigenfunctions $\{\mathrm{e}^{ik_n x},\ n \in \mathbb{Z}\}$ form a complete basis with respect to the L^2 Hilbert structure.

We then define the auxiliary function w as

$$w(x, t) := u(x, t) - v(x, t). \tag{6}$$

The function w is fully determined by the given functions u and v, hence all its boundary values are known. This function encodes information on how the given boundary conditions change the nature of the solution when compared with the solution of the periodic problem.

In each of the following sections, we will select different boundary conditions (2c), and examine the properties of the resulting w, in particular its regularity, with the aim of characterising u as a w-regularity perturbation of v.

Table 1 Summary of problems, examples, and results

u Problem	v Problem	w Problem	Result	Eg
Dirichlet type	Periodic	Forced periodic	Theorem 3	Section 3.1
				Section 3.2
$u(0,t) = u(1,t)$	Periodic	Forced periodic	Theorem 3	Section 3.3
quasiperiodic	Periodic	Forced quasiperiodic	Remark 5	Section 3.4
$u(0,t) = e^{i\theta}u(1,t)$	Quasiperiodic	Forced quasiperiodic	Remark 6	

For most of the u problems considered here, it is appropriate to select v as the solution of a periodic problem and then w can be viewed as the solution of a problem with zero initial condition and with boundary conditions that are, formally, inhomogeneous periodic or quasiperiodic conditions. To emphasize the role of the inhomogeneities in the boundary conditions of w, we describe such problems as "forced (quasi)periodic".

In the last case, we choose v as the solution of a quasiperiodic, rather than periodic, problem.

In Table 1 we summarise the type of boundary value problem given for u and selected for v. Once these are given, the function w is fixed but the boundary value problem for it can be given either in terms of the boundary values of u or of v, resulting in a different problem for w; the third column in this table summarises what particular problem is selected for w for each of the examples we treat in this paper.

3.1 Uncoupled BC of Dirichlet Type

Here we consider the problem (1) for $u(x,t)$. In this case, the function $w(x,t)$ satisfies

$$w_t + w_{xxx} = 0, \quad x \in (0,1), \ t > 0; \qquad w(x,0) = 0, \ x \in (0,1); \tag{7a}$$
$$w(0,t) = w(1,t), \qquad t > 0; \tag{7b}$$
$$\partial_x w(0,t) = \partial_x w(1,t) + h_1(t), \qquad t > 0; \tag{7c}$$
$$\partial_{xx} w(0,t) = \partial_{xx} w(1,t) + h_2(t), \qquad t > 0. \tag{7d}$$

The known, smooth functions $h_1(t)$ and $h_2(t)$ are given by

$$h_1(t) = \partial_x u(0,t),$$
$$h_2(t) = \partial_{xx} u(0,t) - \partial_{xx} u(1,t).$$

Hence w can be regarded as the solution of a forced periodic problem, with zero initial condition.

Lemma 1 *The function $w(x,t)$ that solves (7) is a continuous function of x and admits the representation*

$$w(x,t) = \sum_{n\in\mathbb{Z}} \mathrm{e}^{ik_n x + ik_n^3 t} \left[ik_n H_1(k_n,t) + H_2(k_n,t)\right], \tag{8}$$

with $k_n = 2\pi n$,

$$H_1(k,t) = \int_0^t \mathrm{e}^{-ik^3 s} h_1(s)ds, \qquad H_2(k,t) = \int_0^t \mathrm{e}^{-ik^3 s} h_2(s)ds. \tag{9}$$

Proof Consider the Fourier transform of $w(x,t)$ on $[0,1]$, defined by

$$\hat{w}(k,t) = \int_0^1 \mathrm{e}^{-ikx} w(x,t)dx.$$

Then, using the Fourier transform and integration by parts, the PDE for $w(x,t)$ yields the following ODE for $\hat{w}(k,t)$:

$$\begin{aligned}\left(\mathrm{e}^{-ik^3 t}\hat{w}(k,t)\right)_t = -[k^2 w(0,t) - ikw_x(0,t) - w_{xx}(0,t)] \\ +\mathrm{e}^{-ik}[k^2 w(1,t) - ikw_x(1,t) - w_{xx}(1,t)].\end{aligned}$$

We set

$$F(k,t) = \int_0^t \mathrm{e}^{-ik^3 s}[k^2 w(0,s) - ikw_x(0,s) - w_{xx}(0,s)]ds, \tag{10a}$$

$$G(k,t) = \int_0^t \mathrm{e}^{-ik^3 s}[k^2 w(1,s) - ikw_x(1,s) - w_{xx}(1,s)]ds. \tag{10b}$$

Then, since $\hat{w}(k,0) = 0$, the solution of the ODE is given by

$$\hat{w}(k,t) = \mathrm{e}^{ik^3 t}\left[-F(k,t) + \mathrm{e}^{-ik} G(k,t)\right].$$

Using the boundary conditions, we find

$$F(k,t) = G(k,t) - ikH_1(k,t) - H_2(k,t),$$

with $H_1(k,t)$, $H_2(k,t)$ defined in (9). Hence

$$\hat{w}(k,t) = \mathrm{e}^{ik^3 t}\left[(\mathrm{e}^{-ik} - 1)G(k,t) + ikH_1(k,t) + H_2(k,t)\right].$$

Evaluating this expression at $k = k_n$ for $n \in \mathbb{Z}$, we obtain

$$\hat{w}(k_n, t) = e^{ik_n^3 t}\big[ikH_1(k_n, t) + H_2(k_n, t)\big].$$

Hence, using the Fourier series representation

$$w(x, t) = \sum_{n\in\mathbb{Z}} e^{ik_n x}\hat{w}(k_n, t), \tag{11}$$

we arrive at the representation (8) for $w(x, t)$.

Since the functions $h_1(t)$ and $h_2(t)$ are differentiable, the coefficients in the series (8) decay at least as $1/k_n^2$, which guarantees the continuity of $w(x, t)$ with respect to x. □

Remark 2 It is crucial in the argument above that the first boundary term in the Fourier transform of the PDE, namely the two terms $k^2 w(0, t)$ and $k^2 w(1, t)$, vanish. Indeed the presence of either of these terms would make the decay in k of the coefficients in the Fourier series (11) too slow to guarantee that the solution is continuous.

From all this we infer that the solution $u(x, t)$ of the original Dirichlet type problem has the formal representation

$$u(x, t) = v(x, t) + w(x, t) = \sum_{n\in\mathbb{Z}} e^{ik_n x + ik_n^3 t}\hat{f}(k_n)$$
$$+\sum_{n\in\mathbb{Z}} e^{ik_n x+ik_n^3 t}\int_0^t e^{-ik_n^3 s}(ik_n u_x(0, s) + u_{xx}(0, s) - u_{xx}(1, s))ds.$$

If we did not have the a priori knowledge that the function $u(x, t)$ exists and is unique, the above would be purely a formal expression, with nothing new to offer; it would not yield a way to represent $u(x, t)$ effectively. However, because we do have wellposedness of (1), expressing $u(x, t)$ in this way gives information on how its regularity properties depend on the regularity of initial and boundary conditions.

For the solution $v(x, t)$ of the purely periodic problem, the regularity depends only on the functional class of $f(x)$. It is less known that if $f(x)$ is only of bounded variation, but not continuous, the regularity of the solution remains in the same class at certain values of the time in a dense set of measure 0, but improves for almost all t. This is known in the context of the periodic problem as the phenomenon of *revivals*.

The second sum on the right hand side of the expression above for $u(x, t)$ conveys information on how the regularity is affected by the (homogeneous) boundary conditions. Lemma 1 implies that the second term is always continuous as a function of x. Therefore, u itself is a continuous perturbation of v.

3.2 Mixed BC of Dirichlet Type

If the boundary conditions for $u(x,t)$ include the Dirichlet-type condition $u(0,t) = u(1,t) = 0$, plus another condition possibly coupling the ends of the interval [0, 1], the analysis of the previous examples remain essentially unaltered. For example, if the third condition is $u_x(0,t) = \gamma u_x(1,t)$, for some $\gamma \in (0,1)$, the argument detailed above follows through with

$$H_1(k,t) = \int_0^t (\gamma - 1)u_x(1,s)e^{-ik^3 s}ds,$$

and the value of $H_2(k,t)$ given by the latter of Eq. (9). Therefore, the same conclusion can be drawn: regardless of the regularity (or lack of regularity) of $u(x,t)$ as a function of x, the function $w(x,t)$ is continuous in x, so u is a continuous perturbation of v.

Note that, unlike the previous example, in this case the spatial operator admits an L^2 basis of eigenfunctions, even though the eigenvalues cannot be determined explicitly other than as roots of a transcendental equation. Using the Fokas transform approach, the associated generalised Fourier series can be determined by a contour deformation technique [6, 11].

3.3 Coupled BC: Pseudo-periodic

We now turn to boundary conditions that couple the endpoints of the interval [0, 1], and assume that the given boundary conditions for $u(x,t)$ are the pseudo-periodic conditions

$$\beta_j \partial_x^j u(0,t) = \partial_x^j u(1,t), \quad j = 0, 1, 2, \qquad \beta_j \in \mathbb{C}. \tag{12}$$

Conditions need to be imposed on the β_j's to ensure the problem is well-posed, see [12]. We assume this to be the case.

The function $w(x,t)$ satisfies, along with a zero initial condition, the boundary conditions

$$\partial_x^j w(0,t) = \partial_x^j w(1,t) + h_j(t), \quad j = 0, 1, 2, \tag{13}$$

where, in this case,

$$h_j(t) = (1 - \beta_j)\partial_x^j u(0,t), \qquad j = 0, 1, 2.$$

A lemma entirely analogous to Lemma 1 yields for $w(x,t)$ the representation

$$w(x,t) = \sum_{n\in\mathbb{Z}} e^{ik_n x + ik_n^3 t}\left[-k_n^2 H_0(k_n,t) + ik_n H_1(k_n,t) + H_2(k_n,t)\right], \tag{14}$$

where $k_n = 2n\pi$ and

$$H_j(k,t) = \int_0^t \mathrm{e}^{-ik^3 s}(1-\beta_j)\partial_x^j u(0,s)ds.$$

As noted in Remark 2, the function $w(x,t)$ now has coefficients that can be guaranteed to decay only as n^{-1}, and therefore it may have lower regularity than the given initial datum $f(x)$, assumed Hölder continuous. However, if it so happens that $\beta_0 = 1$, then $H_0 = 0$, so the coefficients in Eq. (14) decay like n^{-2}, w is continuous and, as before, u is a continuous perturbation of v.

The results presented in the previous sections can be summarised and generalised as the following theorem.

Theorem 3 *Suppose problem* (2) *is wellposed and u is its solution. If the given linearly independent boundary conditions* (2c) *are either*

$$u(0,t) = 0, \qquad u(1,t) = 0, \qquad \textit{one other boundary condition on } u$$

or

$$u(0,t) = u(1,t), \qquad \textit{two other boundary conditions on } u,$$

then $u(x,t) = v(x,t) + w(x,t)$, for v the solution (5) *of periodic problem* (3) *and w a continuous function of x.*

3.4 An Outlier: Quasi-periodic BC

For the particular case that $\beta_0 = \beta_1 = \beta_2$ in example (12), known as the quasi-periodic case, one can pursue an alternative argument to give some interesting qualitative information about the function $u(x,t)$. This information is consistent with the fact that quasi-periodic problems for the Airy equation do not in general exhibit the phenomenon of weak revivals [2]. It is also consistent with the fact that the spectral structure of the quasi-periodic spatial operator can be easily derived by a shift on the structure of the periodic operator, and is well known.

Assume that the given boundary conditions for $u(x,t)$ are the quasi-periodic conditions

$$\partial_x^j u(0,t) = \mathrm{e}^{i\theta}\partial_x^j u(1,t), \quad j = 0,1,2, \quad \theta \in \mathbb{R}. \tag{15}$$

The function $w(x,t)$ satisfies, along with zero initial conditions, the boundary conditions

$$\partial_x^j w(0,t) = \mathrm{e}^{i\theta}\partial_x^j w(1,t) + (1-\mathrm{e}^{-i\theta})\partial_x^j v(1,t), \quad t > 0, \tag{16}$$

where $v(x,t)$ is the solution of the purely periodic problem (3). It is still true that w obeys boundary conditions (13) with $\beta_0 = \beta_1 = \beta_2 = e^{-i\theta}$, but we shall make use of the alternative characterisation (16) in the following argument.

Lemma 4 *The function $w(x,t)$ solution of (7a), (16) admits the representation*

$$w(x,t) = \sum_{n\in\mathbb{Z}} e^{i(k_n-\theta)x+i(k_n-\theta)^3 t} \int_0^t (1-e^{i\theta})V(k_n-\theta,s)e^{-i(k_n-\theta)^3 s}ds, \tag{17}$$

with $k_n = 2n\pi$ and

$$V(k,t) = k^2 v(0,t) - ikv_x(0,t) - v_{xx}(0,t). \tag{18}$$

Proof Consider the Fourier transform of $w(x,t)$, which satisfies as before the following ODE:

$$\left(e^{-ik^3 t}\hat{w}(k,t)\right)_t = -F(k,t) + e^{-ik}G(k,t),$$

with $F(k,t)$ and $G(k,t)$ given by Eq. (10). Then, since $\hat{w}(k,0) = 0$, the solution of the ODE is given by

$$\hat{w}(k,t) = e^{ik^3 t}\left[-F(k,t) + e^{-ik}G(k,t)\right].$$

Using the boundary conditions, we find

$$e^{-i\theta}F(k,t) = G(k,t) + (1-e^{-i\theta})\int_0^t e^{-ik^3 s}V(k,s)ds,$$

with $V(t,k)$ defined in (18). Hence

$$\hat{w}(k,t) = e^{ik^3 t}\left[(e^{-i(k+\theta)}-1)F(k,t) + (e^{-i(k+\theta)} - e^{-ik})\int_0^t e^{-ik^3 s}V(k,s)ds\right].$$

Evaluating this expression at $k = k_n - \theta$ for $n \in \mathbb{Z}$, we obtain

$$\hat{w}(k_n-\theta,t) = e^{i(k_n-\theta)^3 t}(1-e^{i\theta})\int_0^t e^{-i(k_n-\theta)^3 s}V(k_n-\theta,s)ds.$$

We now invert this to obtain the generalised Fourier series expression

$$w(x,t) = \sum_{n\in\mathbb{Z}} e^{i(k_n-\theta)x}\hat{w}(k_n,t), \tag{19}$$

which is the representation (17) for $w(x,t)$. □

From all this, we infer that the solution $u(x,t)$ of the original quasi-periodic problem has the formal representation

$$
\begin{aligned}
u(x,t) =& v(x,t) + w(x,t) = \sum_{n\in\mathbb{Z}} \mathrm{e}^{ik_n x + ik_n^3 t}\hat{f}(k_n) \\
&+ \sum_{n\in\mathbb{Z}} \mathrm{e}^{i(k_n-\theta)x}\mathrm{e}^{i(k_n-\theta)^3 t}\int_0^t (1-\mathrm{e}^{i\theta})\mathrm{e}^{-i(k_n-\theta)^3 s} V(k_n-\theta, s)ds. \quad (20)
\end{aligned}
$$

The function $V(k,t)$ is made up of the boundary values of the x-periodic function $v(x,t)$. Knowledge of this function as well as the characterisation of the eigenvalues of the spatial operator, is enough to represent $u(x,t)$ effectively, and $V(k,t)$ can easily be calculated from representation (5).

Remark 5 Note that the presence of the term $k_n^2 v(0,t)$ in the definition of the generalised Fourier coefficient $\widehat{w}(k,t)$ implies that the convergence of the series for $w(x,t)$ is slow and not uniform; unless $v(0,t)=0$, this term implies a no better regularity of w than that of the solution of the purely periodic problem.

Note also that the second term on the right of representation (20) contains the exponential $\mathrm{e}^{i(k_n-\theta)x+i(k_n-\theta)^3 t}$ which is both space- and time-periodic with a period congruent to $\theta+\mathbb{Q}$, while $v(x,t)$ is periodic with period in $\mathbb{Q}$. Therefore if $\theta\notin\mathbb{Q}$, the function $u(x,t)$ cannot have any periodicity property. This confirms the result of [2], namely the fact that this quasi-periodic problem, surprisingly, does not exhibit revivals if $\theta\notin\mathbb{Q}$.

Remark 6 Suppose the boundary conditions for u are

$$
u(0,t) = \mathrm{e}^{i\theta}u(1,t), \qquad \text{two other boundary conditions on } u, \qquad (21)
$$

with $\theta\in\mathbb{R}$ but, to avoid the regime already covered by Theorem 3, suppose θ is not an even integer multiple of π. Suppose this problem for u is wellposed. Note that this includes certain pseudoperiodic problems (12), but not all wellposed such problems.

We can make the decomposition $u(x,t) = v(x,t) + w(x,t)$ with v satisfying the quasiperiodic problem

$$
\begin{aligned}
&v_t + v_{xxx} = 0, \quad x\in(0,1),\ t>0; \qquad v(x,0) = f(x), \quad x\in(0,1); \\
&\partial_x^j v(0,t) = \mathrm{e}^{i\theta}\partial_x^j v(1,t), \quad j=0,1,2, \quad t>0,
\end{aligned}
$$

and w satisfying the boundary forced quasiperiodic problem

$$
\begin{aligned}
&w_t + w_{xxx} = 0, \quad x\in(0,1),\ t>0; \qquad w(x,0) = 0, \quad x\in(0,1); \\
&\partial_x^j w(0,t) = \mathrm{e}^{i\theta}\partial_x^j w(1,t) + h_j(t), \quad j=0,1,2, \quad t>0,
\end{aligned}
$$

in which

$$
h_0(t)=0, \qquad h_1(t) = u_x(0,t) - \mathrm{e}^{\mathrm{i}\theta}u_x(1,t), \qquad h_2(t) = u_{xx}(0,t) - \mathrm{e}^{\mathrm{i}\theta}u_{xx}(1,t).
$$

Then, as discussed above, the (non)existence of revivals for v is determined by the (ir)rationality of θ and, using an argument exactly paralleling the proof of Lemma 1, w

is continuous. Therefore, u, being a continuous perturbation of v, exhibits continuous perturbations of revivals if and only if $\theta \in \mathbb{Q}$. This is the analogue of Theorem 3 for boundary conditions (21).

Conclusion We have embedded the solution of the periodic problem in the solution of certain classes of homogeneous boundary value problems to determine how the boundary conditions perturb the qualitative properties of the periodic solution.

For homogeneous Dirichlet type separated boundary conditions, and for boundary conditions that describe continuous extension from $[0, 1]$ to $(-\infty, \infty)$, we found that the remaining one or two boundary conditions add a component that superimposes a continuous function of x onto the periodic solution, irrespective of the overall x regularity of the full solution.

On the other hand, in the case of some particular quasi-periodic problems, which in the case of second order problems can always be recast in terms of periodic boundary conditions, this approach confirms that while the solution depends only on the boundary values of the periodic solutions, the boundary conditions not only add a less regular component to the purely periodic solution, but also that the interaction between the periodic and the non-periodic part of the solution can silence completely the echo of periodicity.

These remarks have particularly significant consequences in case of low-regularity initial data and the phenomenon of weak revivals. This is explored further in [3].

Acknowledgements The authors thank the Isaac Newton Institute for Mathematical Sciences for support and hospitality during programme *Dispersive hydrodynamics: mathematics, simulation and experiments, with applications in nonlinear waves*, when work on this paper was undertaken. This work was supported by EPSRC Grant Number EP/R014604/1.

References

1. Berry, M.V.: Quantum fractals in boxes. J. Phys. Math. Gen. **29**(20), 6617 (1996)
2. Boulton, L., Farmakis, G., Pelloni, B.: Beyond periodic revivals for linear dispersive PDEs. Proc. R. Soc. A. **477**(2251), 20210241 (2021)
3. Boulton, L., Farmakis, G., Pelloni, B., Smith, D.A.: Weak revivals for linear time-evolution equations, in preparation (2022)
4. Erdoğan, M.B., Tzirakis, N.: Dispersive Partial Differential Equations: Wellposedness and Applications, vol. 86. Cambridge University Press (2016)
5. AS Fokas and Beatrice Pelloni: A transform method for linear evolution PDEs on a finite interval. IMA J. Appl. Math. **70**(4), 564–587 (2005)
6. Fokas, A.S., Smith, D.A.: Evolution PDEs and augmented eigenfunctions. Adv. Differ. Equ.**21**(7/8), 735–766 (2016)
7. Fokas, A.S.: A unified transform method for solving linear and certain nonlinear PDEs. Proc. R. Soc. Lond. Ser. Math. Phys. Eng. Sci. **453**, 1411–1443 (1997)
8. Jackson, D.: Expansion problems with irregular boundary conditions. Proc. Am. Acad. Arts Sci. **51**
9. Olver, P.J.: Dispersive quantization. Am. Math. Mon. **117**(7), 599–610 (2010)
10. Olver, P.J., Sheils, N.E., Smith, D.A.: Revivals and fractalisation in the linear free space Schrödinger equation. Q. Appl. Math. **78**(2), 161–192 (2020)

11. Pelloni, Beatrice: The spectral representation of two-point boundary-value problems for third-order linear evolution partial differential equations. Proc. R. Soc. Math. Phys. Eng. Sci. **461**(2061), 2965–2984 (2005)
12. Smith, D.A.: Well-posed two-point initial-boundary value problems with arbitrary boundary conditions. **152**, 473–496 (2012)
13. Smith, D.A.: Revivals and fractalization. Dyn. Syst. Web. 1–8 (2020)
14. Talbot, H.F.: Facts related to optical science. No. IV. Philos. Mag. **9** (1836), 401–407
15. Trogdon, Thomas, Biondini, Gino: Evolution partial differential equations with discontinuous data. Q. Appl. Math. **77**(4), 689–726 (2019)

The Fokas Method for the Well-posedness of Nonlinear Dispersive Equations in Domains with a Boundary

Dionyssios Mantzavinos

Dedicated to Professor Athanassios S. Fokas

Abstract The Fokas method, also known as the unified transform, is one of the most remarkable breakthroughs noted in the study of linear and integrable nonlinear partial differential equations at the turn of the new millennium. Its numerous implications, along with the elegance of the ideas forming its foundation, led the great Israel Gelfand to once describe it as one of the most exciting developments in the area of partial differential equations since the time of Fourier. In this article, we offer further evidence in support of that statement by elucidating the analogy between the Fokas method and the celebrated Fourier transform in the context of both linear and nonlinear dispersive equations. First, we review the Fokas-Gelfand derivation of the Fourier transform pair via the technique of inverse scattering but applied to linear (as opposed to nonlinear) equations—an idea that subsequently contributed to the development of the linear component of the Fokas method. Then, we discuss a novel approach for proving the well-posedness of initial-boundary value problems for general nonlinear (i.e. not necessarily integrable) dispersive equations. This approach utilizes the Fokas method in analogy with the way that the Fourier transform is used in the classical harmonic analysis-based approach for proving the well-posedness of the initial value problem for these nonlinear equations. In this regard, the new approach further establishes the Fokas method as the direct analogue of the Fourier transform in domains with a boundary.

Keywords Fokas method · Unified transform · Dispersive equations · NLS · KdV · Well-posedness · Initial-boundary value problems · Fourier transform · Inverse scattering transform

D. Mantzavinos (✉)
Department of Mathematics, University of Kansas, Lawrence, KS 66045, USA
e-mail: mantzavinos@ku.edu

T. Bountis et al. (eds.), *Chaos, Fractals and Complexity*, Springer Proceedings in Complexity, https://doi.org/10.1007/978-3-031-37404-3_24

1 Introduction

Dispersive partial differential equations describe phenomena in which waves of different wavelengths propagate at different speeds. Two prominent examples are the Korteweg-de Vries (KdV) equation[1]

$$u_t + u_{xxx} + uu_x = 0 \tag{1}$$

and the (cubic) nonlinear Schrödinger (NLS) equation

$$iu_t + u_{xx} \pm |u|^2 u = 0. \tag{2}$$

In the above equations, $u = u(x, t)$ is a function of space x and time t, with the various indices denoting partial derivatives with respect to the relevant variable. Moreover, in Eq. (2), the positive sign in front of the nonlinearity corresponds to the focusing NLS and the negative sign to the defocusing NLS. Both KdV and NLS are nonlinear evolution equations that arise as approximations, under certain regimes, of the fundamental Euler equations for incompressible and inviscid flow. Furthermore, NLS has a ubiquitous presence in mathematical physics, being a central model in such diverse areas as optics, plasmas, and Bose-Einstein condensates.

When considered on the infinite line $-\infty < x < \infty$, Eqs. (1) and (2) must be supplemented with an initial condition of the form

$$u(x, 0) = u_0(x) \tag{3}$$

for some given function $u_0(x)$. This is known as the initial value problem (IVP) or Cauchy problem. One can then ask whether such an IVP can be solved and, if so, how and in what sense. In particular, one can also ask whether the choice of initial data $u_0(x)$ affects the solvability of the IVP and the various properties of its solution.

A key element in regard to the above questions is the fact that both KdV and NLS have a remarkably rich structure as completely integrable systems. For KdV, this feature was discovered by Gardner et al. [18], and for NLS it was established by Zakharov and Shabat [43], while the seminal 1968 work of Lax [36] provided a solid theoretical framework for studying completely integrable equations with the help of what are nowadays known as Lax pairs. These are systems of linear equations that allow integrable nonlinear equations to be "linearized" by means of expressing them as compatibility conditions of these linear systems. For example, a Lax pair for the KdV Eq. (1) is given by the 2 × 2 linear system

$$\begin{aligned} &\mu_{xx} + \left(\tfrac{1}{6}u - k\right)\mu = 0, \\ &\mu_t + \left(\tfrac{1}{3}u + 4k\right)\mu_x - \tfrac{1}{6}u_x\mu = 0, \end{aligned} \qquad \mu = \mu(x, t, k), \ k \in \mathbb{C}, \tag{4}$$

[1] Although KdV also contains the linear term u_x, for our purposes it suffices to consider the simplified form (1).

Fig. 1 Outline of the inverse scattering transform method

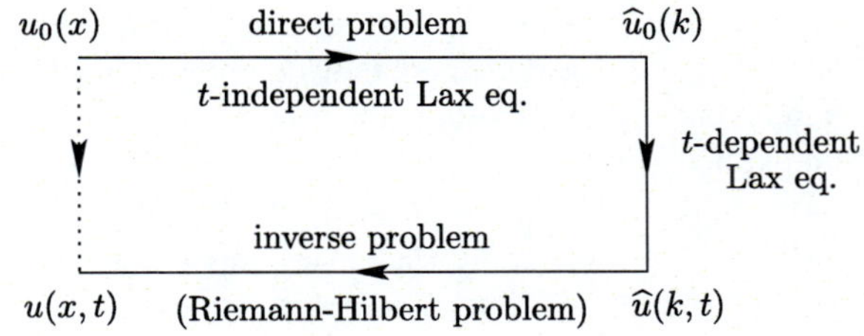

since KdV follows from that system under the simple symmetry condition $\mu_{xxt} = \mu_{txx}$ (which for continuous mixed derivatives follows by the Clairaut/Schwarz theorem). In [18], the authors studied the IVP (1), (3) for KdV on the infinite line by introducing the inverse scattering transform method. For the NLS equation and the IVP (2), (3), the corresponding formalism was developed in [43]. The method consists of three main steps that can be outlined as follows (see also the diagram of Fig. 1):

- the spectral analysis of the t-independent component of the Lax pair (first equation in (4)), thus mapping the initial data $u_0(x)$ to spectral data $\widehat{u}_0(k)$;
- the evolution of the spectral data $\widehat{u}_0(k)$ via the t-dependent component of the Lax pair (second equation in (4));
- the inversion of the resulting time-dependent spectral function $\widehat{u}(k,t)$ from the spectral kt-space to the physical xt-space, in order to recover the solution $u(x,t)$ of the IVP. This step typically involves the formulation and analysis of a Riemann-Hilbert problem.

The above procedure is conceptually identical to the well-known Fourier transform method for solving the IVP of linear evolution equations. In this sense, the inverse scattering transform can be regarded as a nonlinear analogue of the Fourier transform.

When available, the inverse scattering transform is a truly powerful method. Nevertheless, from the broader perspective of the analysis of nonlinear dispersive equations (and, more generally, nonlinear evolution equations), the method has some important limitations in its applicability. First and foremost, it can only be employed for integrable equations.[2] In addition, even then it comes with certain restrictions on the smoothness and decay at infinity of the initial data, e.g. on the infinite line these must belong in the class of "rapidly decaying" functions satisfying $\int_{-\infty}^{\infty} (1+|x|)\,|u_0(x)|\,dx < \infty$. These limitations rule out the vast majority of nonlinear evolution equations and, importantly, any such equation in space dimension three or higher. Moreover, even when studying integrable equations like KdV and NLS, conditions like the one above exclude large and significant classes of initial data. Indeed, as noted on page 257 of [33], the inverse scattering transform machinery seems to break down "even under very mild relaxations" of the "rapidly decaying" condition (see also [7]).

[2] There does not exist a universally accepted definition of complete integrability. Here, we identify an integrable equation by its ability to be "linearized" via a Lax pair.

Although the above limitations cannot be overcome in the context of the inverse scattering transform method, they do not pose a problem if one changes perspective and revisits the IVPs with a different goal, i.e. without the ambition of constructing an explicit solution map like the one produced via inverse scattering. In fact, the most fundamental question for the KdV and NLS IVPs is that of *well-posedness*. Originally formulated by Hadamard, this notion refers to the existence and uniqueness of solution of a given equation, as well as to the continuous dependence of that solution on the data. In the absence of well-posedness, the analysis of a model becomes pointless, regardless of the other features that this may have. For example, the "bad" Boussinesq equation, the first equation for which a soliton solution was written down, is not particularly useful otherwise since it is ill-posed. Through the years, various techniques have been developed for proving the well-posedness of IVPs that involve evolution equations. In the case of dispersive equations, a very effective such technique combines the powerful tools of harmonic analysis and the Fourier transform with the contraction mapping theorem for studying these equations in suitable Banach spaces. We hereafter refer to this technique as the Fourier transform approach.

It is widely known that well-posedness is affected by a number of factors, including the nature of the equation and the regularity and decay of the data. However, it is often less emphasized that it is also affected by the nature of the associated physical domain. In the case of a fully unbounded domain like the infinite line, one has an IVP; on the other hand, when the spatial domain involves a boundary (e.g. in one dimension, the half-line $0 < x < \infty$ or the finite interval $[0, 1]$), one instead has an initial-boundary value problem (IBVP). For any given equation, these two types of problems are generally very different, and this is also reflected in the analysis of their well-posedness. In fact, the well-posedness of nonlinear dispersive equations in the context of IBVPs is much less studied (and understood) than their IVP well-posedness.

Through a systematic effort that began in 2012, Alex Himonas and the author introduced a new approach for the well-posedness of IBVPs for nonlinear dispersive equations which takes advantage of the Fokas method in analogy to the way that the classical Fourier transform approach utilizes the Fourier transform. In that sense, the Fokas method can be regarded as the natural analogue of the Fourier transform in the IBVP setting. This novel well-posedness approach is reviewed in Sect. 3. Of course, the connection of the Fokas method with the Fourier transform dates back a lot further—specifically, its origins can be traced back to the 1994 paper of Fokas and Gelfand [14], where the Fourier transform pair is rediscovered through an inverse scattering analysis of the linear Schrödinger equation on the infinite line. This understanding later contributed to the realization that the Fokas method has significant implications at the level of IBVPs for *linear* equations, despite the fact that it had originally been motivated through the study of IBVPs for integrable *nonlinear* equations. For that reason, and due to the fact that the linear component of the Fokas method plays a fundamental role in the new well-posedness approach discussed in Sect. 3, in Sect. 2 we review the derivation of the Fourier transform pair in the style of [14], using some of the ideas that later led to the integrable nonlinear component of the Fokas method.

2 Inverse Scattering for Linear Equations: Rediscovering the Fourier Transform

The motivation behind the discovery of Lax pairs and the introduction and subsequent development of the inverse scattering transform method had to do with the study of integrable *nonlinear* equations; *linear* equations were not part of that motivation, since their IVP could be easily solved via the Fourier transform. Nevertheless, Fokas and Gelfand [14] came to the realization that every *linear* evolution equation can also be expressed as the compatibility condition of a Lax pair. Let us, for example, consider the Airy equation

$$u_t + u_{xxx} = 0, \tag{5}$$

which corresponds to the linear part of the KdV Eq. (1). With the help of the formal adjoint equation $-\widetilde{u}_t - \widetilde{u}_{xxx} = 0$, which is obtained by replacing ∂^j with $(-1)^j \partial^j$ in the x and t partial derivatives, we can write (5) in the divergence form $(e^{-ikx-ik^3t}u)_t + (e^{-ikx-ik^3t}[u_{xx} + iku_x - k^2u])_x = 0$, $k \in \mathbb{C}$. Seeking $M = M(x, t, k)$ such that $M_x = e^{-ikx-ik^3t}u$ and $M_t = -e^{-ikx-ik^3t}(u_{xx} + iku_x - k^2u)$, we see that the above divergence form (which is equivalent to (5)) is nothing but the symmetry requirement $M_{xt} = M_{tx}$. That is, the Airy Eq. (5) is the compatibility condition of the linear system for M, which is therefore a Lax pair for that equation. In fact, the exponential term can be absorbed by letting $M(x, t, k) = e^{-ikx-ik^3t}\mu(x, t, k)$, giving rise to the Lax pair

$$\mu_x - ik\mu = u, \quad \mu_t - ik^3\mu = -\left(u_{xx} + iku_x - k^2u\right). \tag{6}$$

Following the above realization, Fokas and Gelfand applied the inverse scattering transform formalism to Lax pairs like (6) in order to solve the IVP of linear evolution equations analogously to their integrable nonlinear counterparts, i.e. as if Fourier transform were not known/available. This direction was especially motivated by a long-standing open problem, namely the advancement of the inverse scattering transform method from the IVP to the IBVP setting, e.g. for solving the KdV equation on the half-line with nonzero Dirichlet data. Indeed, as noted on page 1 of [13], when this problem was first suggested to Ablowitz and Fokas by Julian Cole in 1982, they first attempted to solve the corresponding linear problem, namely the Airy Eq. (5) on the half-line, by using an appropriate *spatial* transform. The reason for first seeking a spatial transform for the linear IBVP had to do with the observation that, in the case of the IVP, in the linear limit the inverse scattering transform reduces to the Fourier transform [2]. Thus, knowledge of the relevant spatial transform in the case of the linear IBVP could provide the basis for developing the analogue of the inverse scattering transform method for integrable nonlinear IBVPs. To their surprise, Fokas and Ablowitz could not find an appropriate spatial transform for solving the linear Airy equation on the half-line; in fact, such a transform does *not* exist for any linear evolution of spatial order higher than two[3] [13]. Taking into account that even the "simple" task of solving *linear* IBVPs via spatial transforms was an open problem, it becomes evident that any progress made in the study of linear equations via inverse scattering ideas, like the one pursued in [14] as mentioned above, could have far-reaching implications also for integrable nonlinear equations.

Let us now follow the approach of [14] in order to integrate the Lax pair (6) and hence solve the IVP for the Airy equation on the infinite line.[4] As usual in the inverse scattering transform method, we work under the assumption of existence of solution and, in particular, we assume sufficient smoothness and decay at infinity as necessary. As noted earlier (see diagram of Fig. 1), there are three main steps: the direct problem, the inverse problem, and the time evolution of the spectral data.

Direct problem. Treating t, k as parameters—and thus suppressing them from the arguments of μ, u—we integrate the t-independent part of the Lax pair (6) to obtain the following expressions for the particular solutions $\mu^{\pm}$ that correspond to zero "boundary" conditions at $\pm\infty$, i.e. $\lim_{x\to\pm\infty} \mu^{\pm}(x) = 0$:

$$\mu^{+}(x) = \int_{-\infty}^{x} e^{ik(x-y)} u(y)dy, \quad \mu^{-}(x) = -\int_{x}^{\infty} e^{ik(x-y)} u(y)dy. \tag{7}$$

Inverse problem. Changing our perspective, we use the expressions (7) in order to define μ as a piecewise function of k (this time, we suppress the dependence on x, t) by $\mu(k) = \mu^{+}(k)$ for $\mathrm{Im}(k) > 0$ and $\mu(k) = \mu^{-}(k)$ for $\mathrm{Im}(k) < 0$. Then, introducing the *notation*

$$\widehat{u}(k) := \int_{-\infty}^{\infty} e^{-iky}\, u(y)dy, \quad k \in \mathbb{R}, \tag{8}$$

we observe that $\mu(k)$ satisfies the following scalar *Riemann-Hilbert problem*:

- $\mu(k)$ is analytic in $\mathbb{C} \setminus \mathbb{R}$ (by the form of (7) and a Paley-Wiener theorem like Theorem 7.2.4 in [40]);
- along $\mathbb{R}$, $\mu(k)$ satisfies the jump condition $\mu^{+}(k) - \mu^{-}(k) = e^{ikx}\,\widehat{u}(k)$, $k \in \mathbb{R}$;
- integration by parts in (7) implies $\mu(k) = O(1/k)$ as $|k| \to \infty$.

The solution of this scalar Riemann-Hilbert problem is readily obtained via the Plemelj formulae (Lemma 7.2.1 in [1]) as

[3] Although a temporal Laplace transform is available, it comes with certain disadvantages, most notably its inability to generalize to the integrable nonlinear equations.

[4] In [14], the authors illustrated their approach via the linear Schrödinger equation; the analysis is essentially the same in both cases.

$$\mu(x,k) = \frac{1}{2i\pi}\int_{-\infty}^{\infty} \frac{e^{i\lambda x}\,\widehat{u}(\lambda)}{\lambda - k} d\lambda, \quad k \notin \mathbb{R}. \tag{9}$$

Inserting this expression into the t-independent part of the Lax pair (6) and taking $|k| \to \infty$ yields the following representation for $u(x)$ in terms of the notation $\widehat{u}(k)$ introduced by (8):

$$u(x) = \frac{1}{2\pi}\int_{-\infty}^{\infty} e^{ikx}\,\widehat{u}(k)dk, \quad x \in \mathbb{R}. \tag{10}$$

Time evolution. Observe that the expressions (7) satisfy $\lim_{x\to\pm\infty}(e^{-ikx}\mu^{\pm}) = \pm\widehat{u}$. Therefore, restoring the time variable t and taking the two limits $x \to \pm\infty$ of the t-dependent part of the Lax pair (6) while assuming that $u, u_x, u_{xx} \to 0$ in those limits, we obtain the equation $\widehat{u}_t - ik^3\,\widehat{u} = 0$. In view of the initial condition (3) and the notation (8), this equation implies $\widehat{u}(k,t) = e^{ik^3t}\,\widehat{u}_0(k)$, which can be combined with the representation (10) to yield the solution to the IVP (6), (3) in the explicit form

$$u(x,t) = \frac{1}{2\pi}\int_{-\infty}^{\infty} e^{ikx+ik^3t}\,\widehat{u}_0(k)dk, \quad x \in \mathbb{R},\ t \geq 0. \tag{11}$$

What is truly remarkable is the fact that the spectral transform (8), which arises *spontaneously* in the above analysis, is nothing but the celebrated *Fourier transform*! Furthermore, the solution of the relevant inverse (Riemann-Hilbert) problem readily yields the inversion of this spontaneously emerging transform, namely the *inverse Fourier transform* (10)! That is, in addition to providing the explicit solution formula to the Airy equation IVP (6), (3) (which is, of course, the well-known Fourier transform solution of this problem), the analysis of [14] leads to the *rediscovery* of the Fourier transform itself, and also to an elegant proof of its inversion!

In this regard, as noted at the beginning of this section, the contribution of [14] was of crucial importance because it suggested that devising a method for the spectral analysis of the Lax pairs of *linear* equations in the IBVP setting could provide the correct way of generalizing the inverse scattering transform method from the IVP to the IBVP setting. Soon after, this turned out to be indeed the case with the introduction and subsequent development of the Fokas method for *both* linear and integrable nonlinear equations in the IBVP setting.

3 A Novel Approach for the Well-posedness of IBVPs

The Fokas method, also known as the unified transform, was introduced by Fokas in 1997 [12] and subsequently developed by him and numerous collaborators (see [13, 17] and the references therein). The method has groundbreaking implications not only for integrable nonlinear equations, but also for linear equations. In the nonlinear case, it provides the extension of the inverse scattering transform method to the IBVP setting. In the linear case, it produces novel solution formulae for IBVPs formulated in various physical domains, with different types of nonzero boundary conditions, and in any number of spatial dimensions; as such, *the linear component of the Fokas method is the direct analogue of the Fourier transform in the IBVP setting*.

As noted in the introduction, a fundamental question for any given nonlinear dispersive equation is the one of (Hadamard) well-posedness, i.e. existence, uniqueness, and continuous dependence of the data-to-solution map. Although this topic has been studied extensively in the direction of the IVP (see, for example, the books [6, 8, 34, 37, 42] and the vast number of references therein), until recently it had remained largely unexplored in the case of IBVPs (essentially, the works [3, 9–11, 29, 30, 38]), despite the fact that this latter class of problems is very significant with regard to applications.

The main reason for this disproportion is the absence of the Fourier transform from the IBVP setting. Indeed, in the case of dispersive equations, the proof of well-posedness for the IVP relies heavily on the rich and powerful collection of harmonic analysis techniques that surround the Fourier transform. Importantly, the solution via the Fourier transform of the associated forced linear IBVP provides the starting point for defining the iteration map used for proving existence and uniqueness of solution via a fixed point argument (contraction mapping approach). Hence, in the case of IBVPs, without even a way of solving the linearized equations (recall discussion in Sect. 2), it is not surprising that very little progress had been made towards a *general approach* for establishing well-posedness of these problems in the case of (dispersive) nonlinear equations.

A systematic effort towards this goal began in 2012, when the author arrived at the University of Notre Dame to work under the mentorship of Professor Alex Himonas. The main idea had been proposed to Himonas by Fokas a few years earlier, in 2008, and consisted in employing the explicit solution formulae produced by the Fokas method in the case of (forced) linear IBVPs in order to set up the iterations for proving the well-posedness of the corresponding nonlinear problems via contraction mapping. The main source of optimism in regard to this suggestion was that, as mentioned earlier, for linear equations, the Fokas method is the analogue of the Fourier transform in the IBVP setting. Hence, it seemed reasonable to expect that the Fokas solution formulae could fulfill the role of generating iteration maps for nonlinear IBVPs in the same way that the Fourier transform formulae do in the case of nonlinear IVPs.

Regardless of how natural this idea may at first seem, however, when attempting to implement it one is quickly met with important challenges. For example, one must figure out how to obtain estimates in those function spaces that are natural to dispersive equations—such as Sobolev spaces or Bourgain spaces, which are typically studied (and even defined) with the help of the Fourier transform—when the Fokas solution formulae involve integrals along *complex* contours of the spectral k-plane (as opposed to the Fourier transform (8), which is defined only for $k \in \mathbb{R}$).

Another challenge has to do with the correct function space for the boundary data. For example, in the case of the IVP (2), (3) for the NLS equation, the initial datum $u_0(x)$ is typically placed in Sobolev spaces H^s and the solution is obtained in the associated Hadamard-type spaces $C_t H_x^s$ (at least for smooth enough data, i.e.

high enough s). However, on the half-line, one must *additionally* prescribe data at the boundary $x = 0$, e.g. via the Dirichlet boundary condition $u(0, t) = g_0(t)$ and so one must determine a suitable function space also for $g_0(t)$. Whether or not this space depends on the space H^s for $u_0(x)$ and if so, the precise relationship between the two spaces, is a question that adds to the complexity of IBVPs when compared to the IVP.[5]

A combination of ideas inspired by aspects of the Fokas method, together with suitably adapted results from the classical harmonic analysis toolbox used for the IVP, made it possible to pursue Fokas's suggestion and introduce an approach for establishing the well-posedness of IBVPs for nonlinear dispersive equations in a way conceptually analogous to the Fourier transform approach used for the IVP. This new approach has been employed for various problems involving the NLS, KdV, "good" Boussinesq, and biharmonic Schrödinger equations [15, 16, 20, 22, 24, 25, 35, 39], while it has also proved effective outside the dispersive class, for a nonlinear reaction-diffusion model [26]. In the new approach, the key to overcoming the challenges described above was the study of what we refer to as the *pure linear IBVP*. This problem consists of the homogeneous linearized version of the equation under study, supplemented with zero initial data and *nonzero but compactly supported* boundary data. The pure linear IBVP can be thought of as the simplest *genuine* IBVP, since it incorporates the challenges of an IBVP without the "distractions" caused by the initial data and the nonlinearity/forcing.

In the case of the Dirichlet half-line problem for the NLS Eq. (2), the pure linear IBVP is given by

$$\begin{aligned} &iu_t + u_{xx} = 0, \quad 0 < x < \infty, \ 0 < t < T, \\ &u(x, 0) = 0, \quad u(0, t) = g(t), \quad \text{supp}(g) \subset (0, T), \end{aligned} \tag{12}$$

where $T > 0$ is fixed (since we are interested in local well-posedness). Using the Fokas method, the solution of problem (12) is found to be

$$u(x, t) = \frac{1}{\pi} \int_C e^{ikx - ik^2 t} k \widehat{g}(-k^2) dk, \tag{13}$$

where $\widehat{g}(-k^2)$ is the Fourier transform (8) of $g(t)$ evaluated at $-k^2$ and the complex contour C is the positively oriented boundary of the first quadrant of the complex k-plane. Below, we illustrate how the Fokas formula (13) can be used in order to estimate the solution of (12) for each $t \in [0, T]$ as a function in the Sobolev space $H^s(0, \infty)$, $s \geq 0$, on the half-line. Note that this space can be defined either as a restriction of the infinite-line space $H^s(\mathbb{R})$ or, directly, via the norm equal to the sum of the $L^2(0, \infty)$-norms of the derivatives up to order s (using the Slobodeckij

[5] In some cases, there exist results on the time regularity of the IVP solution that can provide helpful insights about the regularity of the boundary data [32]. In general, however, such results may not be available.

seminorm if s is fractional). We shall only provide the details for the case $s = 0$, which corresponds to $L^2(0, \infty)$; the full estimation can be found in [15].

The contour C comprises the positive halves of the real and imaginary axes. Denoting the respective parts of the solution by u_{re} and u_{im}, we have $u = u_{\text{re}} + u_{\text{im}}$ with

$$u_{\text{re}}(x,t) = \frac{1}{\pi}\int_0^\infty e^{ikx} \cdot e^{-ik^2t}\, k\widehat{g}(-k^2)dk, \quad u_{\text{im}}(x,t) = \frac{1}{\pi}\int_0^\infty e^{-kx} \cdot e^{ik^2t}\, k\widehat{g}(k^2)dk.$$

Since the expression for u_{re} also makes sense for $x < 0$, it can be regarded as a function on the infinite line. Thus, by the Plancherel theorem,

$$\sup_{t\in[0,T]} \|u_{\text{re}}(t)\|_{L_x^2(0,\infty)} \lesssim \left\|e^{-ik^2t}\, k\widehat{g}(-k^2)\right\|_{L_k^2(0,\infty)} \simeq \|g\|_{H_t^{\frac{1}{4}}(\mathbb{R})}. \tag{14}$$

On the other hand, the expression for u_{im} does not make sense for $x < 0$, thus a different idea is needed. In particular, observe that, up to a constant, the $L_x^2(0, \infty)$-norm of u_{im} is just the $L_x^2(0, \infty)$-norm of the Laplace transform with respect to k of the quantity $e^{ik^2t}\, k\widehat{g}(k^2)$. Hence, by the boundedness of the Laplace transform in $L^2(0, \infty)$ [19],

$$\sup_{t\in[0,T]} \|u_{\text{im}}(t)\|_{L_x^2(0,\infty)} \lesssim \left\|e^{ik^2t}\, k\widehat{g}(k^2)\right\|_{L_k^2(0,\infty)} \simeq \|g\|_{H_t^{\frac{1}{4}}(\mathbb{R})}. \tag{15}$$

Together, estimates (14) and (15) imply that if the boundary datum of the pure linear IBVP (12) belongs to $H_t^{1/4}$ then the solution of this problem belongs to $C_t L_x^2(0, \infty)$. Furthermore, through the generalizations of these estimates for $s \geq 0$, the Sobolev space $H_t^{(2s+1)/4}$ spontaneously emerges as the correct space for the Dirichlet boundary datum $g_0(t)$. This fact is corroborated via a separate analysis of the time regularity of the homogeneous and forced linear Schrödinger IVPs, which actually shows that the above choice of space for the boundary datum is sharp. Eventually, via a contraction mapping argument, the various linear estimates derived with the help of the Fokas method solution formula (12) imply the Hadamard well-posedness of the Dirichlet problem for NLS on the half-line. More precisely:

Theorem *([15]). Suppose $1/2 < s \leqslant 3/2$. Then, the IBVP for the cubic NLS Eq. (2) on the half-line with initial data $u_0 \in H^s(0, \infty)$ and Dirichlet boundary data $g_0 \in H^{(2s+1)/4}(0, T)$ is well-posed in the sense of Hadamard. In particular, there exists a unique solution $u \in C([0, T^*]; H^s(0, \infty))$, which satisfies*

$$\sup_{t\in[0,T^*]} \|u(t)\|_{H^s(0,\infty)} \leqslant c_s\Big(\|u_0\|_{H^s(0,\infty)} + \|g_0\|_{H^{\frac{2s+1}{4}}(0,T)}\Big)$$

with $c_s = c(s) > 0$ and $0 < T^* \leqslant \min\Big\{T, c_s\Big(\|u_0\|_{H^s(0,\infty)} + \|g_0\|_{H^{\frac{2s+1}{4}}(0,T)}\Big)^{-4}\Big\}$, and the data-to-solution map $\{u_0, g_0\} \mapsto u$ is locally Lipschitz continuous.

The above result can also be established for the general semilinear Schrödinger equation of nonlinearity $\alpha > 1$. Moreover, by adapting the proof of the famous Strichartz estimates [41] that are used for sharp well-posedness of the NLS IVP, it is possible to extend the above result to the interval $0 \leq s < 1/2$ and hence obtain *sharp* well-posedness on the half-line (like for the IVP, the solution will now belong in a finer space motivated by the Strichartz estimates). Indeed, a sharp result of this kind was proved in [21], where the approach introduced in [15] was advanced *for the first time to higher than one spatial dimensions* for the NLS equation on the half-plane $\mathbb{R} \times \mathbb{R}^+$.

In fact, the analysis carried out in [21] and, more recently, in [23], led to a remarkable and perhaps unexpected discovery, namely that the celebrated $X^{s,b}$ spaces, which were introduced by Bourgain [4, 5] as *solution* spaces for proving the sharp well-posedness of the periodic and non-periodic NLS and KdV IVPs, now arise spontaneously as *boundary data* spaces in the estimation of the Fokas method solution for the pure linear IBVP associated with NLS on the half-plane. More precisely, for initial data $u_0 \in H^s(\mathbb{R} \times \mathbb{R}^+)$, it is shown in [21] that the Dirichlet boundary data must belong to a certain restriction of the space $X^{s,1/4} \cap X^{0,(2s+1)/4}$. In the case of the Neumann and Robin problems studied in [23], the corresponding space is a restriction of $X^{s,-1/4} \cap X^{0,(2s-1)/4}$.

In lieu of an epilogue, we emphasize that, despite the substantial progress made during the last decade on the well-posedness of nonlinear IBVPs via the novel Fokas-method-inspired approach outlined above, a plethora of important problems remain open. For example, recently the new approach was further extended in the direction of Bourgain spaces [27, 28], improving the result of [16] for the KdV equation on the half-line from H^s with $3/4 < s < 1$ (which is consistent with the IVP result of [31]) down to $s > -3/4$, matching the IVP result of [32]. Nevertheless, although the results of [27, 28, 32] are optimal with respect to contraction mapping techniques, they are not sharp in general, since it was recently shown in [33] without using a contraction mapping technique that the KdV IVP is well-posed in H^{-1}. Whether or not this result also holds on the half-line is currently unknown. The adaptation of the new approach to other higher-dimensional equations and/or domains such as the quarter-plane is another interesting direction that should be explored.

In conclusion, the Fokas method has provided the key to developing an effective, universal approach for the rigorous well-posedness of IBVPs that involve nonlinear dispersive (and non-dispersive) equations. This is yet another aspect of the remarkable impact that the method has had on the analysis of linear and nonlinear IBVPs since its introduction in 1997. Furthermore, it is also indicative of the influence that the method will continue to have on the field for the years to come.

Acknowledgements The author is thankful to the anonymous referees of the manuscript and gratefully acknowledges partial support via a grant from the National Science Foundation (NSF-DMS 2206270).

References

1. Ablowitz, M., Fokas, A.: Complex Variables: Introduction and Applications. Cambridge University Press (2003)
2. Ablowitz, M., Kaup, D., Newell, A., Segur, H.: Inverse scattering transform—Fourier analysis for nonlinear problems. Stud. Appl. Math. **53**, 249–315 (1974)
3. Bona, J., Sun, S., Zhang, B.-Y.: A non-homogeneous boundary-value problem for the Korteweg-de Vries equation in a quarter plane. Trans. Amer. Math. Soc. **354**, 427–490 (2002)
4. Bourgain, J.: Fourier transform restriction phenomena for certain lattice subsets and applications to nonlinear evolution equations. I. Schrödinger equations. Geom. Funct. Anal. **3**, 107–156 (1993)
5. Bourgain, J.: Fourier transform restriction phenomena for certain lattice subsets and applications to nonlinear evolution equations. II: The KdV equation. Geom. Funct. Anal. **3**, 209–262 (1993)
6. Bourgain, J.: Global Solutions of Nonlinear Schrödinger Equations. AMS (1999)
7. Degasperis, A., Sabatier, P.C.: Extension of the one-dimensional scattering theory, and ambiguities. Inverse Probl. **3**, 73–109 (1987)
8. Cazenave, T.: Semilinear Schrödinger Equations. Courant Lecture Notes In Mathematics. AMS (2003)
9. Colliander, J., Kenig, C.: The generalized Korteweg-de Vries equation on the half-line. Commun. PDE **27**, 2187–2266 (2002)
10. Faminskii, A.: A mixed problem in a semistrip for the Korteweg-de Vries equation and its generalizations. Dinamika Sploshn. Sredy **258**, 54–94 (1988)
11. Faminskii, A.: An initial boundary-value problem in a half-strip for the Korteweg-de Vries equation in fractional-order Sobolev spaces. Commun. Part. Differ. Equ. **29**, 1653–1695 (2004)
12. Fokas, A.: A unified transform method for solving linear and certain nonlinear PDEs. Proc. R. Soc. A **453**, 1411–1443 (1997)
13. Fokas, A.: A unified approach to boundary value problems. In: CBMS-NSF Regional Conference Series in Applied Mathematics, vol. 78. SIAM (2008)
14. Fokas, A., Gelfand, I.: Integrability of linear and nonlinear evolution equations and the associated nonlinear Fourier transforms. Lett. Math. Phys. **32**, 189–210 (1994)
15. Fokas, A., Himonas, A., Mantzavinos, D.: The nonlinear Schrödinger equation on the half-line. Trans. Amer. Math. Soc. **369**, 681–709 (2017)
16. Fokas, A., Himonas, A., Mantzavinos, D.: The Korteweg-de Vries equation on the half-line. Nonlinearity. **29**, 489–527 (2016)
17. Fokas, A., Pelloni, B.: Unified Transform for Boundary Value Problems: Applications and Advances. SIAM (2015)
18. Gardner, C., Greene, J., Kruskal, M., Miura, R.: Method for solving the Korteweg-de Vries equation. Phys. Rev. Lett. **19**, 1095–1097 (1967)
19. Hardy, G.H.: Remarks in addition to Dr. Widder's note on inequalities. J. Lond. Math. Soc. **4**, 199–202 (1929)
20. Himonas, A., Mantzavinos, D.: The "good" Boussinesq equation on the half-line. J. Differ. Equ. **258**, 3107–3160 (2015)
21. Himonas, A., Mantzavinos, D.: Well-posedness of the nonlinear Schrödinger equation on the half-plane. Nonlinearity. **33**, 5567–5609 (2020)
22. Himonas, A., Mantzavinos, D.: The nonlinear Schrödinger equation on the half-line with a Robin boundary condition. Anal. Math. Phys. **11**, 1–25 (2021)
23. Himonas, A., Mantzavinos, D.: The Robin and Neumann problems for the nonlinear Schrödinger equation on the half-plane. Proc. R. Soc. A. **478**, 20220279 (2022)
24. Himonas, A., Mantzavinos, D., Yan, F.: The nonlinear Schrödinger equation on the half-line with Neumann boundary conditions. Appl. Num. Math. **141**, 2–18 (2019)
25. Himonas, A., Mantzavinos, D., Yan, F.: The Korteweg-de Vries equation on an interval. J. Math. Phys. **60**, 051507 (2019)

26. Himonas, A., Mantzavinos, D., Yan, F.: Initial-boundary value problems for a reaction-diffusion equation. J. Math. Phys. **60**, 081509 (2019)
27. Himonas, A., Yan, F.: The Korteweg-de Vries equation on the half-line with Robin and Neumann data in low regularity spaces. Nonlinear Anal. **222**, 113008 (2022)
28. Himonas, A., Yan, F.: A higher dispersion KdV equation on the half-line. J. Differ. Equ. **333**, 55–102 (2022)
29. Holmer, J.: The initial-boundary-value problem for the 1D nonlinear Schrödinger equation on the half-line. Diff. Int. Eq. **18**, 647–668 (2005)
30. Holmer, J.: The initial-boundary-value problem for the Korteweg-de Vries equation. Commun. Part. Differ. Equ. **31**, 1151–1190 (2006)
31. Kenig, C., Ponce, G., Vega, L.: Well-posedness of the initial value problem for the Korteweg-de Vries equation. J. Am. Math. Soc. **4**, 323–347 (1991)
32. Kenig, C., Ponce, G., Vega, L.: A bilinear estimate with applications to the KdV equation. J. Am. Math. Soc. **9**(2), 571–603 (1996)
33. Killip, R., Visan, M.: KdV is well-posed in H^{-1}. Ann. Math. **190**, 249–305 (2019)
34. Klein, C., Saut, J.-C.: Nonlinear Dispersive Equations: Inverse Scattering and PDE Methods. Springer (2021)
35. Köksal, B., Özsari, T.: The interior-boundary Strichartz estimate for the Schrödinger equation on the half line revisited. Turk. J. Math. **46**, 3323–3351 (2022)
36. Lax, P.: Integrals of nonlinear equations of evolution and solitary waves. Commun. Pure Appl. Math. **21**, 467–490 (1968)
37. Linares, F., Ponce, G.: Introduction to Nonlinear Dispersive Equations. Springer (2009)
38. Özsari, T.: Weakly-damped focusing nonlinear Schrödinger equations with Dirichlet control. J. Math. Anal. Appl. **389**, 84–97 (2012)
39. Özsari, T., Yolcu, N.: The initial-boundary value problem for the biharmonic Schrödinger equation on the half-line. Commun. Pure Appl. Anal. **18**, 3285–3316 (2019)
40. Strichartz, R.: A Guide to Distribution Theory and Fourier Transforms. CRC Press (1994)
41. Strichartz, R.: Restrictions of Fourier transforms to quadratic surfaces and decay of solutions of wave equations. Duke Math. J. **44**, 705–714 (1977)
42. Tao, T.: Nonlinear dispersive equations: local and global analysis. In: CBMS Regional Conference Series in Mathematics, vol. 106. AMS (2006)
43. Zakharov, V., Shabat, A.: Exact theory of two-dimensional self-focusing and one-dimensional self-modulation of waves in nonlinear media. Sov. Phys. JETP. **34**, 63–69 (1972)

Athanassios S. Fokas: A Renaissance Scientist

Athanassios S. Fokas: A Renaissance Scientist

George Dassios

Abstract In what follows, some of the seminal contributions of Thanassis (short version of Athanassios) Fokas are reviewed. The goal is to justify the characterization by Israel Gelfand that "Fokas is a rare example of a scientist of the style of Renaissance". Gelfand's statement reflects the fact that Thanassis' career follows parallel academic paths involving Mathematics, Physics, Engineering, Biology and Medicine. In this talk, I will try to justify Gelfand's characterization, utilizing my memories from the nearly half a century that I know Thanassis, as a collaborator and a close friend. I will focus my presentation on his fundamental scientific contributions. Additional achievements, including awards and prizes, can be easily found on the web.

Thanassis was born at Argostoli, Kefalonia, in 1952, where he attended primary and secondary education. He was a very good goalkeeper and for some time during his early youth he had to decide between science and soccer. He studied Aeronautics at Imperial College, London. Why Aeronautics? Perhaps, because during the 70's Aeronautics was an exciting scientific area, but more importantly because it is very mathematical. For his undergraduate thesis, he studied the Wiener-Hopf technique, which is a special case of a more general method known as the Riemann-Hilbert (RH) formalism. This formalism, which belong to Complex Analysis, later played a crucial role in many of his research efforts.

The glamour of aeronautics was not enough to drag him away from mathematics, and in 1975 he moved to USA where he started graduate studies in Applied Mathematics at the California Institute of Technology. He received his Ph.D. in 1979, submitting a dissertation on the investigation of a new type of non-geometrical symmetries, called Lie-B*ä*cklund symmetries. At Caltech, he began his research by studying ordinary differential equations (odes) which possess additional non-geometric symmetries. Such odes are distinguished and can be solved analytically. This led Thanassis to pose the question of whether there also exist nonlinear partial differential equations (pdes) possessing non-geometric symmetries. He understood that

G. Dassios (✉)
Department of Chemical Engineering, University of Patras, Patras 26500, Greece
e-mail: gdassios@otenet.gr

T. Bountis et al. (eds.), *Chaos, Fractals and Complexity*, Springer Proceedings in Complexity, https://doi.org/10.1007/978-3-031-37404-3_25

nonlinear pdes with additional symmetries are precisely the particular class of pdes known as integrable, which at that time were studied by several different groups worldwide. These pdes can also be solved analytically via the Inverse Scattering Transform. Importantly, they possess certain remarkable solutions, called solitons, with many applications, from nonlinear optics to particle physics. In a joint work with Yannis Yortsos, they identified such a nonlinear pde which appears in the modeling of water flooding used for oil extraction.

At that time, one of the main international centers in the field of integrable systems was Clarkson University. In 1979, Thanassis met the Chairman of the Department of Mathematics and Computer Science of Clarkson University, Mark Ablowitz. It immediately became clear that they had a lot to share in mathematics and thus Mark became Thanassis' de facto mentor.

While still at Caltech, Thanassis tried to contact the great mathematician Israel Gelfand, in order to send him his joint paper with Fuchssteiner on symmetries (this important paper has nearly 2000 citations). Because Gelfand's address was in a department of biology, he assumed that Gelfand was involved with mathematical biology. This was the main motivation for his decision to leave mathematics for a few years to study biology and medicine. Meanwhile, he needed a mathematics position for a short period of time, and because of his relationship with Ablowitz, he accepted in 1980 a position at Clarkson University.

There, he concentrated on the solution of nonlinear integrable evolution pdes. At that time, the solution of the initial value problem of such pdes in one space dimension was well understood, but the analysis of the analogous problem in two space dimensions was "terra incognita". Ablowitz realized that the next breakthrough would be in this area. Thus, they studied a particular pde, called the Benjamin-Ono equation, which due to the involvement of the Hilbert transform is, in a sense, between one and two space dimensions. They hoped it could provide a bridge from one to two dimensions. The Lax pair of this equation involves a Riemann Hilbert (RH) problem, for which, as already mentioned, Thanassis had earlier expertise. It turned out that this work was of pivotal importance: Fokas and Ablowitz, not only solved this equation by a slight generalization of a RH problem [2], but also solved with the same method several integrable pdes in two space dimensions [3]. Furthermore, they showed that the remaining nonlinear integrable pdes in two spatial dimensions could be solved by the so-called d-bar technique, which provides a major generalization of the RH method [4].

The very fruitful collaboration of Fokas and Ablowitz led them, within three most productive years, to achieving the following remarkable unification: (a) Integrable ODEs, as well as evolution pdes in one spatial dimension, can be analyzed via the RH method. (b) Certain integrable evolution pdes in two spatial dimensions can be analyzed via a non-local version of the RH method. (c) The remaining integrable evolution pdes in two spatial dimensions can be analyzed via the d-bar method. In this case one goes beyond the realm of analytic functions and considers functions which are nowhere analytic.

The satisfaction of achieving the above unification, together with the disappointment that boundary value problems for very simple linear pdes could not be solved analytically (such problems had to wait for the emergence of the Fokas method), provided additional motivation for Thanassis to attempt to swim in different waters: It was time for medicine.

As he admits, studying medicine was the most difficult and most rewarding experience of his entire academic life. He started at the lowest step of the ladder: out of the 32 students admitted to the two-year Ph.D.-MD program of the University of Miami in 1983, he was the only person with no background in biology or medicine. His fellow students had a Ph.D. in anatomy, physiology, pharmacology, etc., whereas he had a Ph.D. in mathematics. The first six months, when he had to cover all relevant basic sciences, were extremely hard. However, by the time he moved to clinical rotations, it was a different story, and things became much better.

He managed to complete his medical degree within the top three of his class, which he considers one of his most important achievements. Unable to decide if he should return to mathematics, he stayed a year longer in medical school where he took many electives in neurology. Subsequently, he was offered a position at the Medical School of Stanford University for his specialization, and although he did start his internship in medicine, sleep deprivation and the loss of excitement, led Thanassis to his final decision to return to mathematics. His stay in Stanford University, however, gave him the opportunity to meet and work with the preeminent applied mathematician of that period, the great Joe Keller. The paper they wrote at that time on chronic myelogenous leukemia continuous to be cited to this day.

Upon his return to Clarkson University, he formed a strong group in the area of integrable systems. This group included the distinguished mathematician from the former Soviet Union, Alexander Its, who became one of his closest friends and most valuable collaborators. Its is a world expert on the Painlevé equations, and Fokas together with Its and two of Its' former students wrote a comprehensive book on these classical odes [5]. Related to this work, they visited a famous physicist at Princeton University who suggested that they analyze a particular double limit of one of the Painlevé equations. According to this physicist, the computation of this limit was needed in one-dimensional quantum gravity. Thanassis considers this problem as the most challenging problem in asymptotics he has worked on, until his recent analysis on the Riemann zeta function. Together with Its and a former student of Its they managed to solve it, but much to their disappointment, when they returned to Princeton to announce their success, they were informed that this problem was not physically important after all!

Fortunately, mathematical efforts are often rewarded proportionally, and in this case, the associated formalism introduced in that work led to a huge development in the important area of orthogonal polynomials and random matrices: their work implies that these mathematical entities can be re-cast in the framework of the RH formalism [6]. A few years earlier, Percy Deift and Xin Zhou had developed a powerful technique for computing the asymptotics of RH problems. Thus, combining the new formulation with the Deift-Zhou technique it became possible to obtain results in this area which at that time were beyond any expectation. A particular application of the work in the asymptotics of the Painlevé equations is contained in the talk of Rogers at this Conference [1].

Meanwhile, Thanassis felt that he had some unfinished business in the field of symmetries: in collaboration with Fuchssteiner they had achieved a complete understanding of the symmetry structure of integrable evolution pdes in one space dimension. However, the analogous problem in two spatial dimensions remained open. In an unexpected breakthrough, Thanassis, in collaboration with Paolo Santini, solved this problem in 1987 [7].

In order to announce his return to integrable systems, Thanassis asked the great Russian mathematical physicist Vladimir Zakharov to co-edit with him a book summarizing important developments in integrable systems. In this connection, he contacted Gelfand, asking him to write the chapter on symmetries. Gelfand, who knew of the recent breakthrough of Fokas and Santini, suggested that they write this chapter jointly. This was the beginning of a long and close collaboration with Gelfand, which produced twelve papers. More importantly, this collaboration had an enormous impact on Thanassis' career. In his own words: "Mark Ablowitz was my early mentor, but no one has influenced me more than Israel Gelfand".

One of the reasons that Gelfand liked to interact with Thanassis was Thanassis' medical background. Interestingly, in his entire life Gelfand refused to mix mathematics and biology; indeed, he led two parallel lives. Thanassis managed to convince Gelfand to collaborate with him on a problem which was strictly mathematical, but at the same time it was important in Medicine. In this way they could both be happy. This was an inverse problem arising in magnetoencephalography (MEG). It involves the reconstruction of the electric current inside the brain from the knowledge of the magnetic flux outside the head measured with the extremely sensitive apparatus of the SQUID (Superconductive QUantum Interference Device). As it was already known since 1853 to Helmholtz, this problem does not have a unique solution. Actually, no one knew until the work of Thanassis which part of the current can be determined from the data. This is a hard problem, since the conductivity of the brain generates induction currents which "hide" the primary neuronal current that needs to be identified. Fokas and Gelfand, in collaboration with the late Yaroslav Kurylev, obtained in 1994 the first important result for MEG: they characterized explicitly the part of the current that can be reconstructed from the data in the very special (and physically unrealistic) case of a spherical, homogeneous conductor.

In the summer of 1991, Gelfand spent one month at the island of Kefalonia as a guest of Thanassis. There, a conceptually important breakthrough was achieved. They were able to show that within the unification scheme of Fokas and Ablowitz of the early 1980s, they could also include the most well-known pdes, namely, the linear ones! Indeed, the classical way of solving linear evolution pdes in one and two spatial dimensions involves using the Fourier transform in one and two dimensions. Remarkably, these transforms can be constructed via a RH and a d-bar formalism respectively, which are precisely the tools used in the analysis of integrable evolution pdes in one and two spatial dimensions.

In 1995, 20 years after his graduation from the Department of Aeronautics, Thanassis returned to Imperial College in a Chair in Applied Mathematics. Since 1983, when a good understanding of the solution of the initial value problem of evolution pdes in one and two spatial dimensions was achieved, Thanassis had developed a true obsession: solving, for these nonlinear pdes, the class of physically more important and mathematically much more demanding boundary value problems.

His first attempt to study such problems, just before going to medical school, was a failure. Actually, until 1997, the solution of boundary value problems was considered the most important open problem in the analysis of integrable systems. Finally, after 15 years of hard work he was able to solve a typical such problem [8]. Interestingly, in this process he obtained much more than he ever expected: he realized that his new method provided a completely new way of solving boundary value problems for linear pdes. Earlier on, he was skeptical regarding the novelty of his approach for linear pdes. After all, the relevant classical techniques were introduced by some of the giants of the 18th century, including Fourier and Laplace. In addition, he never forgot an advice of Gelfand's: "There is no problem competing with your contemporaries but be extremely careful when you compete with the classics!"

After the publication of hundreds of papers by experts in many different groups, including the groups of Bernard Deconinck and Beatrice Pelloni, it became clear that the new Unified Transform introduced by Thanassis, also referred to as the Fokas method, is indeed new and much superior to anything achieved earlier. Importantly, it has the advantage that it is computationally friendly, so that even undergraduates can use it for the computation of the solution of physically important boundary value problems. In fact, it is already taught at the undergraduate level at several universities. Recently, his new method has allowed him to return to his roots: Fokas, together with collaborators at the University of Cambridge, were able to demonstrate that the Fokas method makes the classical Wiener-Hopf technique obsolete.

Many talks at the 28th Summer School–Conference on "Dynamical Systems and Complexity" of 18–26 July, 2022, (see [1]) were dedicated to this seminal development, starting with the introductory talk by Kaxiras. I note that Fokas and Kaxiras just published a remarkable book containing a variety of techniques in applied mathematics [9]; in this book, for the first time, the Fokas method is presented in a pedagogical way. In their talks, Himonas and Manzavinos review yet another application of the Fokas method: it provides a new powerful approach for establishing the well-posedness of boundary value problems for nonlinear pdes. In [1], Turker presented unexpected results in the important area of control theory; Colbrook reviewed the numerical implication of the Fokas method to elliptic pdes; Smith and Pelloni [1] discussed remarkable implications to the classical area of spectral theory; Fernandez presented interesting implications in the area of fractional calculus, and Saridakis discussed a problem arising in the modeling of glioblastoma; Lenells discussed the most important so far applications of the Fokas method to integrable nonlinear pdes, presenting the recently obtained complete solution of the x-periodic problem, which had remained open since the mid 1970s despite the involvement of several outstanding mathematical figures [10].

In 2002, Thanassis was appointed to the newly inaugurated Chair of Nonlinear Mathematical Science at the Department of Applied Mathematics and Theoretical Physics of the University of Cambridge. As I mentioned earlier, Fokas and Gelfand had shown that the RH and d-bar formalisms provide a completely new way of rederiving the classical Fourier transform in one and two dimensions. Actually, this approach provides a new and powerful algorithm for deriving a variety of integral transforms. In 2002, Roman Novikov used this method to derive the attenuated Radon transform, which plays the same crucial role in the important medical imaging technique of Single Positron Emission Computerized Tomography (SPECT), that the classical Radon Transform plays in Computerized Tomography. After studying Roman's paper, Thanassis realized that this result could be easily derived via a small modification of the results of his earlier paper with Roman, where their method with Gelfand was used for the rederivation of the Radon Transform. At that time, he was elected a member of the Academy of Athens and was able to establish at the Academy a Center of Mathematics, with George Kastis as its Director. He made the decision that this Center should concentrate on the imaging techniques of Positron Emission Tomography (PET) and of SPECT. The goal was to implement numerically the analytical formulae obtained via the Fokas-Gelfand approach. Actually, it took 15 years to achieve this goal [11]. The relevant results are reviewed in the presentations of Kastis and Protonotarios [1].

Thanassis continued his efforts on MEG. The basic question was clear: is it possible to extend the earlier result of Fokas-Gelfand-Kurylev, which is mathematically beautiful but physiologically useless, to a result which is both beautiful and useful? In 2005, the European Union announced the funding of 15 Honorary Marie Curie Chairs of Excellence for 15 scientists from all over the world to spend three years at a European university. Thanassis and I submitted a proposal to collaborate on "the electromagnetic activity of the human brain", and we were successful. In the period of 2005 to 2008 we worked together in Cambridge trying to eliminate from the existing results on Electroencephalography (EEG) and MEG the restriction of spherical geometry. In this project, I had once more the opportunity to use the theory of ellipsoidal harmonics which is a most important component of my work of the last 30 years. After a series of 15 papers, which includes works with collaborators in my group in the University of Patras, the problem has been completely solved.

In 2009, 156 years after the fundamental result of Helmholtz, Thanassis published a paper announcing the final analytical result [12]. It states that, within the EEG modality, no more than one of the three scalar functions specifying the neuronal current can be recovered, whereas within the MEG modality two of these scalar functions are visible from outside the head (one of these two functions being the same with the one obtained via EEG). Remarkably, this result is true for any smooth geometry of the brain-head system, as well as a shell type inhomogeneous brain model. This result gave rise to a beautiful mathematical formula for the part of the current that could be determined from the data. However, the question of whether this formula could be implemented numerically, so that finally a useful result could be obtained, remained open. This was indeed achieved in the period 2009–2019, through continuous support from the UK and Europe, in collaboration with Parham Hashemzadeh. The EEG-MEG results are presented in our joint book with Thanassis [13]. Recent further developments are presented in the talk of Paraskevopoulou, at this School [1].

In 2010, Thanassis taught at Cambridge a course that included a short summary of the Riemann Hypothesis, which remains the most famous open problem in the history of Mathematics. This hypothesis involves a certain function, the Riemann zeta function, which is defined in the complex z-plane. The Riemann Hypothesis states that the Riemann zeta-function has all its roots on the negative even integers (trivial roots) and on the line Rez = 1/2 (non-trivial roots). The difficult problem is to show that there are no roots for points which satisfy $0 < Rez < 1/2$ and have a large imaginary part. Indeed, using today's powerful computers, it has been verified that the Riemann zeta function does not have any zeros for Imz up to approximately 10 to the power 13, which is a large number, but still very small in comparison with infinity! This discussion implies that it is crucial to understand the large Imz-asymptotics of the Riemann zeta function.

Lindelöf postulated the relevant asymptotic behavior. In terms of importance, the Lindelöf hypothesis is considered second only to the Riemann Hypothesis. Thanassis decided to use his expertise in complex analysis and asymptotics to attack this hypothesis. In this connection, together with the outstanding analyst Jonatan Lenells, they published a major contribution in the Memoirs of the American Mathematical Society [14]. After many unsuccessful attempts and with the crucial input of Lenells and his remarkably original Ph.D. student Anthony Ashton, Thanassis finally derived a novel integral equation satisfied by the Riemann zeta function. The computation of the large Imz-asymptotics of this equation gave the proof of the Lindelöf Hypothesis for a slight variant of the Riemann zeta function [15]. New progress has been achieved by representing the Riemann zeta function in terms of its finite Fourier transform. This expression was motivated by the application of the Fokas method to the solution of pdes in a finite domain. A summary of recent results is presented in the talk of Kalimeris [1].

In the last ten years of his life, Israel Gelfand concentrated on the solution of one of the most important open problems in biology, namely, protein folding. Since a protein is uniquely determined by its amino acid sequence, given this sequence, one should be able to predict its three-dimensional structure, which is responsible for the biological properties of the given protein. However, this problem remained open, in spite of the efforts of many brilliant scientists. Gelfand tried to convince Thanassis to collaborate with him on protein folding. However, for several years Thanassis refused, for several reasons. First, he knew about the difficulty of this problem since he was a medical student. Second, he was aware that several powerful groups were dedicated to the solution of this problem. Third, Gelfand refused dogmatically to use mathematics in biology, believing that the mathematics needed to solve biological problems had not yet been invented. Finally, Thanassis agreed to look at this problem, only because of his love and respect for Gelfand.

Somehow, together with the late Theodore Papatheodorou, they managed to obtain an important result: they discovered certain topological rules which limit enormously the possible associated three-dimensional arrangements. Gelfand was very excited with this result but refused Thanassis' suggestion to supplement it with the mathematical technique of optimization. After Gelfand's death Thanassis stopped working in this area, and since their three associated papers in the Proceedings of the National Academy of Sciences were not getting many citations, he thought for several years that this was the only project that he had wasted his time. However, one of the most active groups in this area, the group of the late Christodoulos Floudas, at Princeton University, made a crucial use of this result. Floudas wrote just before his untimely death that, "We are now able to predict the 3D structure of beta proteins with a success rate of 80%. This is based on the seminal works of Fokas and Gelfand who discovered unexpected topological properties of these proteins, which we have incorporated in our optimization mathematical model".

In September 2015, Thanassis was awarded a five-year Senior Fellowship by the Engineering and Physical Science Research Council of United Kingdom. This relieved him from any teaching and administrative obligations at Cambridge, allowing him to live what he called a "dream life". This fellowship was awarded for three projects: the Fokas method, medical imaging, and the asymptotics of the Riemann zeta function. At the same time, he was honored by the University of Southern California with an Adjunct Professorship at the Viterbi School of Engineering.

More than ten years ago, Costas Vayenas, a foreign member of the USA National Academy of Engineering, proposed an iconoclastic model for Particle Physics. He suggested that the three quarks within a neutron are not confined as a result of the strong interactions, but because of super-relativistic gravity. Furthermore, he claimed that these light quarks are, actually, electron neutrinos. Using an ad hoc simple formula for the relativistic gravitational force, as well as a simple Bohr type model, Vayenas was able to compute from first principles the mass of the neutron and to obtain its well-known value. Thanassis, in order to justify this model, decided to derive the relativistic gravitational law from first principles, i.e., from the theory of general relativity. It should be noted that the problem of computing the force between even two relativistic particles, the so-called two-body problem of general relativity, remains open despite the efforts of many brilliant physicists, beginning with Einstein. However, the above case, due to the occurrence of the small masses, corresponds to the so-called Minkowskian approximation, for which it is possible to obtain asymptotically analytical results.

Thanassis, in collaboration with Luc Blanchet (a world expert in this type of computations) obtained an explicit formula for the relevant force [16]. Furthermore, Thanassis also computed the limit of this formula when the speed of the particles approaches the speed of light. Remarkably, this yields a force which possesses the basic characteristics of the strong force, namely confinement and asymptotic freedom! However, a variety of experiments suggest that the mass of the light quarks is much larger than the mass of the electron neutrinos. Using for the mass of the quarks the experimentally verified value, instead of the value of the mass of the electron neutrino used by Vayenas, one finds a force which is quite large, but much smaller than the needed value. Recent developments were presented in the talk of Manolis Floratos, see [1].

Concluding, I must mention Thanassis' very recent contribution to integrable nonlinear evolution pdes in three spatial dimensions. A fundamental open question in the field of integrability has been the question of the existence of integrable nonlinear evolution equations in higher than two spatial dimensions. In 2006, Thanassis obtained 4+2 generalizations of the Kadomtsev-Petviashvilli (KP) and the Davey-Stewartson (DS) equations, by constructing integrable analogues of the KP and DS in four spatial and two temporal dimensions. The solution of the Cauchy problem of these equations was obtained using a non-local d-bar formalism. For this work Thanassis used a nonlinear Fourier transform in four real dimensions.

The question of reducing integrable equations from 4+2 to 3+1, and establishing that the initial value problems of the resulting 3+1 equations are well posed, although discussed in several papers, remained open. This is reviewed in the talk of van der Weele, see [1]. Finally, Thanassis solved this problem in a paper just published in the Proceedings of the Royal Society [17]. The unexpected achievement of the results of this work is the derivation of a formalism capable of dealing with the case that the time dependence of the nonlinear Fourier transform contains an exponential with a non-vanishing real part. Remarkably, this has interesting implications beyond the area of nonlinear integrable pdes: it introduces a new transform for solving a large class of linear pdes. Particular solutions of the new integrable KP in three spatial dimensions are presented in the talk by Jingsong He [1].

It should be noted that Thanassis has made several additional important contributions that time limitations prevent me from discussing. They include, his recent very interesting work on the modeling of the Covid-19 pandemic reviewed in the talk of Dikaios [18], as well as the work on the fractal nature of some of the paintings of Piet Mondrian reviewed by Bountis [19]. In addition, I must mention the remarkable extension of classical results on conformal mappings, obtained jointly with Darren Crowdy [20] (a work that somehow, remains largely unknown).

Perhaps the best way to end my presentation is to return to the title of my talk: Athanasios Fokas: A Renaissance Scientist. Taking into consideration Thanassis' seminal contributions in so many different areas, as well as his remarkable, soon to be published deep book "Ways of Comprehending", which in addition to mathematics, physics, engineering, biology, and medicine, also contains philosophy and painting, I hope you will agree with me that Gelfand was right!

References

1. http://cosa.inn.demokritos.gr/28th-summer-school-dynamical-systems-and-complexity-program/
2. Fokas, A.S., Ablowitz, M.J.: The inverse scattering transform for the Benjamin-Ono equation-a pivot to multidimensional problems. Stud. Appl. Math. **68**, 1 (1983)
3. Fokas, A.S., Ablowitz, M.J.: On the inverse scattering of the time dependent Schrődinger equation and the associated KPI equation. Stud. Appl. Math. **69**, 211 (1983)
4. Fokas, A.S.: Inverse scattering of first-order systems in the plane related to nonlinear multidimensional equations. Phys. Rev. Lett. **51**, 3 (1983)
5. Fokas, A.S., Its, A.R., Kapaev, A.A., Yu Novokshenov, V., Painleve Transcendents: A Riemann-Hilbert Approach. AMS (2006)
6. Fokas, A.S., Its, A.R., Kitaev, A.V.: The isomonodromy approach to matrix models in 2D quantum gravity. Comm. Math. Phys. **147**, 395 (1992)
7. Fokas, A.S.: Symmetries and integrability. Stud. Appl. Math. **77**, 253 (1987)
8. Fokas, A.S.: A unified transform method for solving linear and certain nonlinear PDE's. Proc. R. Soc. Lond. A. **453**, 1411 (1997)

9. Fokas, A.S., Kaxiras, E.: Modern Mathematical Methods for Computational Sciences and Engineering. World Scientific (2022)
10. Fokas, A.S., Lenells, J.: A new approach to integrable evolution equations on the circle. Proc. R. Soc. A. **477**, 20200605 (2021)
11. Fokas, A.S., Kastis, G.A.: Mathematical methods in PET and SPECT imaging. In: Scherzer, O. (ed.) Handbook of Mathematical Methods of Imaging. Springer (2015)
12. Fokas, A.S.: Electromagnetoencephalography for the three-shell model: distributed current in arbitrary, spherical and ellipsoidal geometries. J. R. Soc. Interface. **6**, 479 (2009)
13. Dassios, G., Fokas, A.S.: Electroenchephalography and Magnetoencephalography: An Analytical-Numerical Approach. De Gruyter (2019)
14. Fokas, A.S., Lenells, J.: On the asymptotics to all orders of the Riemann zeta function and of a two-parameter generalization of the Riemann zeta function. Mem. Amer. Math. Soc. **275**(1351). AMS (2022)
15. Fokas, A.S.: A novel approach to the Lindelöf hypothesis. Trans. Math. Appl. **3**, 1 (2019)
16. Blanchet, L., Fokas, A.S.: Equations of motion of self-gravitating N-body systems in the first post-Minkowskian approximation. Phys. Rev. D. **98**, 084005 (2018)
17. Fokas, A.S.: Integrable nonlinear evolution equations in three spatial dimensions. Proc. R. Soc. A. **478**, 20220074 (2022)
18. Fokas, A.S., Dikaios, N., Yortsos, Y.C.: An algebraic formula, deep learning and a novel SEIR-type model for the COVID-19 pandemic (submitted)
19. Bountis, T., Fokas, A.S., Psarakis, E.Z.: Fractal analysis of tree paintings by Piet Mondrian. Int. J. Arts Technol. **10**, 27 (2017)
20. Crowdy, D., Fokas, A.S.: Conformal mappings to a doubly connected polycircular arc domains. Proc. R. Soc. A. **463**, 1885 (2007)

Author Index

A
Anesiadis, K., 221
Aubry, Serge, 3

B
Barmparis, G. D., 213
Basios, Vasileios, 153
Bezerianos, Anastasios, 245
Boukos, N., 125
Bountis, Tassos, 109

C
Cao, Yulei, 265
Charalambopoulos, Antonios, 189
Chatziafratis, A., 289
Chatzigeorgiou, M., 125
Christodoulidi, Helen, 109
Constantoudis, V., 125
Courson, Juliette, 233

D
Dassios, George, 363
De Sanctis, Angela A., 139
Dogkas, T., 213
Dragomir, Andrei, 245

E
Eleftheriou, M., 213

F
Fokas, Athanassios S., 189, 265, 319

G
Grafakos, L., 289
Gunji, Yukio-Pegio, 153

H
Harsoula, Mirella, 57
He, Jingsong, 265
Himonas, A. Alexandrou, 275
Hizanidis, J., 221

K
Kalimeris, K., 319
Kalimeris, Konstantinos, 169
Kaloudis, Konstantinos, 109
Kamvissis, S., 289
Kanellopoulos, Giorgos, 199
Kastis, George A., 169
Katsanikas, Matthaios, 47
Katsiotis, M., 125
Kompatsiaris, Ioannis, 245

L
Lazarou, Ioulietta, 245
Lenells, J., 319

M
Manos, Thanos, 233

T. Bountis et al. (eds.), *Chaos, Fractals and Complexity*, Springer Proceedings in Complexity, https://doi.org/10.1007/978-3-031-37404-3

Mantzavinos, Dionyssios, 347
Minicucci, Francesca, 139
Moretti, Pier-Francesco, 153

N
Nikolopoulos, Spiros, 245

O
Oikonomou, Fotios D., 139

P
Parakevopoulou, Georgia, 189
Patsis, P. A., 37
Pelloni, B., 333
Perantonis, Stavros, 189
Protonotarios, Nicholas E., 169
Provata, A., 221

Q
Quoy, Mathias, 233

R
Razis, Dimitrios, 199
Robnik, Marko, 83

S
Seet, Manuel S., 245
Smith, D. A., 301, 333
Stratis, I. G., 289

T
Tsironis, G. P., 213
Tzemos, Athanasios C., 71

V
Van der Weele, Ko, 199

W
Wiggins, Stephen, 47